T0332348

Nanotechnology
Commercialisation

Nanotechnology Commercialisation

edited by Takuya Tsuzuki

PAN STANFORD PUBLISHING

Published by

Pan Stanford Publishing Pte. Ltd.
Penthouse Level, Suntec Tower 3
8 Temasek Boulevard
Singapore 038988

Email: editorial@panstanford.com
Web: www.panstanford.com

British Library Cataloguing-in-Publication Data
A catalogue record for this book is available from the British Library.

Nanotechnology Commercialisation

ISBN 978-981-4303-28-6 (Hardcover)
ISBN 978-981-4303-29-3 (eBook)

Printed in the USA

Contents

2. Applications and Market Opportunities of Nanoparticulate Materials

Takuya Tsuzuki

Preface

Nanotechnology has the potential to revolutionise all aspects of our lives. For this to happen, commercialisation activities have a critical role to play. They bring technology innovation into realisation whilst creating economic benefits for society. The past 30 years have seen significant progress in nanotechnology. As a result, many innovative and practical applications of nanotechnology have been explored. This book aims to give an overview of the current trends in and the issues associated with the commercialisation of nanotechnology.

Nanotechnology encompasses many disciplines of science and engineering, including nanomaterials, nanomedicine, nano/micro-electromechanical systems, nanofabrication and nano-instrumentation. This book is unique in that it focuses on the nanomaterial sector. Engineered nanomaterials, especially nanoparticulate materials, are regarded as the leader in nanotechnology commercialisation. This is owing to the wide range of their unique properties, relative ease of fabrication, significant market opportunities and a short product development timeframe. In addition, the nanomaterial sector has attracted much more heated debate than any other areas of nanotechnology with regard to safety, regulation, standardisation and ethics. As such, the nanomaterials sector occupies a unique position in nanotechnology commercialisation.

Another unique aspect of this book is that it fills the existing gap between academic research and commercial production of nanomaterials. As shown in Fig. 1, nanotechnology commercialisation is the culmination of a broad spectrum of collective activities, from laboratory-scale investigations through production scale-up, to the non-technological issues surrounding commercialisation. Understanding the continuum of the spectrum is critical not only for the product development but also for regulatory and risk management purposes. Many stakeholders, including consumers, government officials, corporate managers and university researchers, are involved in the process of nanomaterial commercialisation. The interaction between the stakeholders from different segments of the spectrum is essential for it to be successful. Hence, instead of focusing solely on the business side of nanotechnology commercialisation, a special effort has been made to capture and review each segment of the commercialisation spectrum. The chapters in this book are placed in the order roughly aligned with the spectrum.

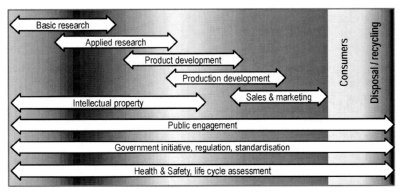

Figure 1   Spectrum of nanotechnology commercialisation activity.

The applications of nanomaterials in biomedical/medicinal areas are largely excluded. Nanotechnology commercialisation in the medical field has unique characteristics in terms of business models, funding sources, regulatory issues and other aspects of commercialisation. Hence, these topics should be covered in a separate review volume. Also excluded are the aspects of organic polymeric nanoparticles. Although polymer nanoparticles have been widely used in paint and plastic industries and will have many important medical applications, they have technological, as well as safety and regulatory issues different from those for inorganic nanomaterials and hence deserve a separate volume of their own. Nonetheless, some aspects of carbon-related nanomaterials such as carbon nanotubes and nano-diamonds are covered in this volume.

This book is the collective achievement of many passionate and dedicated individuals. First of all, I would like to express my sincere gratitude to the chapter authors, who provided wonderful contributions despite their extremely busy schedules. Their presence in this book as renowned experts in their field was critical to the project. I would also like to thank Mr. Stanford Chong, director of Pan Stanford Publishing, for making the publication possible. Finally, I am deeply grateful to my wife, Savitri, who has been a constant source of help and encouragement throughout the project.

Takuya Tsuzuki
Canberra, Australia
2013

Chapter 1

Properties of Nanoparticulate Materials

Takuya Tsuzuki

Research School of Engineering, College of Engineering and Computer Science, Australian National University, Ian Ross Building 31, North Road, Canberra ACT 0200, Australia

takuya.tsuzuki@anu.edu.au

Nanomaterials are a new class of industrial materials. Owing to their unique properties and the recent developments in synthesis methods, current and potential applications of nanomaterials are rapidly expanding into many industries and markets. For the successful development of nano-enabled commercial products, it is critical to understand the unique properties of nanoparticles and how the desired properties can be manifested in the end products. This chapter gives an overview of the unique properties and characteristics of nanoparticulate materials.

Nanotechnology Commercialisation
Edited by Takuya Tsuzuki
Copyright © 2013 Pan Stanford Publishing Pte. Ltd.
ISBN 978-981-4303-28-6 (Hardcover), 978-981-4303-29-3 (eBook)
www.panstanford.com

1.1 Introduction

The word "*nano*" originated from the Greek word νᾶνος (nanos), meaning "dwarf". Scientifically, "nano" means one billionth of a unit. One nanometre (nm) is a length scale equivalent to the one billionth of a metre. Thus, "nano"-materials are the materials that possess miniscule dimensions. According to ISO TS 27687, nanomaterials are defined as the materials that have a characteristic scale of 1–100 nm. To put this into perspective, the size ratio between a nanoparticle of 1 nm in diameter and a soccer ball is equivalent to the size ratio between a soccer ball and the Earth.

The association of humans with nanomaterials is not new. In fact, colloid chemistry has been dealing with the synthesis and characterisation of nanoparticles for centuries. Nevertheless, nano-material science is regarded as a relatively young research field. In fact, it is only since the 1980s that we have seen the progressive development of the knowledge, techniques and instruments for imaging, measuring, manipulating and fabricating nanoscale objects [69].

For example, in 1981, a new instrument, the scanning tunnelling microscope, was invented. The instrument enabled scientists to see and manipulate individual atoms for the first time in human history. In addition, since 1980, many new discoveries were made regarding the unique properties of nanomaterials that differ from the properties of their bulk counterparts [11,12,21,64]. Many new types of nanomaterials have also been found. Buckminsterfullerenes, spherical molecules consisting of only 60 carbon atoms (Fig. 1.1a), were discovered in 1985 [29]. Carbon nanotubes (CNTs), another type of nanomaterial consisting only of carbon atoms (Fig. 1.1b) were re-discovered in 1991 [24]. Graphene, yet another carbon-based nanomaterial, comprising a single-atom-thick layer (Fig. 1.1c), was isolated for the first time in 2004 [6]. These new types of carbon-based nanomaterials exhibit unique properties unobtainable from the conventional carbon-based materials. For example, CNTs show tensile strength 300 times higher than steel and can carry an electric current density 1,000 times higher than metals [5].

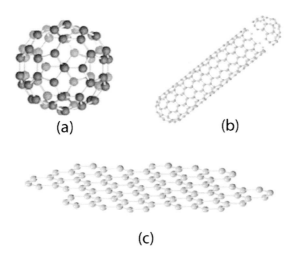

(a) (b)

(c)

Figure 1.1 Carbon-based new nanomaterials; (a) C60 fullerene, (b) carbon nanotube and (c) graphene. Reproduced from Ref. [26], Copyright (2007), with permission from Elsevier.

These developments of new instruments and techniques as well as the discoveries of new properties and materials have given rise to a new wave of materials science and technology in nanoscale dimensions. During its development, nanotechnology has promoted the convergence of many fields of science and has ignited an explosion of multidisciplinary research and development activities in the areas of physics, chemistry, biology, electronics, medical, pharmaceutical, textile, food and so on [56,57].

The development of nanotechnology has also provided tools to analyse the natural materials around us at a nanoscale dimension. For example, it was found that small-scale hairy structures, such as the ones on lotus leaves, create water-repellent (super-hydrophobic) surfaces [10]. It was also found that bone is a natural nanocomposite consisting of ceramic nanoparticles and protein molecules [70]. The mixture of hard-but-brittle ceramic nano-particles and soft-but-flexible protein gives bones special mechanical properties. Mimicking this structure enables the production of lightweight and high-strength materials, which in turn enables the reduction of energy consumption in vehicles and aircraft.

The development of the knowledge, techniques and instruments for imaging, measuring, manipulating and fabricating nanoscale objects has given us not only an insights into Mother Nature but also the opportunities to improve our life in many ways [22,38,49,51,55].

1.2 Nanoparticulate Materials

Nanoparticulate materials consist of nanoscale particles [66]. They do not necessarily have to be spherical in shape. Particles with many different shapes including a hollow-shell, rods, belt, needle, wire, tube, disk or plate, can also be regarded as nanoparticulate materials, as long as one of the dimensions, for instance, the thickness of the plates, is less than 100 nm (Fig. 1.2) [68].

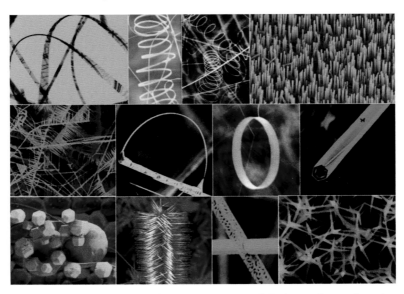

Figure 1.2 A collection of nanostructures of ZnO synthesized under controlled conditions by thermal evaporation of solid powders. Reprinted from Ref. [68], Copyright (2004), with permission from Elsevier.

Nanoparticulate materials are of significant interest in many fields of science because their sizes are similar to the characteristic length scale of key physical and biological parameters.

For example, the electron mean free path in metals is typically <100 nm, exciton-Bohr diameters in semiconductors are 1–100 nm, magnetic domain wall thicknesses are 10–100 nm and the wavelength of ultraviolet light is <350 nm. In biological materials, the diameter of a DNA helix is ~5 nm, the size of a typical virus is 20–300 nm and many other biomolecules fall into the nanoscale range. Because of these similarities in size, many new properties are manifested in nanoscale materials and many new applications in physical, chemical and biological science are realised. In the following sections, we discuss the unique properties of nanoparticles in detail.

1.3 Common Characteristics of All Types of Nanoparticulate Materials

Because of their extremely small size, nanomaterials exhibit new or enhanced properties compared with the same materials of larger dimensions [4]. The unique properties are influenced by the size, shape, crystallinity and other structures of nanoparticles. Since the 1980s, the methods used to control the size and structure of nanoparticles have been considerably improved. As a result, nanoparticles can now be designed to exhibit the desired properties by careful control of their structures. The ability to tailor the unique properties of nanoparticles has caused rapid progress in the commercialisation of nanomaterial-related products. This section discusses the characteristics and unique properties of nanoparticles that are common to all material types including metals, semiconductors and insulators.

1.3.1 High Surface Area

1.3.1.1 Specific surface area

One of the most utilized unique properties of nanoparticles is their high specific surface area. Specific surface area is defined as the surface area per unit weight as expressed in Eq. 1.1:

$$S = \frac{6000}{d \cdot \rho},$$

(1.1)

where S is the specific surface area in m^2/g, d is the particle diameter in nanometres and ρ is the density of the material in g/cm^3. As the particle size decreases, relatively more atoms become exposed on the particle surface. In other words, the number of atoms on the surface increases relative to the total number of atoms. As such, as the particle size is reduced, the specific surface area increases. For example, when the diameter of gold particles is reduced to 1 nm, 1 g of nanoparticles has a surface area of as large as ~300 m^2. On the other hand, if the gold particles have a diameter of 0.1 mm, then 1 g of the powder has only ~0.003 m^2 of surface area (Table 1.1). This means that as much as 100 kg of ~0.1 mm-sized gold particles are required to provide the same surface area as 1 g of ~1 nm-sized gold nanoparticles.

Table 1.1 Surface-to-volume atomic ratio of spherical gold particles (approx. only)

Particle diameter (nm)	Total atom count	Surface atoms (%)	Specific surface area (m^2/g)
1000	~30,000,000,000	~0.2	0.3
100	~30,000,000	~1.6	~3
10	~30,000	~15	~31
1	~30	~90	~310

The high surface area of nanoparticulate materials is useful for many applications that utilize surface-related functionality, including catalytic and photocatalytic activities, gas sensing abilities and solubility.

1.3.1.2 Melting point depression

Nanomaterials melt at lower temperatures than bulk materials. Melting point depression is important in applications that involve high temperatures such as three-way automotive catalysts and ceramic forging.

Melting temperature is associated with the so-called cohesive energy of materials. Cohesive energy is defined as the energy difference between the atoms in the solid and the atoms in a free state. Hence, cohesive energy is related to the thermal energy required to free the atoms from the solid. In general, solids with a higher cohesive energy require a higher temperature to melt.

Chemical bonds between atoms provide positive cohesive energy. Since the atoms on the surface have fewer neighbouring atoms than the atoms inside of the particles, the increase of surface-to-volume atomic ratio in nanoparticles leads to reduced cohesive energy as in Eq. 1.2 [41]:

$$E_{\text{cohesive}} = E_{\text{bulk}}\left(1 - \frac{a}{d}\right),\qquad(1.2)$$

where E_{cohesive} is the cohesive energy of a nanoparticle, E_{bulk} is the cohesive energy of bulk, d is the nanoparticle diameter and a is the atomic diameter. Figure 1.3 shows the example of melting point depression in gold nanoparticles [8].

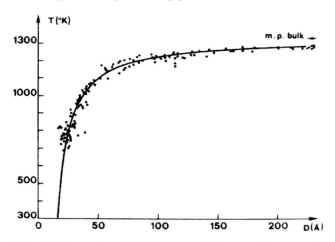

Figure 1.3   Melting point of fold nanoparticles as a function of particle size. Reprinted with permission from Ref [8]. Copyright (1976) by American Physical Society.

1.3.1.3   Solubility enhancement

At the atomic scale, the dissolution of materials by solvents occurs at the surface of materials. As a consequence, smaller particles with a higher specific surface area have higher solubility. The high solubility of nanoparticulate materials is useful for many applications in the food, pharmaceutical, medical as well as agricultural sectors to enhance bioavailability and nutrient delivery.

When the particles have a wide size distribution, the particles in the solution tend to experience the so-called "Ostwald ripening", where smaller particles dissolve and re-deposit onto larger particles, as a result of their size-dependent solubility. This "ripening" results in a narrower particle size distribution with a larger mean particle size than those of the original particle size distribution.

Semiconductor nanoparticles exhibit photo-induced dissolution effects. When semiconductor nanoparticles are irradiated with the light having a high enough energy, electrons and holes are generated via the photo-excitation of valence band electrons to the conduction bands. The photo-generated electrons and holes facilitate redox reactions and ultimately induce dissolution of nanoparticles that are normally thermodynamically hindered. This light-enhanced dissolution, sometimes called photo-corrosion, is more prominent in smaller nanoparticles due to their high surface areas. It may cause problems in the product integrity of particle-suspension systems or in the environmental safety of accidentally released nanoparticles.

1.3.1.4 Reduced sintering temperature

As the particle size is reduced, metal and ceramic powders start sintering at significantly lower temperatures [61]. Some examples of size-dependent sintering onset-temperatures are listed in Table 1.2 [20,61]. The possible reasons for the reduced sintering temperature in nanoparticles are considered to be (i) enhanced diffusibility of surface atoms and (ii) shorter distances (= particle diameter) for the grain boundaries to move. The reduction of particle size into nanoscale also reduces the time required to form a fully dense sintered body [16].

Table 1.2   Size-dependent sintering onset-temperatures, T

Material	Particle diameter (nm)	Sintering onset-temperature, $T(K)$	T/T_m
TiO_2	13	823	0.40
	40	950	0.46
Fe	30	393	0.21
	2,000	900	0.50

Note: T_m is themelting point of bulk material [20,61].

A fine grain size and full densification are two of the most important factors for obtaining reproducible and improved

properties of consolidated ceramics. The unique sintering characteristics of nanoparticles can be utilized to decrease process cost
in the production of ceramics in solid oxide fuel cells, bioceramic
implants and dental applications. The reduced sintering temperature helps to cast/consolidate the powder without the addition
of sintering aids, undesired crystal structure transformation and
thermal decomposition.

1.3.1.5 Thermodynamically metastable crystal structures

Surface energy is a product of surface area, A, and free surface energy
per unit surface area, γ:

$$\Delta G = \gamma \cdot A, \tag{1.3}$$

where the surface energy, ΔG, is expressed in the form of excess
free energy created by the formation of a new surface. In nanoparticles, the thermodynamics of crystal structures are altered by
the large surface energy derived from the high surface area [35,42].
The crystal structure is determined so as to minimize the total
energy of the system including both bulk and surface energies.

Different crystals have different bulk and surface energies. Even
if the chemical compositions are the same, different polymorphs
(crystal structures) have different bulk and surface energies. In
some cases, a polymorph that is most thermodynamically stable
at room temperature under atmospheric pressure has a surface
energy higher than that of the other metastable polymorphs.
When particle size is reduced to nanoscale, the high surface area
gives rise to high net surface energies and, as a result, structural
transitions occur.

For example, rutile, a polymorph of TiO_2, has a surface
energy (when anhydrous) of 2.2 J/m^2, whereas anatase, another
polymorph of TiO_2, has an anhydrous surface energy of 0.74 J/m^2,
much lower than that of rutile [42]. In a bulk form, rutile is
thermodynamically more stable than anatase. However, when
the size of TiO_2 particles is reduced to nanoscale, the anatase
structure becomes more favourable than the rutile structure, in
order to reduce the total energy of the system. This transition
occurs around 50 nm.

This size effect on the crystal structure is depicted in Fig. 1.4
for ZrO_2 as a function of specific surface area. In the bulk form,

the monoclinic crystal structure is the thermodynamically most stable structure of ZrO_2. However, ZrO_2 nanoparticles tend to have a tetragonal or amorphous structure rather than the monoclinic structure. This is because the tetragonal and amorphous structures have surface energies lower than that of the monoclinic structure. The preferential formation of metastable crystal structures in TiO_2 and ZrO_2 was experimentally demonstrated [42].

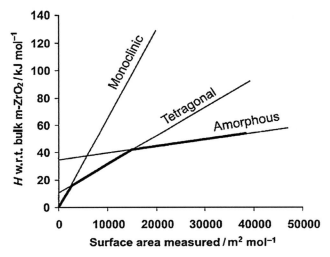

Figure 1.4   Phase stability crossover of nanocrystalline ZrO_2. The thick red line segments indicate the energetically stable phases. Reprinted with permission from Ref. [54]. Copyright (2005) John Wiley and Sons.

Table 1.3 lists materials that show preferential formation of metastable phases in nanoscale. Since the surface energy depends on the crystal facets, the critical size to cause phase transition also depends on particle morphology [3]. It is reported that the critical pressure to cause crystal phase transition in nanoparticles is also influenced by particle size in many materials [60].

The size dependence of crystal structures has a significant implication in many applications of nanoparticles. One example is the production of dense ZrO_2 ceramics for dental or biomedical implants. The sintering of ZrO_2 requires temperatures over 1000°C. At such high temperatures, the tetragonal phase is more stable than the monoclinic phase. Hence, sintering of micron-sized ZrO_2

powder leads to crystal phase transformation from monoclinic to tetragonal. This often results in crack formation in the sintered body because of the volume change associated with the crystal phase transition. The use of ZrO_2 nanoparticles with the tetragonal crystal structure will overcome this problem.

Table 1.3   Metastable crystal phase formation in nanoparticles

Material	Particle diameter	Observed crystal structure in nanoparticles	Thermodynamically stable structure in a bulk form	Ref.
TiO_2	13 nm	Anatase	Rutile	[30]
ZrO_2	40 nm	Tetragonal	Monoclinic	[50]
Al_2O_3	30 nm	Cubic (γ-)	Hexagonal (α-)	[35]
Y_2O_3	10 nm	Monoclinic	Cubic	[18,19]
Fe_2O_3	<30 nm	Cubic (γ-)	Hexagonal (α-)	[43]
Cu_2O	<25 nm	Cubic	Monoclinic	[48]
Ce_2S_3	10–80 nm	Cubic	Orthorhombic	[65]
ZnS	2.8 nm		Hexagonal	[53]
$BaTiO_3$	<30 nm	Cubic	Tetragonal	[23]

Note: References listed are examples only.

Another example is the production of TiO_2 nanoparticles for UV screening applications. Visible transparency is essential to many of UV screening applications. This can be achieved by using nanoparticles as a UV screening agent, owing to their low light-scattering powder. However, the synthesis of nanoscale TiO_2 often results in the anatase phase that has much higher photoactivity than rutile, resulting in unfavourable side effects in UV screening applications.

It is also reported that the luminescence characteristics of rare earth doped Y_2O_3 nano-phosphors depend on the crystal structures of the Y_2O_3 host and thus can show size dependency [13].

1.3.1.6   Luminescent quenching

Luminescence is the phenomenon in which external energy input causes the emission of light from materials. Nanoparticles with luminescence properties have various uses in electronics and biomedical applications. The major drawback of nanoscale phosphors and luminescent materials is their low quantum efficiency (the ratio between input energy and output emission energy) compared to that of larger particles or bulk crystals. This luminescence quenching is mainly due to the surface defects acting as non-radiative recombination sites for the photo-excited electrons and holes, wherein excited charges are trapped by defects and combined to generate heat instead of light. As the particle sizes are reduced, the specific surface area increases, and hence the number of surface defects increases. Thus, a reduction of particle size normally results in reduced luminescence intensity.

In some cases, surface defects give rise to additional luminescence frequency, especially in semiconductor nanoparticles. In metal oxide nanoparticles, surface defects such as oxygen vacancies and cation (metal ion) vacancies can create additional electron energy levels in the bandgap. When the electrons are transferred from a defect energy level to the lower valence band energy level, the energy difference is emitted as light [1]. As the particle size is reduced to gain larger surface areas, this defect luminescence dominates over the bulk luminescence. The surface defect luminescence has been extensively investigated for ZnO and SiO_2 nanoparticles [15]. Surface defect luminescence is sensitive to the environment surrounding the nanoparticles and chemical species on the surfaces, so that it can be used for the detection of gas or biological molecules. For the applications of luminescent nanoparticles in electronic devices, it is critical to suppress the surface-defect-derived luminescence.

1.3.1.7 Surface treatments

The surface chemistry of nanoparticles is extremely sensitive to the following factors:

- synthesis routes
- crystallinity
- crystal structure
- the crystal facet exposed on the particle surface
- molecules attached on the surface (particle growth limiting agents, dispersants, etc.)

These factors affect the surface-related unique characteristics of nanoparticles including solubility, melting and sintering temperatures, luminescence intensity and crystal structure. This fact implies that even if the mean particle size, size distribution and specific surface area are the same, nanoparticles prepared using different techniques may exhibit very different properties. In this regard, it is difficult to predict the properties of commercial nanoparticle products from the material safety data sheet (MSDS) or even from product specification sheets, as these documents rarely contain sufficient information on the surface chemistry of the products. By the same token, tailoring surface chemistry by applying ligands, impurities and other foreign materials enables the control of the unique surface-related properties of nano-particles without changing the particle size and shape [15].

1.3.2 Small Light-Scattering Power

Another widely used property of nanoparticles is the high optical transmission of particle suspension systems. The turbidity of particle dispersion systems results from light scattering by the suspended particles. When the diameter of particles becomes smaller than the optical wavelength, the scattering of light by the particles becomes negligible. As a result, the turbidity of the system decreases and transparency increases.

This effect is described by the Rayleigh approximation of Mie scattering theory as expressed in Eq. 1.4 [7]:

$$I_{\text{scat}} = I_0 \frac{8\pi^4 d^6}{r^2 \lambda^4} \left(\frac{m^2 - 1}{m^2 + 2}\right)^2 (1 + \cos^2\theta), \qquad (1.4)$$

where I_0 is the intensity of incident light, I_{scat} is the intensity of scattered light by a particle, d is the particle diameter, λ is the wavelength of incident light, r is the radial distance, θ is the scattering angle and m is the relative refractive index defined as

$$m = \frac{n_{\text{particle}}}{n_{\text{media}}}, \qquad (1.5)$$

where n_{particle} and n_{media} are the refractive index of the particle and its surrounding medium, respectively. This approximation is valid when the particle diameter and the optical wavelength fulfil the following condition:

$$\frac{2\pi d}{\lambda} << 1. \qquad (1.6)$$

As can be seen in Eq. 1.4, the light scattering efficiency of a particle decreases proportionally to the 6th power of the particle diameter. As such, a slight decrease in particle size leads to drastic change in the turbidity and in turn the transparency of nanoparticle dispersion systems. For instance, when the particle diameter is halved from 100 nm to 50 nm, the light scattering intensity is reduced to 3% of the original value.

Normally, nanoparticles smaller than 100 nm meet the condition in Eq. 1.6 for the scattering of visible light (λ: 400–750 nm). Hence, when nanoparticles of this size range are well dispersed in a transparent medium, the particle suspension system will appear highly transparent. This effect is most useful in many applications of nanoparticles, including phosphorescent panels, UV screening coatings and sunscreens, transparent polymer nanocomposites and diesel fuel additives, where transparency has high commercial value.

1.3.3 Phonon Confinement Effects

Rare earth oxides and oxysulfides make excellent nano-phosphors by the doping with rare earth ions such as Eu^{3+}. The localized electronic states of the doped rare earth ions are influenced by the nanoscale dimension of particles through electron–phonon interactions.

In a small particle, phonons (= sound, the vibration of atoms and molecules) have to form standing waves, or they will be cancelled out by self-interference while moving back and forth within a confine space. The standing waves can have only certain wavelengths or frequencies in a small box. This fact has two important implications:

(i) Since energy is directly related to wavelength, confined phonons can only have certain discrete energy values.

(ii) As the size of the box decreases, the longest wavelength that the standing wave can take, becomes shorter (Fig. 1.5). Hence, the lowest energy associated with the confined phonons increases as the particle size decreases.

To put this in technical terms, phonon confinement in nanoscale particles modifies the phonon density of states from continuous to discrete, resulting in a lack of low-frequency phonon modes [44]. Since low-frequency phonons largely contribute to the non-radiative relaxation between the closely spaced crystal-field energy levels, the phonon-confinement effect gives rise to a significant change in the luminescence dynamics [32]. For example, the fluorescence lifetime of nano-phosphors can be significantly longer than that of bulk materials due to the phonon-confinement effect [31].

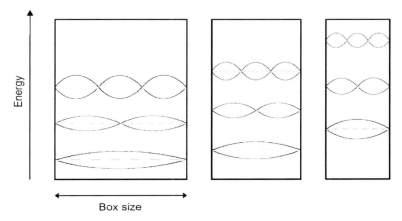

Figure 1.5 Schematic diagram of standing waves in a confined space. It can be noted that the longest allowed wavelength is reduced as the box size decreases.

1.3.4 Nanoparticle Suspension Systems

1.3.4.1 Distance between particles

In many of the applications, nanoparticles are used either in a solid matrix as part of a nanocomposite or in a liquid medium as a nanoparticle dispersion system. When the particle size is reduced to nanoscale, the environment around the nanoparticles in those nanocomposites and nanosuspension systems needs to be considered "nano-scopically".

Of particular importance is the distance between nano-particles. Assuming that spherical nanoparticles form a uniform spatial distribution in a nanocomposite, the surface-to-surface distance between particles can be given as

$$\frac{D}{d} = \left(\frac{A}{f_{\mathrm{v}}}\right)^{1/3} - 1 \qquad (1.7)$$

where D is the mean surface-to-surface distance between spherical particles, d is the particle diameter and f_V is the volume fraction of particles in the media [58]. The A value is the maximum achievable volume fraction of nanoparticles in the system.

Equation 1.7 is plotted in Fig. 1.6 for three hypothetical particle configurations, namely, random close packing (A = 0.64), face-centred cubic packing (hexagonal close packing, A = 0.74), and simple cubic packing (A = 0.52). Note that the y-axis is the distance between particle surfaces (n.b., not a core-to-core distance), D/d, with the particle diameter as a length unit. It is evident that, as the particle's volume fraction increases, the mean surface-to-surface distance between particles quickly becomes comparable to the particle diameter. Even if the particle concentration is as low as 1 vol% (f_V = 0.01), the mean gap distance is only ~3 times larger than the particle diameter. For micron-sized particles, this gap is still large, in micron scale. However, for nanoparticles such as commercial silver colloids of ~10 nm in diameter, the gap between particles is only ~30 nm at this relatively low particle concentration.

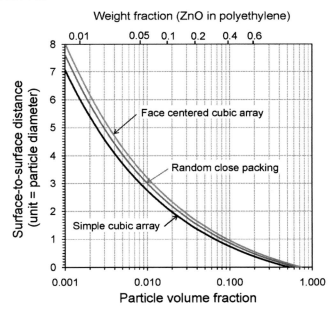

Figure 1.6 Mean gap distance between particles (D/d) as a function of particle volume fraction.

Also presented in Fig. 1.6 as a practical example, is the surface-to-surface particle distance as a function of weight fraction for ZnO nanoparticles dispersed in polyethylene. Some sunscreens or plastic products contain up to 20 wt% (0.2 weight fraction) of ZnO in organic matrices as a UV screening agent. As can be seen in Fig. 1.6, when ZnO nanoparticles of 30 nm in diameter are dispersed in polyethylene at 20 wt% particle concentration, the mean gap distance between particles is only ~45 nm.

This close proximity between nanoparticles in particle suspension systems and nanocomposites can cause serious practical problems in the handling of nanoparticles. Some examples of the implications of short particle-to-particle distances will be discussed below.

1.3.4.2   Particle dispersion

In order to take advantage of the unique properties of nanoparticles, it is critical to assure that particles are well separated from each other. This means that a uniform distribution of the host matrix material between particles is required in nanocomposites and particle-suspension systems. However, when the gap between particles is as small as the nanoparticles themselves, there is a high risk of particle agglomeration. The high surface reactivity and enhanced van der Waals forces of nanoscale particles aggravates the problem. To ensure the high degree of particle dispersion, it is common to introduce polymeric surfactants or dispersants into the system. However, the concentration and molecular weight of those polymeric additives have to be carefully selected. When the particle concentration is too high, a long-chain surfactant with a high molecular weight may bridge particles and cause particle flocculation. In addition, if the host matrix material is a polymer with a high molecular weight, the uniform distribution of the polymer molecules in the nanoscale gap between particles may be challenging to achieve.

1.3.4.3   Rheology

When manufacturers replace conventional micron-sized powder with nanoparticles, they tend to simply swap the raw powder ingredient in the production line, with no change in the process equipment or process parameters. However, this simple-substitution approach often causes serious problems. Even if the same powder concentration is used, nanoparticles give significantly higher viscosity than micron-sized particles, due to the extremely short surface-to-surface distance between particles and a higher number of particles in the unit volume. By reducing the particle size, a normally "watery" particle suspension becomes a thick gel. Hence, a new production facility or processing equipment may be required for the handling of nanoparticles.

Another consideration in handling nanoparticles is that by increasing the concentration of nanoscale particles, the flow behaviour of a particle suspension system changes from Newtonian to shear-thin and finally to shear-thick. The transition of the flow behaviour occurs at much lower particle concentrations than the conventional large particles.

1.3.4.4 Light scattering

Particle sizing: The dynamic light scattering technique is a common method to measure the size distribution of nanoscale particles (see Chapter 9 for more detail). However, at high particle concentration, the technique faces a severe limitation. This technique assumes that the Brownian motion of nanoparticles causes the Doppler shift in the frequency of scattered light. Since the speed of particle diffusion depends on particle size, the detected spectrum of Doppler shift gives the information about the particle size distribution. This measurement principle works well assuming that the light is scattered by a particle only once (single light scattering). However, when the particle concentration becomes high, multiple light scattering effects become non-negligible, which reduces apparent particle size.

Transparency: When both particle size and the distance between particles become much smaller than the wavelength of light, the nanocomposite or nanoparticle suspension system appears as a uniform material to the probe light. As a result, light scattering by particles becomes negligible. In addition, multiple light scattering effects increase the coherent forward light scattering [63]. As such, high particle concentration results in unexpectedly high optical transmittance. This phenomenon is useful in the applications of nanoparticles for transparent functional nanocomposite films, where a large quantity of nanoparticles is required to gain high functionality while retaining high transparency of the composites. Figure 1.7 shows such an example of a refractive-index-engineered nanocomposites film. In the particle concentration range between $f_V = 0.08$ and 0.2, the refractive index of the nanocomposites was continuously modified as the particle concentration increased, while optical transmittance was hardly altered, [63].

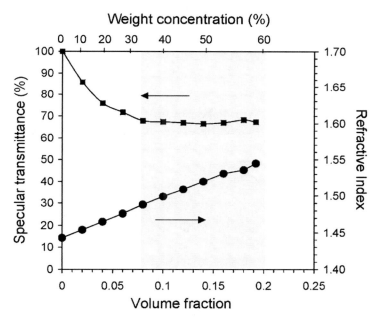

Figure 1.7   Refractive indices and specular transmittance of ZnO/caprylic capric triglyceride hybrid films at 550 nm. Reproduced with permission from Ref. [63]. Copyright (2008) Wiley-VCH.

1.4   Characteristics of Specific Types of Nanoparticulate Materials

In Section 1.3, the characteristics and unique properties of nanoparticles that are common to all material types including metals, semiconductors and insulators were discussed. This section reviews the unique properties of nanoparticles specific to the material types.

1.4.1 Semiconductor Nanoparticles

Quantum size effects that can be observed in semiconductor nanoparticles are one of the most striking properties of nano-particles. Quantum size effects, sometimes referred to as quantum confinement effects, are commonly described as the effects arising from electrons and holes confined in a small space. In quantum mechanics, electrons and holes are treated as "matter waves". Owing to the requirement for matter waves to form standing waves in a confined space, the electrons and holes can have only discrete energy levels, in the same way as confined phonons (see Section 1.3.3). Like the situation with phonons, the lowest energy that the confined electrons can take (ground state energy level) increases as the particle size decreases. Nanoparticles that exhibit quantum size effects are called quantum dots [28].

In semiconductor materials, the energy gap between the valence band (the highest energy band occupied by electrons) and the conduction band (lowest unoccupied energy band) is relatively small so that electrons in the valence band can be readily excited to the conduction band by heat or UV-light. The energy gap between these two bands is called bandgap energy. Any light having energy larger than the bandgap energy can be absorbed by semiconductors and its energy used for the excitation of electrons. The excited electron leaves a hole in the valence band. Since electrons and holes have opposite electric charges, they are bound together via an electrostatic attractive force and form a pair known as an exciton. When the hole and the electron recombine, a photon (light) is emitted, which results in a phenomenon called luminescence. In bulk semiconductors, the lowest energy required to create an exciton is roughly the same as the bandgap energy, and the light energy emitted upon the collapse of an exciton is also nearly the same as the bandgap energy.

When the size of semiconductor nanoparticles becomes smaller than the critical size for exciton, called the exciton-Bohr diameter, the exciton shows quantum confinement effects. In such quantum dots, the minimum energy to create excitons is the sum of the bulk bandgap energy and the ground state energy level of a confined exciton (Fig. 1.8). This minimum energy determines the wavelength of light to be absorbed and emitted. The minimum energy can be increased by reducing the size of the quantum dots.

In this way, the excitation energy and, in turn, emission energy of light can be controlled by changing the size of nanoparticles. Hence, the colour of emitted light can be tailored using the same semiconductor materials, by changing the particle size of the quantum dots. Figure 1.9 shows the example of CdSe quantum dots [11]. As the size of the CdSe quantum dots increases, the colour of the luminescence

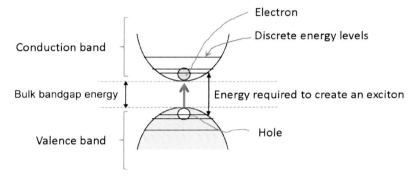

Figure 1.8   Schematic diagram of excitation energy in semiconductor quantum dots.

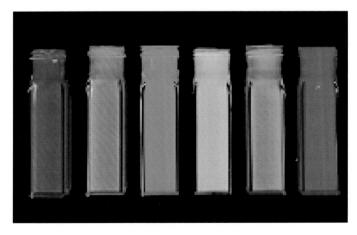

Figure 1.9   Colour photograph demonstrating the wide spectral range of bright fluorescence from different size samples of (CdSe) ZnS ranging from 2.3 to 5.5 nm in diameter. Their photoluminescence peaks occur at (going from left to right) 470, 480, 520, 560, 594 and 620 nm. Reprinted with permission from Ref. [11]. Copyright 1997 American Chemical Society.

changes in a continuous manner from blue through green, yellow, orange, to red. This size-dependent emission can be used in many applications from quantum lasers to biomarkers [9].

Different semiconductor materials have different exciton Bohr diameters and bandgap energies. As such, by using different semiconductor materials in different particle sizes, a wide range of luminescence spectra can be obtained [37].

1.4.2   Metal Nanoparticles

Plasmonics in metal nanoparticles is another often-quoted example of the unique properties of nanoparticles [34,39]. The size reduction of metal particles results in drastic changes in the electronic properties, as the motion of the electrons is restricted in the finite dimension of particles. As a result, metal nanoparticles absorb and scatter light much more strongly than bulk materials. The enhanced optical response of metal nanoparticles is due to the collective oscillation of electrons known as plasmons that are excited on the particle surface by external light. The surface plasmon on nanoparticles has a characteristic resonance frequency. When incoming light has the same frequency as the plasmon resonance frequency, the light is strongly absorbed by metal nanoparticles and then re-emitted in the same frequency.

The frequency of the surface plasmon is influenced by the type of metal, particle size, particle shape, the dielectric environment on the surface and inter-particle spacing [27,45]. For example, spherical silver and gold nanoparticles appear yellow and red, respectively, very different to their bulk colours. The colour of silver nanoparticles can be further tailored by controlling the particle shape (Fig. 1.10) [59]. The colour of gold nanorods changes with the aspect ratio (Fig. 1.11) [40]. The colour of gold nano-shells depends on the shell-thickness [47].

The strong coupling between light and plasmons in nanoparticles gives rise to new phenomena including optical force enhancement and the light-controlled anisotropic growth of nanoparticles [27]. The surface plasmonics of nanoparticles find many applications in optics, opto-electronics, magneto-optics, chemical and biological sensing and tagging, solar cells, diagnostic medical imaging, carriers of quantum bits and so on [25].

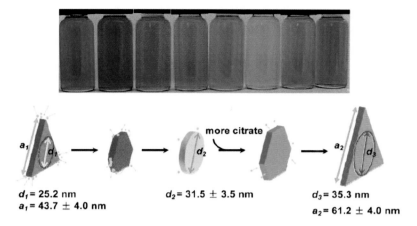

d_1 = 25.2 nm
a_1 = 43.7 ± 4.0 nm

d_2 = 31.5 ± 3.5 nm

d_3 = 35.3 nm
a_2 = 61.2 ± 4.0 nm

Figure 1.10 Colour of silver nanoparticles with different shapes and sizes. The lower illustration describes the photoconversion of nanoprism to nanodisk and reconstruction of silver nanoprism during the modification of a synthesis parameter. Reprinted with permission from Ref [59]. Copyright 2009 American Chemical Society.

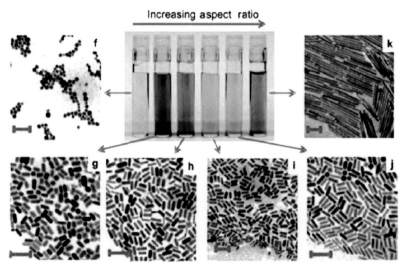

Figure 1.11 Photographs of aqueous solutions of gold nanorods as a function of aspect ratio. Reprinted with per-mission from Ref. [40]. Copyright 2008 American Chemical Society.

1.4.3 Carbon-Based Nanomaterials

1.4.3.1 Fullerenes

Fullerenes are molecules consisting of carbon atoms with a hollow cage-like structure (Fig. 1.1). Sometimes carbon nanotubes and graphenes are included in fullerene families. In this section, only spherical fullerenes also known as Bucky balls or buckminster-fullerenes are discussed. They are named after Richard Buckminster Fuller, as the atomic arrangement in fullerenes resembles the structure of Fuller's famous geodesic dome.

There are many "magic numbers" of carbon atoms that can form stable hollow cage structures. However, the majority of commercially used spherical fullerenes is C60 (buckminsterfullerene), in which a single layer of 60 carbon atoms forms a closed-cage structure in icosahedral-symmetry. The layer contains 20 hexagonal and 12 pentagonal rings and has the same appearance as a soccer ball.

Fullerenes have properties useful to many applications as listed in Table 1.4. Their small size (diameter ~1 nm) is useful for potential medical applications, such as targeted gene or drug delivery and imaging contrast agents [2]. Some commercial products containing C60 have already appeared, especially in the cosmetic, solar cell and polymer composite sectors.

Table 1.4   Key attributes of spherical fullerenes [36]

Properties/ characteristics	Descriptions and applications
n-Type semiconductors	An active component in *pn*-junctions in organic photo-voltaic cells and organic electronic devices such as transistors, light emitting diodes and photo-detectors. C60 inks for solar cell applications are currently on the market.
Anti-oxidant	Estimated to be 100 times more effective than current leading antioxidants such as Vitamin E. Health and personal care applications are considered and skin care product are already commercialised.
Free radical scavenging	Controlling the radical-related neurological damage of diseases as Alzheimer's disease.
Super-conductors	Doped C60 shows superconductivity but only at extremely low temperatures (38K for Cs-doped C60) and hence the commercial application of this property has not been explored.

1.4.3.2 Carbon nanotubes

Carbon nanotubes are cylindrical-shaped molecules consisting of only carbon atoms. The structure is made of single or multiple graphene sheets rolled up into a tube form. Carbon nanotubes can have many structures in terms of diameter, the number of wall layers, length and straightness. Single-wall carbon nanotubes (SWCNTs) have a typical diameter of ~1 nm and multi-wall carbon nanotubes (MWCNTs) can be as large as ~50 nm in diameter. A tube length can be 100,000,000 times of the diameter [67].

The electrical characteristics of SWCNTs depend on the chirality, i.e. the way the graphene sheet is rolled [71]. As shown in Fig. 1.12, there are three types of structures. The "armchair" structures have a metallic nature. The "zigzag" structures can be either semi-metallic or semiconducting, depending on the diameter. The tubes with a chiral angle between 0° and 30° are either semi-metals or semiconductors. Metallic SWCNTs have high electrical conductivity similar to copper. In addition, SWCNTs possess excellent mechanical properties, as listed in Table 1.5 along with other unique attributes. For many electronic applications, the selectivity of chirality is critical [72]. MWCNTs share excellent mechanical and thermal properties with SWCNTs. Their electrical properties are more complex than SWCNTs but are reported to be always electrically conductive.

Table 1.5   Key attributes of single-wall carbon nanotubes [73]

Properties/ characteristics	Descriptions and applications
High stiffness	Young's moduli of SWCNTs or MWCNTs are ~5 times higher than steel.
High strength	Tensile strength of CNTs are 10–40 times higher than steel.
High hardness	The bulk modulus of CNTs can be ~500 GPa, comparable or slightly higher than diamond.
High elasticity	~18% elongation to failure.
High electrical conductivity	Room temperature resistivity is similar to or lower than the in-plane resistivity of graphite. The electric current density can be 1,000 times higher than copper.
High thermal conductivity	CNTs have high thermal conductivity, ~10 times higher than copper, along the tube axis, but significantly low thermal conductivity perpendicular to the tube axis.

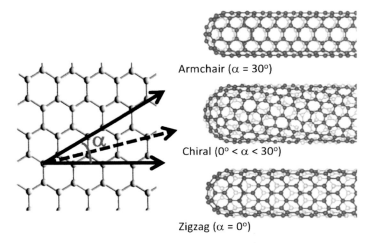

Armchair (α = 30°)

Chiral (0° < α < 30°)

Zigzag (α = 0°)

Figure 1.12 Chirality of SWCNTs, with different wrapping angles, α.

Although CNTs are depicted as straight tubes in many illustrations, in reality, CNTs have different degrees of entanglement and bends. As expected, different structures (diameter, wall thickness, chirality, branching, etc.) result in different properties and toxicity [52]. The degree of agglomeration and the impurity levels stemming from the growth catalysts also largely affect these properties. In many applications where CNTs are embedded in host matrices, surface modification is applied on the tubes, which again alters some of the properties. Owing to the diverse range of unique properties, the applications of CNTs cover many industries, including electronics, automotive, medical, energy and construction.

1.4.3.3 Graphenes

Graphene is a one-atom-thick planar sheet of carbon atoms packed in a honeycomb crystal lattice (Fig. 1.1) [14]. For a long time, it was believed that freestanding graphene was too unstable to exist. However, in 2004, freestanding graphene was successfully prepared in a laboratory [6,46]. The research on the properties, as well as the synthesis methods, of graphene is still in the early stages. Some properties of graphenes are listed in Table 1.6. Investigations on the applications of graphene in conductive polymer nanocomposites for electronics, drug delivery systems and organic pollutant absorption materials have been steadily progressing.

Table 1.6 Key attributes of graphene [14]

Properties/characteristics	Descriptions and applications
High electron mobility	The electron mobility is independent of temperature between 10 K and 100 K. The theoretical electrical resistivity is lower than silver, the lowest resistivity material known at room temperature.
High strength	The tensile strength of graphene is 130 GPa, higher than CNTs. The fracture strength is 100 times higher than that of steel.
High thermal conduction	At near room temperature, the measured thermal conductivity of graphene exceeds that of CNTs.

1.4.4 Magnetic Nanomaterials

Magnetic nanoparticles exhibit unusual properties compared to the bulk magnetic materials [17,33]. The magnetic characteristics of nanoparticles are strongly influenced by finite-size and surface effects.

1.4.4.1 Magnetic materials

Magnetism in materials stems from the spin magnetic moment of electrons. The overall magnetic moment of the material is determined by the distribution of the orientation of spin magnetic moments across the material. Every material has electrons and, hence, can respond to an external magnetic field in one way or another. However, the phrase "magnetic materials" is commonly applied to ferromagnetic, ferrimagnetic and paramagnetic material. Those special materials can be magnetised along the same direction as the external magnetic field (i.e. positive magnetic susceptibility). As a result, they are attracted to external magnetic fields.

Paramagnetic materials have randomly oriented spins. When an external magnetic field is applied, the spins align with the magnetic field to exhibit magnetisation. However, when the external field is removed, the spin orientation becomes random again and paramagnetic materials cease to be "magnets".

On the other hand, ferromagnetic and ferrimagnetic materials can form permanent magnets. In those materials, the spin magnetic moments tend to align spontaneously along easy-magnetisation axes that are dictated by the crystal structure. This parallel alignment of spins occurs in microscopic regions called magnetic domains. Normally many magnetic domains exist in bulk materials. The spins are aligned within each domain, but the spins in separate domains point in different directions. When a ferromagnetic material is magnetized by an external magnetic field, the domains that have the spin direction parallel with the external field, increase their volume at the expense of the volume of other domains (Fig. 1.13). This gives raise to the overall magnetisation in the same direction as the external magnetic field. In ferromagnetic and ferrimagnetic materials, the change in the domain volume requires a certain amount of energy and is somehow restricted. In a technical term, this phenomenon is called domain wall pinning. When the external field is turned off, each domain does not regain the original volume and the overall magnetisation stays in the same direction as the external magnetic field.

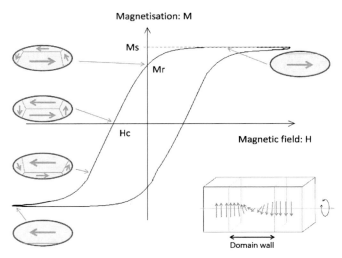

Figure 1.13   Magnetic hysteresis loop of ferromagnetic materials and associated domain structures. Hc: coercivity, Mr: remnant magnetisation, and Ms: saturation magnetisation. The inset shows the spin orientation within a domain wall.

The response of overall magnetisation to the external field is often described using a hysteresis curve (Fig. 1.13). The area that is enclosed in the hysteresis loop represents the quantity of energy that is lost during a cycle in which the domain volume or spin directions in the domains are changed. The lost energy is normally converted to heat. This effect is used for the hyperthermic treatment of cancer cells with nanoscale permanent magnets. In this treatment, magnetic nanoparticles are targeted to cancer cells and an alternating external magnetic field is applied to them. This makes the nano-magnet generate heat in the localised area around the cancer cells, resulting in the death of the cancer cells by overheating. The magnetic field required to cancel the overall magnetisation is called coercivity. The highest magnetisation achievable is called the saturation magnetisation. The higher the coercivity and saturation magnetisations, the more energy loss occurs during the hysteresis cycle.

1.4.4.2 Finite size effect: single domain

A domain wall is a region between two magnetic domains. In a domain wall, the direction of spin changes gradually across the wall thickness (Fig. 1.13). Since the spins are forces to have slightly off-aligned direction to each other, certain energy is required to create domain walls. The thickness of domain walls varies depending on the magnetic materials, but is typically in the range of ~100 nm.

When the size of a magnetic nanoparticle is reduced to the dimension similar to the thickness of domain walls, only one magnetic domain can be formed in the particle, because there is no sufficient room in the particle to create domain walls. The reversal of the magnetisation in a single domain nanoparticle requires higher energy than the multi-domain bulk material, as the reversal needs to rely on the coherent spin-rotation within the domain instead of domain-wall movement. Consequently, as the particle size is reduced to a single domain size, the coercivity of magnetic particles becomes higher (Fig. 1.14). The size range where a ferromagnetic material forms a single domain is normally between 10 and 100 nm.

Single domain magnetic nanoparticles offer the possibility of effective targeted cancer treatment through hyperthermia. Because of their small size, magnetic nanoparticles can be administered directly into the blood stream to circulate around the body. Surface treatment with a certain protein enables targeted accumulation of the magnetic nanoparticles only in pathological areas or organs.

The single domain nature of the magnetic nanoparticles ensures that effective heat generation can be induced by the alternating external magnetic field, only in the targeted region, which, in turn, resulting in the death of tumour cells by localised heat. Other important applications include high-density magnetic recording media.

1.4.4.3   Finite size effect: superparamagnetism

When the size of magnetic nanoparticles is reduced further from the single domain size, thermal energy exceeds the energy required to align spins. As a result, the direction of magnetisation in the nanoparticles fluctuates between two orientations anti-parallel to each other along an easy-magnetisation axis. This causes the magnetic nanoparticles to behave like a paramagnetic material and their coercivity becomes zero (Fig. 1.14). However, their magnetic susceptibility is significantly higher than that of the conventional paramagnetic materials. This means that superparamagnetic materials respond to an external magnetic field much more strongly than normal paramagnetic materials.

Superparamagnetic materials have a combination of strong responsivity to magnetic fields and the ability to become non-magnetic in the absence of an external magnetic field. This unique characteristic offers many applications including ferro-fluid, magnetic separators, magnetic resonance imaging contrast agents and other biomedical applications.

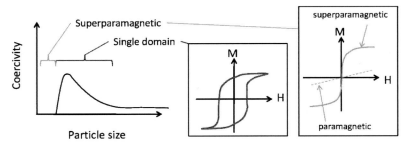

Figure 1.14   Particle size effect on the coercivity of magnetic nanoparticles.

1.4.4.4   Surface effect

When the size of magnetic nanoparticles is further reduced, saturation magnetisation decreases. This effect is caused by the fact that the surface of magnetic materials often has a random spin orientation (spin glass) that does not contribute to magnetisation. When the particle size is reduced, the relative volume of the spin glass layer on the surface increases compared to the total volume of the particle. Hence, the saturation magnetisation varies with particle size, according to Eq. 1.8 [62]:

$$\frac{M_S}{M_{S0}} = \frac{(d-2t)^3}{d^3},$$ (1.8)

where d is the particle diameter, t is the thickness of the surface layer, M_S is the particle's saturation magnetisation and M_{S0} is the bulk saturation magnetisation. The thickness of the spin glass is nearly independent (~1 nm) of the particle diameter. Figure 1.15 shows the effect of particle size on the saturation magnetisation of $MnFe_2O_4$ and γ-Fe_2O_3 nanoparticles. This effect determines the lower size limit of superparamagnetic nanoparticles usable in many applications.

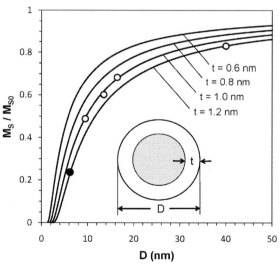

Figure 1.15 Solid curves: particle-size (D) dependence of normalised saturation magnetisation (M_S/M_{S0}) for various values of shell thickness (t), calculated on the basis of Eq. (1.8); M_{S0} denotes the saturation magnetisation for bulk material. Filled circle: data point for γ-Fe_2O_3 nanoparticles. Open circles: data points for $MnFe_2O_4$ nanoparticles. Inset: schematic picture of a particle, consisting of a ferrimagnetic core (shaded region) and a spin-glass shell. Reprinted from Ref. [62]. Copyright (2011) with permission from Elsevier.

1.5 Summary

Nanoparticulate materials exhibit unique properties that are unobtainable from their bulk counterparts. For example, they have very high specific surface areas that give rise to enhanced reactivity and solubility, reduced melting and sintering temperatures, as well as altered crystal structures. Their small physical size induces quantum confinement effects in phonons and electrons to create many new optical and electronic properties. Magnetic nanoparticles exhibit unusual properties resulting from their high surface areas and small diameter. An understanding of these unique properties of nanomaterials is not only critical to the successful development of nano-enabled commercial products but also essential to the standardisation, regulation and safety assessment of nanomaterials.

References

1. Allan, G., Delerue, C., and Lannoo, M. (2006). Nature of luminescent surface starts of semiconductor nanocrystallites, *Phys. Rev. Lett.*, **76**, pp. 2961–2964.

2. Bakry, R., Vallant, R. M., Najam-ul-Haq, M., Rainer, M., Szabo, Z., Huck, C. W., and Bonn, G. K. (2007). Medical applications of fullerenes, *Int. J. Nanomed.*, **2**, pp. 639–649.

3. Barnard, A. S., and Zapol, P. (2004). A model for the phase stability of arbitrary nanoparticles as a function of size and shape, *J. Chem. Phys.*, **121**, 1775770.

4. Batsanov, S. S. (2011). Size effect in the structure and properties of condensed matter, *J. Struct. Chem.*, **52**, pp. 602–615.

5. Bellucci, S. (2005). Carbon nanotubes: physics and applications, *Phys. Status Solidi (c)*, **2**, pp. 34–47.

6. Berger, C., Song, Z., Li, T., Li, X., Ogbazghi, A. Y., Feng, R., Dai, Z., Marchenkov, A. N., Conrad, E. H., First, P. N., and de Heer, W. A. (2004). Ultrathin epitaxial graphene: 2D electron gas properties and a route toward graphene-based nanoelectronics, *J. Phys. Chem. B*, **108**, pp. 19912–19916.

7. Bohren, C. F., and Huffman, D. R. (1998). *Absorption and Scattering of Light by Small Particles* (Wiley-VCH, Weinheim).

8. Buffat, P., and Borel, J. P. (1976). Size effect on the melting temperature of gold particles, *Phys. Rev. A*, **13**, pp. 2287–2298.

9. Chan, W. C. W., Maxwell, D. J., Gao, X., Bailey, R. E., Han, M., and Nie, S. (2002). Luminescent quantum dots for multiplexed biological detection and imaging. *Curr. Opin. Biotechnol.*, **13**, pp. 40–46.

10. Cheng, Y. T., Rodak, D. E., Wong, C. A., and Hayden, C. A. (2006). Effects of micro- and nano-structures on the self-cleaning behaviour of lotus leaves, *Nanotechnology*, **17**, pp. 1359–1362.

11. Dabbousi, B. O., Rodriguez-Viejo, J., Mikulec, F. V., Heine, J. R., Mattoussi, H., Ober, R., Jensen, K. F., and Bawendi, M. G. (1997). (CdSe) ZnS core–shell quantum dots: synthesis and characterization of a size series of highly luminescent nanocrystallites, *J. Phys. Chem. B*, **101**, pp. 9463–9475.

12. Daniel, M. C., and Astruc, D. (2004). Gold nanoparticles: assembly, supramolecular chemistry, quantum-size-related properties, and applications toward biology, catalysis, and nanotechnology, *Chem. Rev.*, **104**, pp. 293–346.

13. Dosev, D., Guo, B., and Kennedy, I. M. (2006). Photoluminescence of Eu^{3+}: Y_2O_3 as an indication of crystal structure and particle size in nanoparticles synthesized by flame spray pyrolysis, *Aerosol. Sci.*, **37**, pp. 402–421.

14. Geim, A. K., and Novoselov, K. S. (2007). The rise of graphene, *Nat. Mater.*, **6**, pp. 183–191.

15. Gong, Y., Andelman, T., Neumark, G. F., O'Brien, S., and Kuskovsky, I. L. (2002). Origin of defect-related green emission from ZnO nano-particles: effect of surface modification, *Nanoscale Res. Lett.*, **2**, pp. 297–302.

16. Groza, J. R. (2007). *Nanostructured Materials, Processing, Properties and Applications*, 2nd ed. Koch, C. C. (ed.), Chapter 5, "Nanocrystalline powder consolidation methods" (William Andrew Publishing, Norwich, New York) pp. 173–234.

17. Gubin, S. P. (2009). *Magnetic Nanoparticles* (Wiley-VCH, Weinheim).

18. Guo, B., Harvey, A. S., Neil, J., Kennedy, I. M., Navrotsky, A., and Risbud, S. H. (2007). Atmospheric pressure synthesis of heavy rare earth sesquioxides nanoparticles of the uncommon monoclinic phase, *J. Am. Ceram. Soc.*, **90**, pp. 3683–3686.

19. Guo, B., Harvey, A., Risbud, S. H., and Kennedy, I. M. (2006). The formation of cubic and monoclinic Y_2O_3 nanoparticles in a gas-phase flame process, *Philos. Mag. Lett.*, **86**, pp. 457–467.

20. Hahn, H., Logas, J., and Averhack, R. S. (1990). Sintering characteristics of nanocrystalline TiO_2, *J. Mater. Res.*, **5**, pp. 609–614.

21. Hayashi, T., Tanaka, K., and Haruta M. (1987). Selective vapour phase epoxidation of propylene over Au/TiO_2 catalysts in the presence of oxygen and hydrogen, *J. Catal.*, **178**, pp. 566–575.

22. Holister, P., Weener, J. W., Román-Vas, C., and Harper, T. (2003). *Nanoparticles: Technology White Paper Nr 3* (Cientifica, UK).

23. Hoshina, T., Kakemoto, H., Tsurumi, T., Wada, S., and Yashima, M. (2006). Size and temperature induced phase transition behaviors of barium titanate nanoparticles, *J. Appl. Phys.*, **99**, article number 054311.

24. Iijima, S. (1991). Helical microtubules of graphitic carbon, *Nature*, **354**, pp. 56–58.

25. Jain, P. K., Lee, K. S., El-Sayed, I. H., and El-Sayed, M. A. (2006). Calculated absorption and scattering properties of gold nanoparticles of different size, shape, and composition: applications in biological imaging and biomedicine. *J. Phys. Chem. B*, **110**, pp. 7238–7248.

26. Katsnelson, M. I. (2007). Graphene: carbon in two dimensions, *Mater. Today*, **10**, pp. 20–27.

27. Kelly, K. L., Coronado, E., Zhao, L. L., and Schatz, G. C. (2002). The optical properties of metal nanoparticles: the influence of size, shape, and dielectric environment, *J. Phys. Chem. B*, **107**, pp. 668–677.

28. Klimov, V. I. (2010). *Nanocrystal Quantum Dots*, 2nd ed. (CRC Press, Boca Raton, USA).

29. Kroto, H. W., Heath, J. R., O'Brien, S. C., Curl, R. F., and Smalley, R. E. (1985). C60: Buckminsterfullerene, *Nature*, **318**, pp. 162–163.

30. Levchenko, A. A., Li, G., Boerio-Goates, J., Woodfield, B. F., and Navrotsky, A. (2006). TiO_2 stability landscape: polymorphism, surface energy, and bound water energetic, *Chem. Mater.*, **18**, pp. 6324–6332.

31. Liu, L., Ma, E., Li, R., Liu, G., and Chen, X. (2007). Effects of phonon confinement on the luminescence dynamics of Eu^{3+} doped Gd_2O_3 nanotubes, *Nanotechnology*, **18**, article number 015403.

32. Liu, G. K., Chen, X. Y., Zhuang, H. Z., Li S., and Niedbala, R. S. (2003). Confinement of electron–phonon interaction on luminescence dynamics in nanophosphors of Er^{3+}: Y_2O_2S, *J. Solid State Chem.*, **171**, pp. 123–132.

33. Lu, A. H., Salabas, E. L., and Schüth, F. (2007). Magnetic nanoparticles: synthesis, protection, functionalization and applications, *Angew. Chem. Int. Ed.*, **46**, pp. 1222–1244.

34. Lu, X., Rycenga, M., Skrabalak, S. E., Wiley, B., and Xia, Y. (2009). Chemical synthesis of novel plasmonic nanoparticles, *Ann. Rev. Phys. Chem.*, **60**, pp. 167–192.

35. McHale, J. M., Auroux, A., Perrotta, A. J., and Navrotsky, A. (1997). Surface energies and thermodynamic phase stability in nanocrystalline aluminas, *Science*, **277**, pp. 788–791.

36. Mélinon, P., and Masenelli, B. (2011). *From Small Fullerenes to Superlattices—Science and Applications* (Pan Stanford, Singapore).

37. Michalet, X., Pinaud, F. F., Bentolila, L. A., Tsay, J. M., Doose, S., Li, J. J., Sundaresan, G., Wu, A. M., Gambhir, S. S., and Weiss, S. (2005). Quantum dots for live cells, *in vivo* imaging and diagnostics, *Science*, **307**, pp. 538–544.

38. Mirkin, C. A. (2005). The beginning of a small revolution, *Small*, **1**, pp. 14–16.

39. Mody, V. V., Siwale, R., Singh, A., and Mody, H. R. (2010). Introduction to metallic nanoparticles, *J. Pharm. Bioall. Sci.*, **2**, pp. 282–289.

40. Murphy, C. J., Gole, A. M., Stone, J. W., Sisco, P. N., Alkilany, A. M., Goldsmith, E. C., and Baxter, S. C. (2008). Gold nanoparticles in biology: beyond toxicity to cellular imaging, *Acc. Chem. Res.*, **41**, pp. 1721–1730.

41. Nanda, K. K., Sahu S. N., and Behera, S. N. (2002). Liquid-drop model for the size-dependent melting of low-dimensional systems, *Phys. Rev. A*, **66,** article number 013208.

42. Navrotsky, A. (2011). Nanoscale effects on thermodynamics and phase equilibria in oxide systems, *Chem. Phs. Chem.*, **12**, pp. 2207–2215.

43. Navrotsky, A., Mazeina, L., and Majzlan, J. (2008). Size-driven structural and thermodynamic complexity in iron oxides, *Science*, **319**, pp. 1635–1638.

44. Nirmal, M., and Brus, L. (1998). Luminescence photophysics in semiconductor nanocrystals, *Acc. Chem. Res.*, **32**, pp. 407–414.

45. Noguez, C. (2007). Surface plasmons on metal nanoparticles: the influence of shape and physical environment, *J. Phys. Chem. C*, **111**, pp. 3806–3819.

46. Novoselov, K. S., Geim, A. K., Morozov, S. V., Jiang, D., Zhang, Y., Dubonos, S. V., Grigorieva, I. V., and Firsov, A. A. (2004). Electric field effect in atomically thin carbon films. *Science*, **306,** pp. 666–669.

47. Oldenburg, S. J., Averitt, R. D., Westcott, S. L., and Halas, N. J. (1998). Nanoengineering of optical resonances, *Chem. Phys. Lett.*, **288**, pp. 243–247.

48. Palkar, V. R., Ayyub, P., Chattopadhyay, S., and Multani, M. (1996). Size-induced structural transitions in the Cu-O and Ce-O systems, *Phys. Rev. B*, **53**, pp. 2167–2170.

49. Pérez, J., Bax, L., and Escolano, C. (2005). *Roadmap Report on Nanoparticles* (Willems & van den Wildenberg).

50. Pitcher, M. W., Ushakov, S. V., Navrotsky, A., Woodfield, B. F., Li, G., Boerio-Goates, J., and Tissue, B. M. (2005). Energy crossovers in nanocrystalline zirconia, *J. Am. Ceram. Soc.*, **88**, pp. 160–167.

51. Pitkethly, M. J. (2004). Nanomaterials—the driving force, *Mater. Today*, **7**, Suppl. 1, pp. 20–29.

52. Poland, C. A., Duffin, R., Kinloch, I., Maynard, A., Wallace, W. A. H., Seaton, A., Stone, V., Brown, S., MacNee, W., and Donaldson, K. (2008). Carbon nanotubes introduced into the abdominal cavity of mice show asbestos-like pathogenicity in a pilot study, *Nat. Nanotechnol.*, **3**, pp. 423–428.

53. Qadri, S. B., Skelton, E. F., Dinsmore, A. D., Hu, J. Z., Kim, W. J., Nelson, C., and Ratna, B. R. (2001). The effect of particle size on the structural transition in zinc sulfide, *J. Appl. Phys.*, **89**, 1328066.

54. Ranade, M. R., Navrotsky, A., Zhang, H. Z., Banfield, J. F., Elder, S. H., Zaban, A., Borse, P. H., Kulkarni, S. K., Doran, G. S., and Whitfield, H. J. (2002). Energetics of nanocrystalline TiO$_2$, *Proc. Natl. Acad. Sci.*, **99**, pp. 6476–6481.

55. Roco, M. C. (1999). Nanoparticles and nanotechnology research, *J. Nanopart. Res.*, **1**, pp. 1–6.

56. Rotello, V. M. (2004). *Nanoparticles: Building Block for Nanotechnology* (Springer, New York).

57. Schmid, G. (2004). *Nanoparticles: From Theory to Application* (Wiley-VCH Verlag, Weinheim).

58. Wolfgang M., Sigmund, W. M., Bell, N. S., Bergström, L. (2000). Novel powder-processing methods for advanced ceramics, *J. Am. Ceram. Soc.*, **83**, pp. 1557–1574.

59. Tang, B., Xu, S., An, J., Zhao, B., and Xu, W. (2009). Photoinduced shape conversion and reconstruction of silver nanoprisms, *J. Phys. Chem. C*, **113**, pp. 7025–7030.

60. Tolbert, S. H., and Alivisatos, A. P. (1994). Size dependence of a first order solid-solid phase transition: wurtzite to rock salt transformation in CdSe nanocrystals. *Science*, **265**, pp. 373–376.

61. Trusov, L. I., Lapovok, V. N., and Novikov, V. I. (1989). *Science of Sintering*, eds. Uskokovic, D. P., Plamour III, H., and Spriggs, R. M.,

"Problems of sintering in ultrafine powders" (Plenum Press, New York). pp. 185–192.

62. Tsuzuki, T., Schäffel, F., Muroi, M., and McCormick, P. G. (2011). Magnetic properties of mechanochemically synthesised γ-Fe$_2$O$_3$ nanoparticles, *J. Alloy. Compd.*, **509**, pp. 5420–5425.

63. Tsuzuki, T. (2008). Abnormal transmittance of refractive-index modified ZnO-organic hybrid films, *Macromol. Mater. Eng.*, **293**, pp. 109–113.

64. Tsuzuki, T., Robinson, J. S., and McCormick, P. G. (2002). UV-shielding ceramic nanoparticles synthesised by mechanochemical processing, *J. Aus. Ceram. Soc.*, **38**, pp. 15–19.

65. Tsuzuki, T., and McCormick, P. G. (1999). Synthesis of ultrafine Ce$_2$S$_3$ powder by mechanochemical processing, *Mater. Sci. Forum*, **315–317**, pp. 586–591.

66. Yokoyama, T., Naito, M., Nogi, K., and Hosokawa M. (2007). *Nanoparticle Technology Handbook* (Elsevier, UK).

67. Wang, X., Li, Q., Xie, J., Jin, Z., Wang, J., Li, Y., Jiang, K., and Fan, S. (2009). Fabrication of ultralong and electrically uniform single-walled carbon nanotubes on clean substrates, *Nano Lett.*, **9**, pp. 3137–3141.

68. Wang, Z. L. (2004). Nanostructures of zinc oxide, *Mater. Today*, **7**, pp. 26–33.

69. Wee, A. T. S., Sow, C. H., and Chin, W. S. (2009). *Science at the Nanoscale— An Introductory Textbook* (Pan Stanford, Singapore).

70. Weiner, S., and Wagner, H. D. (1998). The material bone: structure–mechanical function relations, *Ann. Rev. Mater. Sci.*, **28**, pp. 271–298.

71. Weisman, R. B. (2004). Simplifying carbon nanotube identification, *Ind. Phys.*, February/March 2004, pp. 24–27.

72. Wilder, J. W. G., Venema, L. C., Rinzler, A. G., Smalley, R. E., and Dekker, C. (1998). Electronic structure of atomically resolved carbon nanotubes, *Nature*, **391,** pp. 59–62.

73. Zhang, Q. (2011). *Carbon Nanotubes and Their Applications* (Pan Stanford, Singapore).

Chapter 2

Applications and Market Opportunities of Nanoparticulate Materials

Takuya Tsuzuki

Research School of Engineering, College of Engineering and Computer Science, Australian National University, Ian Ross Building 31, North Road, Canberra ACT 0200, Australia

takuya.tsuzuki@anu.edu.au

2.1 Introduction

The current and potential applications for nanoparticles are growing rapidly [1,28]. These applications cover an extremely broad range of markets and industries including biomedical and cancer treatment, renewable energy, environmental protection, pharmaceuticals, personal care, surface coatings, plastics, textiles, food, building materials, electronics and the automotive industries (Fig. 2.1) [41,43,50]. In fact, a number of nanoparticle-based consumer products are already available [40]. For example, personal care products containing TiO_2 and ZnO nanoparticles are sold extensively to protect human skin from UV rays. Silver

Nanotechnology Commercialisation
Edited by Takuya Tsuzuki
Copyright © 2013 Pan Stanford Publishing Pte. Ltd.
ISBN 978-981-4303-28-6 (Hardcover), 978-981-4303-29-3 (eBook)
www.panstanford.com

nanoparticles are used as an antibacterial agent in many consumables ranging from surgical instruments to pet food bowls. The Woodrow Wilson International Centre for Scholars, Projects on Emerging Nanotechnology, counted nanomaterial-based consumer goods in 2010 and found that there were 1,317. The number is steadily increasing every year [55].

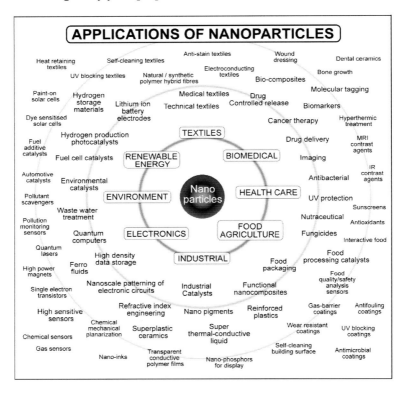

Figure 2.1   Major applications of nanoparticles. Reprinted with permission from Ref [51]. Copyright 2009 Inderscience Publishers.

However, the use of nanoparticles in man-made objects is not new. The Lycurgus cup, a Roman drinking vessel from the 4th century AD, contains 50–100 nm-sized silver and gold nanoparticles embedded in glass, displaying a remarkable optical effect due to the surface plasmon resonance of the nanoparticles. In direct light, the cup appears an opaque green-yellow colour, but when the light shines through the cup, it exhibits a translucent reddish colour [14]. Similar techniques were used to obtain unique colouring effects in the stained glass windows of old churches [44]. Ancient Egyptians used nanoscale carbon particles and multi-wall carbon nanotubes in inks and paints [33]. TiO_2, Fe_2O_3 and NiO nanoparticles were also used in inks in ancient China [33]. However, those early applications of nanoparticulate materials were more or less accidental with no understanding of nanoscale science.

Even in more recent times, nanoparticles have been used for many years in automotive exhaust gas convertors [25], magnetic recording media such as video tapes [30] and car tyres (>1960s) [37]. IC chips in many electronic devices such as iPods, cell phones and computers are fabricated utilising nanoparticles as a polishing agent to form ultra-thin layers [38]. As such, almost everyone in developed countries owns more than one product that utilises nanoparticle technology today.

In this chapter, we review the current and future applications of nanoparticles and the potential markets for commercial nano-enabled products.

2.2   Nanotechnology Market

There are more than 10 market research companies actively analyzing the size of potential nanotechnology markets (Table 2.1). A recent BCC Research report states that [5]:

> The total worldwide sales revenues for nanotechnology were $11,671.3 million in 2009, and are expected to increase to more than $26,000 million in 2015, a compound annual growth rate (CAGR) of 11.1%. The largest nanotechnology segments in 2009 were Nanomaterials. All Nanomaterials will increase from $9,027.2 million in 2009 to nearly $19,621.7 million in 2015, CAGR of 14.7%.

If the figures prove to be accurate, nanotechnology will emerge as a larger economic force than the combined telecommunications and information technology industries seen at the beginning of the technology boom of the late 1990s [35]. The nanotech markets include the energy, healthcare, consumer goods, textile, automotive, aerospace and many other industries. Interestingly, Lux Research insists that the presence of the nanotechnology market is an illusion; there is no nanotechnology market and instead there is a

Table 2.1 Examples of market reports on nanotechnology and nano-materials (2008–2011)

Research company	Title	Publication date
BCC research	Nanotechnology: A realistic market assessment	July 2010
Bizacumen	Nanomaterials—Worldwide market challenges and opportunities	November 2009
Bharat Book	Nanotechnology Market forecast for 2013	January 2011
Cientifica	Nanotechnology Opportunity Report, 3rd ed	June 2008
Espicom Business Intelligence	Nanotechnology: Players, products and prospects to 2018	June 2009
Future Market	World market for nanotechnology and nanomaterials in consumer products, market share, 2010–2015	May 2010
Freedonia	World nanomaterials to 2013-Demand and sales forecasts, market share, market size, market leaders	March 2010
Global Industry Analysis	Nanotechnology—Global strategic business report	October 2010
Global Industry Analysis	Nanomaterials—Global strategic business report	October 2010
ICON Group International	2011–2016 World Outlook for Nanotechnology	January 2011
Lux Research	Ranking the Nations on Nanotech; Hidden havens and false threats	August 2010
Nanoposts	The Global markets and applications for nanotechnology in consumer goods to 2015	February 2010
RNCOS	Nanotechnology market forecast to 2013	January 2011

nanotechnology value chain in the existing market [28]. Regardless of the definition, a large number of small and large companies saw significant business opportunities in nanotechnology commercialisation. Many start-up companies used the word "nano" in their names to take advantage of the hype. In 2008, Credit Suisse developed the Global Nanotechnology Index to focus on companies offering nanotechnology products. Other financial institutions followed this move [13].

In 2002, an analysis by Holister and Harper listed the four major markets for nanoparticles by volume as (i) automotive catalysts (Pt, Pd, CeO_2, ZrO_2, 11,500 tonnes), (ii) chemical mechanical planarisation slurry (Al_2O_3, SiO_2, CeO_2, 9,400 tonnes), (iii) magnetic recording media (Co, Fe_2O_3 and other ferrites, 3,100 tonnes) and (iv) sunscreens (ZnO and TiO_2, 1,500 tonnes) [21]. Although nanoparticle sunscreens are often regarded as the most widely known nanotech products, their volume is estimated to be only ~6% of the combined volume of those four major markets. In terms of business profit, the volume does not necessarily equate with value. In general, cosmetic, pharmaceutical and medical applications can fetch significantly higher profit margins than industrial applications.

In this chapter, the detailed analysis of the nanotechnology market in each sector is left to the commercial market reports (Table 2.1). Instead, the overviews of current nano-enabled consumer goods and future applications in key industrial sectors are presented.

2.3 Opportunities and Challenges

Nanoparticles can be used in many ways at the various stages of manufacturing processes. For example, they can be the raw ingredients which are transformed into non-nano materials during the process. They can be used to assist manufacturing processes as process catalysts or processing tools, which are not incorporated in the final products. For example, chemical mechanical planarisation slurry is a processing tool and the nanomaterials do not remain in the final products, i.e., computer chips. Nanoparticles can also be the vital components incorporated in the final products to give unique functionalities to the products. As such, the consumer products that contain nanoparticles are rather a small part of the examples of nanoparticle technology applications. Many of the applications may be invisible to most of the consumers. In this section, a brief overview of the applications of nanomaterials is given for some key market sectors.

2.3.1 Energy Sector

Owing to the population explosion and global industrialisation, energy demand in the world is expected to double over the next 25 years [34]. Currently, fossil fuels provide ~85% of energy production worldwide, but the resources are limited. In 2002, the reserves of fossil fuels throughout the whole world were projected to last 40 years for oil, 60 years for natural gas and 200 years for coal [34]. Once the demand for fossil fuels exceeds the supply, our lifestyle will be largely compromised and geo-political restrictions/conflict will escalate. Hence, renewable energy sources are urgently needed to fill the demand-supply gap expected in the near future. Nanotechnology presents significant potential to assist the development of clean energy technology in the segments of energy production, storage and conservation [24,34,47].

2.3.1.1 Energy production

The major applications of nanomaterials in this sector are listed in Table 2.2. Solar cells generate electricity directly from sunlight and are one of the most recognised clean energy technologies. The production of fuels such as methanol and diesel from renewable biological sources is an attractive alternative to the direct use of fossil fuels. Fuel cells are another clean energy technology where the chemical energy stored in the fuel (methanol, hydrogen, methane, etc.) is transformed into H_2O and electrical energy with high efficiency, no toxic emission and low environmental impact. Wind power is a major source of clean energy in Europe. In Denmark, ~20% of electricity demand is met by wind power technology. In all of those clean technologies, nanomaterials are expected to have significant influence on improving the efficiencies.

Table 2.2 Examples of nanomaterial applications in the energy generation technology

Application	Description
Solar cells	• Metal nanoparticles embedded in solar cells to increase photon harvest by plasmonics
	• Carbon nanotubes and graphenes for new transparent conducting electrode layers to replace costly indium or rare-earth based transparent electrodes
	• TiO_2 nanoparticles as a working electrode in dye sensitized solar cells (DSSCs). Modification of morphology, porosity and surface chemistry to increase the cell efficiency
	• Quantum dots embedded in polymer or silicon solar cells to increase the cell efficiency
	• Printable solar cells based on CuInGaS nanoparticles for the reduction of cell production cost
Biofuel	• Nanoparticulate catalysts for the biofuel production enhances the production rate
Fuel cells	• Nanoparticles of platinum metal catalysts in polymer electrolyte membrane fuel cells and phosphoric acid fuel cells
	• SiO_2 nanoparticles are considered to increase the efficiency to separate/retain water in the cell
	• Ceramic nanoparticles such as CeO_2 to reduce the sintering temperature for the production of solid oxide fuel cell electrodes
Wind generation	• Polymer nanocomposites that are reinforced with nanoparticles, carbon nanotubes and nanofibres as light weight and high strength structural materials
Hydrogen generation	• Water splitting by photocatalysis of semiconductor nanoparticles, utilising sunlight as the energy source

2.3.1.2 Energy storage

The generated energy needs to be stored for many reasons. For example, energy storage is required for providing mobility and portability to electronic products. Storing energy is also necessary for meeting future electricity demand. In nanomaterial science, the most actively investigated areas in energy storage applications are lithium ion batteries, hydrogen storage materials and super-capacitors. Lithium ion batteries have become the most widely used rechargeable batteries due to their excellent energy density, robust recharging cycle properties, high energy-retention and light weight. In 2007, 60% of lithium ion batteries already utilised nanofibres [11]. Super-capacitors can store a large amount of energy with a very short discharge–recharge time and remarkably long recharge-cycle lifetime. The development of a safe high-density hydrogen-storage system is a key to using hydrogen as an environmentally clean fuel. Nanomaterials play a critical role in improving the performance of those energy storage devices/ materials as listed in Table 2.3.

Table 2.3   Examples of nanomaterial applications in the energy storage technology

Application	Description
Lithium ion battery	• Carbon nanotubes as a novel anode to enhance energy density • Manganese compound nanoparticles for improved discharge-recharge cycle characteristics and reduced recharge time • Novel cathode nanomaterials with nanoscale layered structures • Nanofibre separators for improved efficiency
Super-capacitors	• Electrodes made of nanoparticles and carbon nanotubes, for higher energy density derived from high surface area
Hydrogen storage material	• Nanograined magnesium alloys • Carbon nanotubes, boron nitride, metal-organic nano-framework, etc.

2.3.1.3   Energy conservation

Reducing energy consumption is an important and effective strategy in meeting the ever-increasing energy demand. This can be achieved by (i) increasing the efficiency of energy usage in combustion engines, electric motors and lighting apparatus, and (ii) reducing the demand for energy usage in buildings and transport. Nanomaterials can assist in achieving these objectives through the applications listed in Table 2.4.

Table 2.4   Examples of nanomaterial applications in the energy conservation technology

Application	Description
Fuel borne catalysts	• Nanoparticle catalysts to enhance diesel fuel combustion efficiency and to reduce harmful exhaust gas/particulates
Nanocomposite magnets	• Generation of stronger force from less electric energy by smaller and lighter electric motors
Structural materials	• Nanoparticle/nanotube/nanofibre reinforced polymer composites as light weight and high strength structural materials, to reduce energy consumptions in vehicles and airplanes
Thermal insulation	• Infrared reflecting/absorbing nanocomposite window films • Nanoporous thermal insulators for buildings • Electrochromic windows made with tungsten oxide nanoparticles
Light emitting diode	• Casings made of refractive-index-engineered optical nanocomposites to enhance the energy extraction rate
Nanofluids	• Cooling liquid with improved thermal exchange properties for motors and transformers

2.3.1.4   Market for energy nanotechnology

Owing to the escalating threat of climate change and depletion of fossil fuels, the market for alternative clean energy technologies has grown significantly in the last 20 years. This trend is expected to continue in the foreseeable future and reliance on nanotechnology for technological breakthrough will increase steadily. There are a few published market reports that analyse the impact of nanotechnology in the energy sector (Table 2.5). In January 2011, Future Market Inc. estimated the world market for nanotech-enabled solar cells as US$68 million in 2010 and estimated to rise to $820 million by 2017 [16]. Future Market Inc. stated that [15]:

> Due to their unusual mechanical, electrical and optical properties nanomaterials are being incorporated into new, low-cost and environmentally friendly energy conversion and storage systems, promising greater efficiency than current technologies. The world market for these systems was US$203.7 million in 2010 and is estimated to rise to US$2,280 million by 2017.

Table 2.5 Market reports on nanotechnology and nanomaterials in the energy sector (2007–2011)

Research company	Title	Publication date
BCC Research	Nanotechnology in Energy Applications	March 2007
Cientifica	Nanotechnology for the Energy Market	June 2009
Cientifica	Nanotechnologies for the Portable Electronics Energy Market	February 2007
Cientifica	Nanotechnologies for Sustainable Energy: Reducing Carbon Emissions Through Clean Technologies and Renewable Energy Sources	May 2007
Cientifica	Nanotechnologies for the Energy Market—Residential & Commercial Use	February 2007
Energy Business Reports	Nanotechnology and Photovoltaics Trends & Market Potential	September 2009
Energy Business Reports	Nanotechnology in the Energy Industry: Applications and Market Potential	September 2007
Future Market	Nanotechnology and Photovoltaics: Market, Companies and Products	February 2011
Future Market	Nanotechnology and Nanomaterials for the World Energy Market: Applications, Products, End User Markets, Companies and Revenues	January 2011
Future Market	Nanomaterials and the World Lithium Ion Battery Market: Applications, Products, End User Markets, Companies and Revenues	February 2011
Global Markets Direct	Global Nanotechnology Market for Energy Storage: Analysis and forecast to 2015	December 2009
WinterGreen Research	Energy Harvesting Market Shares, Strategies, and Forecasts, Worldwide, Nanotechnology, 2012 to 2018	November 2011
WinterGreen Research	Worldwide Nanotechnology Thin Film Lithium-Ion Battery Market Shares, Strategies, and Forecasts, 2009–2015	January 2009
WinterGreen Research	Worldwide Nanotechnology Portable Fuel Cell Market Shares Strategies, and Forecasts, 2009–2015	December 2008

WinterGreen Research Inc. predicted a similar market size for energy nanotechnology by 2018 [51]. On the other hand, Cientifica estimated that energy conservation applications will see the highest growth in the energy sector. Cientifica predicted a larger market size than the other reports, stating that [12]:

> The most immediate opportunities lie in saving energy through the use of advanced materials and this is already a $1.6 billion dollar market (2007), predicted to rise to $51 billion by 2014. Despite advances in battery technology, hydrogen storage and fuel cells, energy saving technologies will exhibit faster growth, accounting for 75% of the market for nanotechnologies in 2014, up from 62% in 2007. Solid state lighting, nanocomposite materials, aerogel and fuel borne catalysts will have the greatest impact between now and 2014. Compound annual growth rates are 64% for energy saving technologies and 90% for energy generation, while energy storage applications show a comparatively lowly 30%. Applications in transportation will increase to $50 billion by 2014 with a CAGR of 72%.

2.3.2 Medical, Personal Care and Pharmaceutical Sector

2.3.2.1 Applications for medical nanotechnology

Nanotechnology in the health, medical, pharmaceutical and personal care sectors is referred to as bionanotechnology. These sectors enjoy one of the largest arrays of nanotechnology applications [7,22,26,39,45,49]. Many types of nanomaterials are used in biomedical and personal care applications, including nano-emulsion, nanofibres, metal oxide nanopowders, carbon nanotubes, dendrimers, fullerenes, nanocapsules, silver nanoparticles and quantum dots. For example, nanoparticles less than 100 nm in diameter can go through the smallest blood vessels to carry drugs or medical-imaging agents to designated diseased areas in the body. The similarity in size between biomolecules and nanoparticles makes the bio-marking and bio-sensing applications promising. The unique properties of nanoparticles such as quantum size effects, superparamagnetism, low sintering temperatures, high surface areas and low light-scattering ability are also actively utilised. The key application areas of nanomaterials in biomedical and personal care sectors are listed in Table 2.6.

Table 2.6   Examples of nanomaterial applications in the medical and personal care sectors

Application	Description
Drug delivery	• Delivery of specific drugs by nanoparticles and dendrimers that are targeted to only pathological areas, so as to minimise the side effect and increase the administration efficacy
Medical imaging	• Delivery of nanoparticles only to the pathological areas so as to enhance the image contrast in X-ray, infra-red and magnetic resonance imaging for more accurate diagnosis
Targeted physical cancer treatment	• Hyperthermia treatment to kill tumour cells by magnetic nanoparticles targeted to pathological areas. Localised heat is generated by applying an external alternating magnetic field • Destruction of tumour cells by nano-blast of targeted gold nano-shells by externally applying infrared lasers
Biomarkers Biosensors	• Quantum dots, plasmonic metal nanoparticles, nano-phosphors to use as markers to detect the location and/or presence of specific molecules by optical means in order to examine organ functions or pathological cells
Tissue scaffolds	• Nanofibre and nanoporous implants to regenerate tissue, bone, nerve cells, skin and organs, by utilising the fact that cells grow quicker on nanoscale surfaces
Bio implants	• Calcium hydroxyl apatite nanoparticles to surface treat ceramic and metal implants for enhanced bio-compatibility
Dental implants	• Nanograined dental ceramics for improved strength and plasticity
Fast-effect pills	• Fast solubility of orally administered drugs, by utilising the high surface areas of nanoscale drug particles
Wound dressing	• Nanofibre membrane to block bacteria • Silver nanoparticles as antibacterial agents • Dendrimers and nanoporous materials for controlled drug release
Antimicrobial agents	• Long lifetime antibacterial agents such as silver and zinc oxide nanoparticles, in wound dressing, for sanitisation of medical apparatus and hospital walls, in consumer goods
UV protection	• ZnO and TiO_2 nanoparticles for transparent UV screening applications with high chemical-, photo- and thermal stability
Antioxidant	• Fullerenes as antioxidants for skin care

2.3.2.2 Market for medical nanotechnology

The biomedical applications of nanomaterials are regarded by the general public as one of the most needed and accepted applications areas of nanotechnology. According to Nanoposts.com's analysis, nano-enabled life sciences and healthcare products represent a $1 billion plus market in 2009 [36]. Business Insight estimated a rapid growth in the market size of medical nanotechnology, by stating that [9]:

> The healthcare nanotechnology market growth is largest in North America, at $4.75 billion in 2009, followed by Europe at $3.65 billion. The nanotechnology drug delivery market is expected to grow at a compound annual growth rate of 21.7% for the period 2009–14, to reach almost $16 billion by 2014. Biocompatible implants and coatings and diagnostics comprise some of the major applications which are estimated to experience a high growth between 2009 and 2014 of 42% and 21.8% respectively.

The markets for dental implants ($3.4 billion in 2008) are anticipated to reach $8.1 billion [52]. Other recent market reports are listed in Table 2.7.

Table 2.7   Examples of market reports on nanotechnology in the medical and personal care sector

Research company	Title	Publication date
BCC Research	Nanobiotechnology: Applications and Global Markets	January 2011
Business Insights	Nanotechnology in Healthcare: Market Outlook for Applications, Tools and Materials and 40 Company Profiles	January 2010
Business Insights	Innovation in Drug Delivery: The Future of Nanotechnology and Non-invasive Protein Delivery	November 2006
Cientifica	Nanoparticle Drug Delivery Market Report	September 2007
Espicom Business Intelligence	Emerging Opportunities in Cancer Nanomedicine	August 2006
Frost & Sullivan	The Role of Nanotechnology in European Drug Discovery	May 2006

(Continued)

Table 2.7   (*Continued*)

Research company	Title	Publication date
Frost & Sullivan	Nanomedicine—Global Technology Developments and Growth Opportunities (Technical Insights)	July 2004
Future Markets	The World Market for Nanotechnology in Healthcare, Medicine and Biotech 2011	October 2011
Global Industry Research	Nanotechnology in Drug Delivery—Technological Improvements and Novel Approaches to Fulfil High Potential	December 2010
Global Industry Research	Nanomedicine—Global Strategic Business Report	July 2009
ICON Group	The 2011–2016 World Outlook for Nanomedicine	January 2011
Jain Pharma-Biotech	Nanobiotechnology Applications, Markets and Companies	December 2011
Nanoposts	Global Market & Applications for Nanotechnology in the Life Sciences and Healthcare Industries	February 2009
Nanoposts	Key Products and Players, Global Markets and Applications for Nanotechnology in Personal Care, Cosmetics, Household Care, Packaging and Leisure Wear & Equipment	February 2010
Nanoposts	Nanotechnologies for Household and Personal Care	February 2007
Roc Search	Nanobiotechnology: Trends and Technological Developments	March 2006
Science-Metrix	Nanotechnology—World R&D Report 2008—Research in Medicine and Biology Nanoscience and Nanotechnology	April 2008
Technology for Industry	Microsystems and Nanotechnology in the Medical and Biomedical Market	February 2005
Trimark	Medical Nanotechnology Markets	September 2008
WinterGreen Research	Worldwide Nanotechnology Dental Implant Market Shares, Strategies, and Forecasts	April 2009

There are already a number of commercial nano-enabled medical products on the market, including the MagForce® nanoparticle drug delivery system and the Hydrofiber® wound dressing. The major hurdles in the applications of nanomaterials in biomedical fields are (i) the high cost and long lead time associated with clinical trials and regulatory approvals, (ii) the high production cost and low production volume of nanomaterials, and (iii) consumer's risk perception on nanomaterials. These challenges will be discussed in later chapters.

2.3.3   Environment Sector

Nanomaterials can contribute to improving environmental conditions in many ways. For example, nanotechnology can be used to provide clean air and water, by means of nano-filters, pollutant scavengers and environmental sensors. Nanotechnology is also beneficial for reducing pollutants during the generation and usage of energy as described earlier in this chapter. ZnO and TiO_2 nanoparticles will increase the durability and product lifetime of organic materials such as plastic, wood, textiles and paints, by shielding them from UV-light. This results in the minimisation of the use of these materials and in turn contributes to the protection of limited resources. The key applications of nanomaterials in this sector are listed in Table 2.8 [20,23].

Table 2.9 lists the nanotechnology market reports in the environment sector. In 2009, BCC Research estimated the nanotechnology market in environmental applications, stating that [6]:

> The global market for nanotechnology in environmental applications generated $1.1 billion in 2008 and an estimated $2.0 billion in 2009. This is expected to increase at a compound annual growth rate (CAGR) of 61.8% to reach $21.8 billion in 2014.

In 2011, BCC Research also reported the scope of water-related nanotechnology markets, stating that [4]:

> The global market for nanostructured products used in water treatment was worth an estimated $1.4 billion in 2010 and will grow at a compound annual growth rate (CAGR) of 9.7% during the next 5 years to reach a value of $2.2 billion in 2015. Most of the current market is made up of well-established treatment products, reverse osmosis, nanofiltration, and ultra filtration membrane modules, which may be categorized as nanotech-based. The market for established products was nearly $1.4

billion in 2010 and is expected to reach $2.1 billion in 2015, a compound annual growth rate (CAGR) of 9.2%. Many emerging products, including nanofiber filters, carbon nanotubes, and a range of nanoparticles, are in the pre-commercial stages. The market for emerging products was $45 million in 2010. This sector will increase at a 20% compound annual growth rate (CAGR) to reach a value of $112 million in 2015.

Table 2.8 Examples of nanomaterial applications in the environment sector

Application	Description
Environmental monitors/sensors	• Nano-sensors for monitoring toxic gases and volatile organic compounds (VOC)
Pollutant scavenging	• Decomposition of organic pollutant in air and water using semiconductor nanoparticles' photocatalytic activity with sun light as an energy source
Wastewater treatment	• Destruction of pathogens with photocatalytic nano-particles • Removal of pollutants and pathogens using magnetic nanoparticles
Air purification	• Silver nanoparticles on air filters for pathogen removal • Nanofibre membrane for the removal of particulate pollutants
UV protection	• ZnO and TiO_2 nanoparticle UV screening agents in paints and plastics for the protection of plastics, paints, dye, wood, lather and other organic industrial materials
Desalination	• Nanoporous materials and nano-filtration systems

Table 2.9 Examples of market reports on nanotechnology and nano-materials in the environment sector (2009–2011)

Research company	Title	Publication date
BCC Research	Nanotechnology in Water Treatment	January 2011
BCC Research	Nanotechnology in Environmental Applications: The Global Markets	July 2009
Frost & Sullivan	Impact of Nanotechnology in Water and Waste-water Treatment	December 2006
Nanoposts	Nanotechnologies for Energy and the Environment	May 2007
Nanoposts	Nanotechnologies for Cleantech	November 2008
Nanoposts	Nanotechnologies for Environmental Applications	October 2006

2.3.4   Electronics Sector

Nanoelectronics is a large research area in its own right [18]. Many of the applications involve the fabrication of complex patterns out of nanometre-thick layers by etching. The fabrication technique is mainly based on lithography and was established in the semiconductor industry in the late 20th century. Examples of nano-patterns are quantum dots and nanowires for new logic devices such as single-electron transistors, resonance tunnelling diodes and quantum lasers. Soft lithography is a new technique to create nanoscale three-dimensional structures on a flat surface. In these techniques, stand-alone nanomaterials are rarely used, except as chemical mechanical planarisation (CMP) polishing agents.

Traditionally, data storage systems utilise fine magnetic particles in magnetic tapes and disks. Theoretically, the reduction of particle size improves the data storage density. However, the majority of current data storage devices are made of thin films of giant magneto-resistance materials (in hard drive) or photochromic ultrafine islands (DVD disks) or solid state drives (SSD) that are fabricated using lithography, without relying on stand-alone nano-materials as a component. Although micro-electro-mechanical systems (MEMS) are an important area in nanoelectronics, it is little to do with the applications of stand-alone nanomaterials.

Nonetheless, nanoparticles as stand-alone nanomaterials still have many applications in the electronics industry as listed in Table 2.10. The applications include two of the four major markets for nanoparticles by volume, i.e., chemical mechanical planarisation slurry (Al_2O_3, SiO_2, CeO_2, 9,400 tonnes) and magnetic recording media (Co, Fe_2O_3 and other ferrites, 3,100 tonnes) [21]. The applications of nanomaterials in batteries and solar cells also belong to this sector.

The use of carbon nanotubes in fabricating logic devices such as field emission transistors has been investigated at a laboratory scale. However, it is extremely challenging to place a single carbon nanotube in a precise location in the nano-patterned integrated circuit, let alone reproducing the precision over a million times in a single chip (*n.b.*, in 2011, a single central processing unit of a computer chip contained nearly 2 billion transistors).

Table 2.10   Examples of the applications of stand-alone nanomaterials in the electronics sectors

Application	Description
Micro-processors	• Chemical mechanical planarization polishing agents such as SiO_2, Al_2O_3 and CeO_2 nanoparticles to form flat surfaces in each deposited layers of integrated circuits • Carbon nanotube field emission transistors • Silver nanoparticle conducting inks for printed electronics
Data storage	• Iron oxide based magnetic nanoparticles • Nanocrystal memory • Optical memory unit using non-linear optical effects of metal nanoparticles • Magnetic nanoparticles or diluted-magnetic-semiconductor nanoparticles for spintronics
Optoelectronics	• Photonic crystals • Optical switches based on magnetic nanoparticles, phase-changing nanoparticles and plasmonics metal nanoparticles
Displays	• Carbon nanotube field emission displays for improved efficiency and resolution • Nanophosphors for higher resolution
Transparent conducting layers	• Polymer nanocomposite films containing antimony doped tin oxide or tin doped indium oxide nanoparticles for solar cells, touch screens and antistatic displays
Sensors	• Nanoparticle-based gas sensors

Table 2.11 lists the nanotechnology market report in this sector. In 2009, Global Industry Analysis announced that the global market for nanoelectronics was expected to reach $409.6 billion by 2015 [51]. In 2007, Cientifica estimated that hard disk drives that are enabled by giant magnetoresistance (GMR) accounted for US$25 billion in 2006 with 450 million units shipped [11]. WinterGreen Research has predicted that the market for mid-infrared sensors at US$70.2 million in 2008 will reach US$865.4 million by 2015 [53].

Table 2.11 Examples of market reports on nanotechnology in the electronics sector (2003–2011)

Research company	Title	Publication date
BCC Research	Nanostructured Materials; Electronics/Magnetic/Optoelectronic	December 2007
Cientifica	Nanotechnologies for the Electronics Market	November 2007
Future market	Nanotechnology and Nanomaterials in the Electronics Sector 2011	Sepember 2011
Frost & Sullivan	Nanoelectronics: Markets; Applications; and Technology Developments (Technical Insights)	January 2003
Global Industry Analysis	Nanoelectronics—Global strategic Business Report	July 2009
Global Industry Analysis	Nano-Photonics—Global Strategic Business Report	October 2006
ICON Group	The 2011 Report on Nanoelectronics: World Market Segmentation by City	January 2011
ICON Group	The 2011–2016 World Outlook for Nanoelectronics	January 2011
Science Metrix	Nanotechnology—World R&D Report 2008—Research in Electronic and Computing Nanoscience and Nanotechnology	April 2008
WinterGreen Research	Worldwide Nanotechnology Mid IR Sensor Market Shares Strategies, and Forecasts, 2009 to 2015	June 2009
Science-Metrix	Nanotechnology—World R&D Report 2008—Research in Optics and Photonics Nanoscience and Nanotechnology	April 2008

2.3.5 Textile Sector

The textile sector is more than the clothing industry. It encompasses medical textiles and fibres, military fabrics, construction textiles, filtration materials (non-woven) and other high-tech applications. A common way to use nanomaterials is to apply them onto the surface of fibres and textiles to provide additional functionality. The next generation of nano-enabled textiles consists of smart and intelligent fabrics. In homeland security applications, the functionality of smart textiles is often combined with sensing, actuation, wireless communication of the data, and self-power-generation functions. Common applications of nanomaterials in the textile sector are listed in Table 2.12 and available market reports are given in Table 2.13.

Table 2.12 Examples of nanomaterial applications in the textile sector

Application	Description
Clothing	• Added functionality stemming from the incorporation of nanoparticles inside or on the surface of fibres. The added functions include UV blocking, infrared reflecting, anti-bacterial, stain-resistant, water-repellent, moisture-controlled, frame-retardant, odour removing, anti-static, electric conductivity, heat retaining, temperature regulating and winkle resistance properties, as well as high mechanical strength
	• Nanoparticles to provide the controlled release of fragrances, biocides and anti-fungal agents on textiles
	• Wearable electronics with conducting fibres incorporating carbon nanotubes
Filters and membranes	• Biological filters made with nanofibres
	• Waste water treatment by nanofibre membrane or filters coated with silver/ZnO nanoparticles
	• Air purification with nanofibre membranes
	• Desalination with nano-membranes
	• Nanofibre fuel cell membranes
	• Nanofibre battery separators
Others	• Textile-based nano-sensors where textile is used as a substrate, for the monitoring of environment (toxic gas, pathogen, etc.) and body functions (temperature, heart-beat, etc.)

Table 2.13 Examples of market reports on nanotechnology in the textile sector (2006–2011)

Research company	Title	Publication date
Cientifica	Nanotechnology for—Textile Market	April 2006
Cientifica	Nanotechnologies for the Sport Textile Market	June 2006
Cientifica	Nanotechnologies for the Clothing Textile Market	June 2006
Cientifica	Nanotechnologies for the Military Textile Market	June 2006
Cientifica	Nanotechnologies for the Home Textile Market	June 2006
Cientifica	Nanotechnologies for the Medical Textile Market	June 2006
Cientifica	Nanotechnologies for the Non-Conventional Technical Textile Market	June 2006
Nanoposts	Nanotechnologies for Smart and Responsive Textiles	February 2007
Nanoposts	Global Market for Nanotechnology in Textiles	March 2010
Nanoposts	Nanotechnologies for Wearable and Non-Wearable Textiles	November 2006
Textile Intelligence	Nanotechnology in Technical Textiles and Apparel	April 2011
ABOUT	Nanotechnology and the Apparel Industry	March 2007

Nanofibre research has been active since 2000, but the diameter of most commercial or laboratory-bench-top nanofibres is often larger than 100 nm, and, hence, it is debatable if those nanofibres can be classified as engineered nanomaterials. Nonetheless, nanofibres have opened up new applications in many areas, including desalination and medical membranes.

2.3.6 Food and Agriculture Sectors

The applications of nanomaterials in the food sector can be divided into four classes, namely, (i) food packaging, (ii) food sensing, (iii) food processing and (iv) food additives including nutrition delivery [42]. In food packaging, improved quality and shelf life of packaged food can be realised using nanocomposite materials containing nanoparticles, for regulating gas and moisture permeability (O_2, CO_2, etc.) and UV-light irradiation. In food sensing applications, nano-sensors can be placed in food packaging to monitor oxygen levels and the presence of pathogens and contaminants, so as to give warnings to consumers and manufacturers about the safety of food products. In food processing, apparatus and utensils can be made antibacterial by silver nanoparticle coatings. Some food processing involves nano-emulsion, though it is without the involvement of solid nanomaterials. In food additive applications, nanoparticles or nano-capsules are used to improve delivery of drugs and nutrients, by utilising the high solubility of nanomaterials to provide improved bioavailability. Many developing countries are researching the possibility of using nanoparticles as a viable food supplement of essential elements such as iron, zinc and calcium [46]. The examples of nanomaterial applications in the food and agriculture sectors are listed in Table 2.14.

Table 2.14   Examples of nanomaterial applications in the food and agriculture sectors

Application	Description
Food packaging	• Nanocomposite films containing clay nanoparticles for reduced gas permeability
	• ZnO containing transparent food wraps for protecting food from UV-induced degradation
	• Silver containing food containers for antibacterial function
	• Nano-tagging for anti-counterfeiting and fraud protection
Food sensors	• Nano-electro-mechanical system to detect virus
	• Oxygen sensors based on TiO_2 nanoparticles that change the colour upon exposure to oxygen
Food processing	• Silver nanoparticle coating to provide sanitization of utensils
	• Nano-filtration
Food additive and nutrition delivery	• Nanoparticles and nanocapsules to improve bio-availability of drugs and nutrients such as fish oil and vitamins, without changing the colour, odour and texture of food
Agriculture	• Nanoporous materials for slow release of water, fertilizers and herbicides for plant
	• Nanosensors to monitor soil quality and plant health
	• Nanoparticles and nanocapsules for effective delivery of nutrition to animals and depleted trace elements to soil

The market reports on nanotechnology in the food and agriculture sector are given in Table 2.15. Helmuth Kaiser Consultancy (Germany) predicted that the market in food nanotechnology will surge to US$20.4 billion by 2010 and that nanotechnology will change 25 per cent of the food packaging business in the next decade leading to a yearly market of about US$30 billion [19].

Table 2.15 Examples of market reports on nanotechnology in the food and agriculture sectors (2004–2011)

Research company	Title	Publication date
Cientifica	Nanotechnologies in the Food Industry	2006
Helmuth Kaiser Consultancy	Nanotechnology in Food and Food Processing Industry Worldwide 2008–2010–2015	2004
Helmuth Kaiser Consultancy	Nanofood 2040: Nanotechnology in Food, Food processing, Agriculture, Packaging and Consumption	2007
Nanoposts	Nanotechnology in the Food and Drink Industries. Applications and Markets to 2015	2011

The food sector is one of the most sensitive areas in terms of consumer sentiment. The commercialisation of nanotechnology in the food and agricultural sectors is likely to be severely affected by the way consumers perceive the risk and safety of nanotechnology [10,31]. However, traditionally food contains nanostructures. For example, vegetable matters consist of nanoscale cellulose fibrils. Milk has nanoscale protein colloidal particles. Ice creams normally contain nanoscale emulsions. These facts are rarely appreciated in the debate of the safety of nanomaterials in food applications [2].

2.3.7 Other Sectors Including Industrial Materials and Paints

Examples of nanomaterial applications in the other industries are listed in Table 2.16.

Table 2.16   Examples of nanomaterial applications in industrial sectors

Application	Description
Coatings and Paints	• Self-cleaning coatings with TiO_2 nanoparticles (photo-catalysis) and SiO_2 nanoparticles (lotus effect)
	• ZnO and TiO_2 nanoparticles for UV protecting, antibacterial, anti-mould, antifungal, anti-fouling functions
	• Smart coatings with sensing functions
	• Anti-fogging coatings with TiO_2 nanoparticles (photo induced superhydrophilic effect)
Building material	• Metal oxide nanoparticles, carbon nanotubes and nano-fibers as additives in concrete for densification and improved mechanical strength
	• TiO_2 nanoparticles in cement, window glass, tiles, grouts, sealants for self-cleaning, antibacterial, anti-mould, antifungal, anti-fouling, anti-fogging and deodorising functions
	• Nanoparticle additives in steel for high Young's modulus and high strength with corrosion resistance properties
	• Nanoparticle/nanotube/nanofibre reinforced polymer composites to achieve light weight and high strength structural materials
Automotives and aerospace	• Nanoparticle additives in engine-construction materials for lighter weight, higher strength, improved temperature/corrosion resistance and superior wear resistance
	• Metal oxide nanoparticles, carbon nanotubes and nano-fibres as additives in polymer nanocomposites for densification and improved mechanical strength and wear resistance (structural material, tyres, interior, etc.)
	• Magnetic fluid (magnetic nanoparticle dispersion systems) for suspension and braking systems
	• Nanoparticle/nanotube/nanofibre reinforced polymer composites to achieve light weight and high strength structural materials
	• Nanoparticle-based three way catalytic conversion catalyst systems
Fraud protection	• Nanophosphors or nanobarcode particles for invisible tagging

The use of nano-enabled structural materials benefits many industries. For example, the use of polymer nanocomposites offers the opportunity to construct light weight and high strength structural materials. Recent research demonstrated that by reducing the size of the particulate fillers to nanoscale, the mechanical properties of polymer composites can become comparable to those of commercial magnesium alloys [54]. The use of these novel structural materials provides aeroplanes and automobiles with enhanced safety, energy efficiency, performance and comfort, as well as reduced emissions, noise and maintenance requirements. The building industry can also benefit from the use of nano-enabled structural materials [29].

The automotive and aerospace industries can also benefit from the use of nanomaterials through energy and electronics applications as described earlier [27,48].

The surface coating industry is deeply associated with the building, automotive and defence industries as these can benefit from the development of new nano-enabled coating products. The coating industry is one of the first sectors where early entry of nano-enabled consumer goods was successfully achieved.

Fraud protection including the protection of brand, authentication, quality and safety is a serious issue in many industries. It is estimated that the cost of counterfeit goods reached ~US$600 billion in 2007. Nanomaterials that are too small to detect with the naked eyes can act as an effective tagging device through the value chains in many products without compromising the quality and functions of the products. Magnetic and phosphorescent properties of nanoparticles as well as their unique chemical compositions and shapes contribute to their use as effective invisible tagging agents.

The market reports on nanotechnology in the coating and other industrial sectors are given in Table 2.17. Future Market Inc. has estimated that the conservative estimate for nano-enabled products in the automotive industry for 2010 was US$246 million and that by 2015 estimates are $888 million (conservative) and $1.852 billion (optimistic) [17]. The analysis of the coating markets for nanotechnology by BCC Research was that [3]:

> The nanocoatings market was nearly $2.1 billion in 2009. This market is projected to reach $3.3 billion in 2010 and $17.9 billion in 2015, for a 5-year CAGR of 39.5%. The nanoadhesives market was $171 million in

2009. This market is projected to reach approx $257 million in 2010 and to reach $1.2 billion in 2015, for a 5-year CAGR of 36.4%.

Table 2.17 Examples of market reports on nanotechnology and nano-materials (2006–2011)

Research company	Title	Publication date
BCC Research	Nanotechnology in Coatings and Adhesive Applications: Global Markets,	January 2010
BCC Research	Nanotechnology in Paper Manufacturing: Global markets	September 2010
Frost & Sullivan	Nanotechnology for Defence and Security (Technical Insights)	March 2006
Future Market	Nanotechnology and Nanomaterials in the Automotive Industry	February 2011
Future Market	The World Market for Nanocoatings	November 2011
IRL	Opportunities for Nanotechnology in the European Coatings Industry	September 2011
Nanoposts	Nanotechnologies for Anti-Microbial and Self-Cleaning Coatings	March 2009
Nanoposts	Nanotechnology in the Automotive and Transportation Industry: Applications and Markets to 2015	March 2011
Nanoposts	Nanostructured Coatings: Applications and Markets to 2015	January 2011
Nanoposts	Nanotechnologies for Packaging & Product Security	February 2007
WinterGreen Research	Worldwide Nanotechnology Electric Vehicle (EV) Market Shares, Strategies, and Forecasts, 2009 to 2015	February 2009

2.4 Nanoparticle Applications in Consumer Products

Nanotech-enabled consumer products are possibly the first example of commercialised nanotechnology that the general public encounters. As such, the risks and benefits of the entire field of nanotechnology may be judged by the consumers, solely on the performance of consumer goods such as silver coated door knobs, stain-resistant trousers and UV-blocking make-ups. Hence, it is worthwhile to have an overview of nanotechnology-related consumer goods.

Around the end of the last millennium, many nano-enabled consumer products started to appear in the market [8,32]. Since then, the number of consumer products has been steadily increasing. In 2010, Nanoposts estimated that the global market for nanotechnology in the consumer goods markets was worth approximately UK£1,435 million in 2009 and was expected to rise to UK£2,740 million by 2015 [35].

As stated in the introduction section, the Woodrow Wilson International Centre for Scholars, Projects on Emerging Nano-technology, counted nanomaterial-based consumer goods in 2010 to be 1,317 [55]. Among them, only 313 products were identified for the use of inorganic nanomaterials. Among the 313 products, over 50% contain silver nanoparticles (Fig. 2.2) [55]. Most of the applications of silver nanoparticles utilise silver's antibacterial property and high electric conductivity. The reason for the large usage of silver nanoparticles is their ease of production and a wide breadth of application opportunity. Because of its long-lasting antibacterial properties, silver can replace many conventional antibacterial chemicals in a variety of consumer goods from refrigerators to hair dryers, providing much easier market entry opportunities than the other nano-applications.

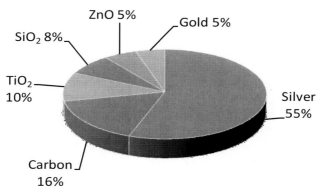

Figure 2.2   Ratio of nanotech-enabled consumer goods counted by Woodrow Wilson International Centre for Scholars, Projects on Emerging Nanotechnology, sorted by the type of inorganic nanomaterials [55].

Carbon and TiO_2 are the second and third most used nanomaterials in the 313 products (Fig. 2.2). Carbon is primarily used in the form of carbon nanotubes to reinforce plastic parts when light weight and high strength are simultaneously required, for example, in bike frames, tennis rackets and other sports goods. TiO_2 is mostly used in the applications of self-cleaning surfaces via superhydrophilicity and antibacterial activity as well as in sunscreens as UV screening agents.

The other remaining 1,004 products out of 1,317 contain either non-specific chemicals or nano-films or nanoscale wax particles. Nano-wax is widely used in household cleaning and polishing products.

The largest market segment (~450) of the listed 1,317 products is the personal care products including cosmetics and sunscreens, followed by textile-related products (~200). This fact does not imply that the use of nanotechnology is largely limited to these two industries, but this is due to the following rather non-technical reasons:

- *Structure of the industry*: Traditionally the cosmetic and textile industries consist of a large number of small players throughout the value chain. Hence, the industries have relatively low market entry thresholds for start-ups and small operators.
- *High value/volume ratio*: The cosmetic industry can normally command high profit margins compared with those of the plastic, paint and other industrial-material-related markets. Hence, it is an attractive industry for start-up companies who need to become fully established financially as quickly as possible.
- *Short product development timeframe*: The time required to develop intermediate and end products is relatively short in the cosmetic and textile industries. The product development does not require years of extensive R&D activity to overcome technological hurdles. In addition, unlike biomedical applications, regulatory barriers for product launch are very low or non-existent.
- *Low capital investment*: The manufacture of intermediate and end products does not require a heavy capital investment in plant, machinery and quality control facilities. Unlike

biomedical applications, there is little need for costly clinical trials.

Hence, it is misleading to judge the general benefit and risks of nanotechnology solely based on the speed at which nanotech-enabled products reach the market or on the current range of nanotech-related consumer products in the market. In the long term, the true value of nanotechnology is more likely to be manifested in the energy, environment and biomedical sectors, rather than in consumer goods.

The real number of nanotechnology-related products is expected to be significantly higher than the count made by the Woodrow Wilson International Centre for Scholars. For example, integrated circuits used in computers, flush memories and mobile phones have nanostructures, and they are manufactured using CMP nanoparticles. Notwithstanding this, computers, flush memories and mobile phones are rarely counted as nano-enabled consumer-products. Also largely excluded in the count is the use of nano-SiO_2 in car tyres and magnetic nanoparticles in magnetic recording media such as video tapes. The European Commission estimated that ~5% of the cosmetic products on the market already contain nanomaterials. Nanopost.com reported that there are more than 2,500 personal care products that contain either nano-titanium dioxide or nano zinc oxide, including moisturisers, eye liners, lip sticks, make-up foundations, soaps, sunscreens, mascara, and nail polish [35].

2.5 Summary

The unique properties of nanomaterials have a broad range of applications in many industries. Already thousands of nano-enabled commercial products are on the market in order to take advantage of the unique properties of nanomaterials. Important applications in the energy, biomedical, environmental and other sectors are rapidly being developed. Through those applications, nanotechnology will demonstrate its practical benefits to the general public and will also provide economic benefits to society.

The current and future market size for nanotechnology has been analysed by many companies. Most market research reports predicted that nanotechnology markets will be tens of billions of dollars in the near future. As the product development and market acceptance continue to progress, the use of nanoparticle technology will spread further, wider and deeper in many industries and market sectors.

References

1. American Ceramic Society (2010). *Progress in Nanotechnology: Applications* (John Wiley & Sons, Hoboken).

2. Barnard, A. S., and Guo, H. (2012). *Nature's Nanostructures* (Pan Stanford, Singapore).

3. BCC Research (2011). *Nanotechnology in Coatings and Adhesive Applications: Global Markets* (http://www.bccresearch.com/report/nanotechnology-coatings-adhesive-nan048a.html/). Last accessed on 15 April 2013.

4. BCC Research (2011). *Nanotechnology in Water Treatment* (http://www.bccresearch.com/report/nanotechnology-water-treatment-nan051a.html/). Last accessed on 15 April 2013.

5. BCC Research (2010). *Nanotechnology: A Realistic Market Assessment* (http://www.bccresearch.com/report/NAN031D.html/). Last accessed on 15 April 2013.

6. BCC Research (2009). *Nanotechnology in Environmental Applications: The Global Markets* (http://www.bccresearch.com/report/nanotechnology-environmental-applications-nan039b.html/). Last accessed on 15 April 2013.

7. Bogedal, M., Gleiche, M., Guibert, J. C., Hoffschlz, H., Locatelli, S., Malsch, I., Morrison, M., Nicollet, C., and Wagner, V. (2003). *Nanotechnology and the Implications for the Health of the EU Citizen*, in *Nanoforum Report*, ed. Morroson, M. (European Nanotechnology Gateway).

8. Brumfiel, G. (2006). Consumer products leap aboard the nano bandwagon. *Nature*, **440**, p. 262.

9. Business Insights (2010). *Nanotechnology in Healthcare: Market Outlook for Applications, Tools and Materials and 40 Company Profiles* (http://www.researchandmarkets.com/reports/1205887/nanotechnology_in_healthcare_market_outlook_for). Last accessed on 15 April 2013.

10. Chun, A. L. (2009). Will public swallow nanofood? *Nat. Nanotechnol.*, **4**, pp. 790–791.

11. Cientifica (2007). *Nanotechnologies for the Electronics Market* (http://www.researchandmarkets.com/reports/570935/nanotechnologies_for_the_electronics_market). Last accessed on 5 December 2011.

12. Cientifica Ltd. (2007). *Nanotechnologies for Sustainable Energy: Reducing Carbon Emissions Through Clean Technologies and Renewable Energy Sources* (http://www.researchandmarkets.com/reports/470560/nanotechnologies_for_sustainable_energy_reducing). Last accessed on 5 December 2011.

13. Credit Suisse (2008). *Press Release: Credit Suisse Global Nanotechnology Index* (https://infocus.credit-suisse.com/app/article/index.cfm?fuseaction=OpenArticle&aoid=223383&lang=en). Last accessed on 15 April 2013.

14. Freestone, I., Meeks, N., Sax, M., and Higgitt, C. (2007). The lycurgus cup–A Roman nanotechnology, *Gold Bull.,* **40**, pp. 270–277.

15. Future Market Inc. (2011). *Nanotechnology and Nanomaterials for the World Energy Market: Applications, Products, End User Markets, Companies and Revenues* (http://www.futuremarketsinc.com/index.php?option=com_content&view=article&id=43:the-world-market-for-nanoenergy&catid=1&Itemid=49/). Last accessed on 15 April 2013.

16. Future Market Inc. (2011). *Nanotechnology and Photovoltaics: Market, Companies and Products* (http://www.futuremarketsinc.com/index.php?option=com_content&view=article&id=46:the-world-market-for-autonano&catid=1&Itemid=53/). Last accessed on 15 April 2013.

17. Future Market (2011). *Nanotechnology and Nanomaterials in the Automotive Industry: Applications, Products, World Markets, Companies and Revenues* (http://www. futuremarketsinc.com/index.php?option=com_content&view=article&id=47&Itemid=54/). Last accessed on 15 April 2013.

18. Global Industry Analysis Inc. (2009). *Nanoelectronics—Global Strategic Business Report* (San Jose, USA).

19. Helmut Kaiser Consultancy (2008). *Study: Nanotechnology in Food and Food Processing Industry Worldwide 2008–2010–2015* (http://www.hkc22.com/Nanofood.html/). Last accessed on 15 April 2013.

20. Hilie, T., Munasinghe, M., Hlope, M., and Deraniyagala, Y. (2006). *Nanotechnology, Water and Development* (Meridian Institute, Washington DC, USA).

21. Holister, P., and Harper, T. E. (2002). *The Nanotechnology Opportunity Report* (CMP Científica, Madrid).

22. Hurst, S. J. (2011). *Biomedical Nanotechnology; Method and Protocol* (Humana Press, Springer, New York).

23. Kim, J. (2011). *Advances in Nanotechnology and the Environment* (Pan Stanford, Singapore).

24. Lambauer, J. (2011). *Nanotechnology and Energy–Science, Promises and Its Limits* (Pan Stanford, Singapore).

25. Liu, W. (2005). Catalyst technology development from macro-, micro- down to nano-scale, *China Particuol.,* **3**, pp. 383–394.

26. Loeffler, J., and Sutter, U. (2006). *Nanomaterial Roadmap 2015: Roadmap Report Concerning the Use of Nanomaterials in the Medical and Health Sector* (Steinbeis-Europa-Zentrum, Karlsruhe).

27. Loeffler, J., Sutter, U., Jourdain, E., and Kristiansen, S. (2006). *Nanomaterial Roadmap 2015: Roadmap Report Concerning the Use of Nanomaterials in the Aeronautics Sector* (Steinbeis-Europa-Zentrum, Karlsruhe).

28. Manalowski, N., Heimer, T., Luther, W., and Werner, M. (2006). *Growth Market Nanotechnology: An Analysis of Technology and Innovation* (Wiley-VCH, Weinheim).

29. Mann, S. (2006). Nanotechnology and Construction, in *Nanoforum Report*, ed. Morroson, M. (European Nanotechnology Gateway).

30. McHenry, M. E. (2000). Nano-scale materials development for future magnetic applications, *Acta Mater.,* **48**, pp. 223–238.

31. Miller, G., and Senjen, R. (2008). *Out of the Laboratory and on to Our Plates: Nanotechnology in Food and Agriculture* (Friends of the Earth).

32. Morroson, M. (2006). Nanotechnology in Consumer Products, in *Nanoforum Report* (European Nanotechnology Gateway).

33. Murr, L. E. (2009). Nanoparticulate materials in antiquity: the good, the bad and the ugly, *Mater. Charact.,* **60**, pp. 261–270.

34. Nanoforum (2004). Nanotechnology helps solve the world's energy problems, in *Nanoforum Report*, ed. Morroson, M. (European Nanotechnology Gateway).

35. Nanoposts.com (2010). *Key Products and Players, Global Markets and Applications for Nanotechnology in Personal Care, Cosmetics, Household Care, Packaging and Leisure Wear & Equipment* (http://www.nanoposts.com/). Last accessed on 15 April 2013.

36. Nanposts.com (2009). *Global Market and Applications for Nanotechnology in the Life Sciences and Healthcare Industries* (http://www.researchandmarkets.com/reports/694620/global_market_and_applications_for_nanotechnology). Last accessed on 15 April 2013.

37. Niedermeier, W., and Schw aiger, B. (2007). Performance enhancement in rubber by modern filler systems, *Kautschuk Gummi* Kunststoffe, April 2007, pp. 184–187.

38. Paik, U., and Park, J. G. (2007). *Nanoparticle Engineering for Chemical-Mechanical Planarization: Fabrication of Next-Generation Nanodevices* (CRC Press, Boca Raton).

39. Pan Stanford series of biomedical nanotechnology (www.panstanford.com/categories/Book-Series/Pan-Stanford-Series-on-Biomedical-Nanotechnology).

40. Perez, J., Bax, L., and Escolano C. (2005). *Roadmap Report on Nanoparticle* (Willems & van den Wildenberg, Spain).

41. Pitkethly, M. J. (2004). Nanomaterials—the driving force, *Mater. Today*, **7**, Suppl. 1, pp. 20–29.

42. Ravichandran, R. (2010). Nanotechnology applications in food and food processing: Innovative green approaches, opportunities and uncertainties for global market, *Int. J. Green Nanotechnol. Phys. Chem.*, **1**, pp. 72–96.

43. Roco, M. C. (1999). Nanoparticles and nanotechnology research, *J. Nanopart. Res.*, **1**, pp. 1–6.

44. Roque, J., Molea, J., Perez-Arntegui, J., Calabuig, C., Portillo, J., and Vendrell-Saz, M. (2007). Luster colour and shine from the olleries Xiques workshop in Paterna (Spain), 13th century AD: nanostructure, chemical composition and annealing conditions, *Archeometry*, **49**, pp. 511–528.

45. Ruch, C. (2004). Nanotechnology in the EU–Bioanalytical and Biodiagnostic Techniques, in *Nanoforum Report*, ed. Morroson, M. (European Nanotechnology Gateway).

46. Salamanca-Buentello, F., Persad, D. L., Court, E. B., Marin, D. K., Daar, A. S., and Singer, P. A. (2005). Nanotechnology and the developing world, *PLos Med.*, **2**, pp. 383–386.

47. Sutter, U., and Loeffler, J. (2006). *Nanomaterial Roadmap 2015: Roadmap Report Concerning the Use of Nanomaterials in the Energy Sector* (Steinbeis-Europa-Zentrum, Karlsruhe).

48. Sutter, U., Loeffler, J., Bidmon, M., Valadon, H., and Ebner, R. (2006). *Nanomaterial Roadmap 2015: Roadmap Report Concerning the Use of Nanomaterials in the Automotive Sector* (Steinbeis-Europa-Zentrum, Karlsruhe).

49. Tibbas, H. F. (2010). *Medical Nanotechnology and Nanomedicine* (CRC Press, Boca Raton).

50. Tsuzuki, T. (2009). Commercial scale production of inorganic nanoparticles, *Int. J. Nanotechnol.,* **6,** pp. 567–568.

51. WinterGreen Research Inc. (2011). *Energy Harvesting Market Shares, Strategies, and Forecasts, Worldwide, Nanotechnology, 2012 to 2018* (http://www.researchandmarkets.com/reports/1958613/energy_harvesting_market_shares_strategies_and/). Last accessed on 15 April 2013.

52. WinterGreen Research Inc. (2009). *Worldwide Nanotechnology Dental Implant Market Shares, Strategies, and Forecasts, 2009 to 2015* (http://www.researchandmarkets.com/reports/941893/worldwide_nanotechnology_dental_implant_market). Last accessed on 15 April 2013.

53. WinterGreen Research (2009). *Worldwide Nanotechnology Mid IR Sensor Market Shares Strategies, and Forecasts, 2009 to 2015* (http://www.researchandmarkets.com/reports/1055867/worldwide_nanotechnology_mid_ir_sensor_market). Last accessed on 15 April 2013.

54. Yano, H., and Nakahara, S. (2004). Bio-composites produced from plant microfiber bundles with a nanometer unit web-like network, *J. Mater. Sci.,* **39,** pp. 1635–1638.

55. Woodraw Wilson International Center for Scholars, Project on Emerging Nanotechnologies (2011). http://www.nanotechproject.org/inventories/consumer/analysis_draft/. Last accessed on 15 April 2013.

Chapter 3

Production Techniques of Nanoparticles on a Laboratory Scale

Putla Sudarsanam and Benjaram M. Reddy

Inorganic and Physical Chemistry Division,
Indian Institute of Chemical Technology,
Uppal Road, Hyderabad 500607, India

bmreddy@iict.res.in, mreddyb@yahoo.com

The research intriguing of nanomaterials synthesis has been growing significantly to meet the demand for future applications. Most of the excitement in this field has arisen due to the novel size- and shape-tunable properties of nanostructures. These fascinating characteristics make them potential candidates in many interesting fields. For instance, nanoparticles have found enormous attention in catalysis, material science, biotechnology, environmental science, pharmacy, electronic industry, etc. This chapter reviews prevalent developments towards facile design of nanoscale materials. Primarily, solution-based methods possess influential routes in a controlled and reproducible manner. In addition, gas- and solid-phase techniques are stimulating greatly into a new branch of nanosynthetic chemistry. A wide range of nano-

Nanotechnology Commercialisation
Edited by Takuya Tsuzuki
Copyright © 2013 Pan Stanford Publishing Pte. Ltd.
ISBN 978-981-4303-28-6 (Hardcover), 978-981-4303-29-3 (eBook)
www.panstanford.com

fabrication methods, including precipitation, deposition-precipitation, sol-gel methods, microemulsion techniques, hydrothermal and solvothermal syntheses, microwave-assisted processes, polyol method, flame spray pyrolysis, template-directed synthetic methods, mechanochemical processing, ionic-liquid assisted methods, and so on, are discussed elaborately. A meticulous study of nanoconfinement variation with reaction parameters is also elucidated in all methodologies.

3.1 Introduction

Over the past decades, the nano-size revolution has intensified tremendously in the science and engineering communities. It has been accomplished by the frontier character of nanoscale materials between bulk and atomic components. Different classes of nanomaterials have been advanced considerably and showed their valuable applications in several provocative domains. Particularly, inorganic nanoparticles (NPs) have rapidly emerged as new appealing materials owing to their unique size- and shape-dependable properties. Plentiful studies on these nanostructures have revealed the supremacy of nano-size in view of their properties and applications. A well-known example of synergism between nanoscale discovery and scientific evolution is the nanocatalysis, which is a phenomenon of significant basic research and remarkable practical use in many industrial areas. Also, their promising use in materials science, pharmacy, biotechnology, environmental science, and electronic industry is well appreciated. Accordingly, now all kinds of inorganic NPs have been realised, such as metals, metal oxides, hydroxides, carbonates and semiconductors. It is very exciting to note at the outset that there are multiple production techniques to synthesise functional nanomaterials [32,46,100,148,191]. However, it is worthwhile to know the distinct characteristics of NPs prior to the discussion of nanosynthetic methods.

The word "nanoparticle" is typically used to describe the broad spectrum of materials in the nanometre scale (1 to 100 nm) as depicted in Fig. 3.1.

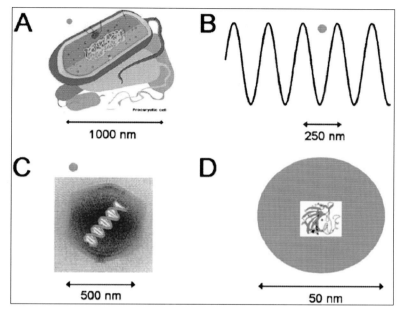

Figure 3.1 Depiction of the size regime of nanoparticles related to common nano-sized objects: (A) prokaryotic cell, (B) ultraviolet wave, (C) virus, and (D) enzyme. Blue spheres signify a 50 nm metal nanoparticle. Reprinted from ref. [178], with permission from the Royal Society of Chemistry.

A nanometre is a billionth of metre, which is approximately 1/80,000 and 10 times of the diameter of a human hair and a hydrogen atom, respectively [178]. In general, bulk materials comprise invariable properties regardless of its size. However, properties of materials vary remarkably as their size approaches a few nanometres. Consequently, they exhibit very small size and high specific surface area with most of the atoms lying near or at the surface of particles. Additionally, an unpredicted change occurs in colour, melting point, boiling point, conductivity, magnetism, light absorptivity, and chemical reactivity. NPs display substantial divergence in surface energy, surface area to volume ratio, and ensemble properties. Moreover, they possess larger percentage of active surface sites (e.g. edge, corner, and kink) than those in the bulk material [124]. These attractive properties could dominate the contributions made by the small bulk of the material. A classic example is the pronounced reduction of melting point of gold (Au) with decrease in the particle size. Au particles in the 5, 2 and 1 nm size range would be melt at 1103, 623, and 473 K, respectively. In contrast, bulk gold melts at higher temperature (1337 K) [25].

A great deal of efforts has been directed towards the fruitful synthesis of nanoparticles. However, there are still many more factors that need to be considered in the perspective of nanomaterials production. If we desire to speak ideally, the chosen method must be simple, less expensive, eco-friendly, and commercially viable [15]. The simultaneous control of particle size and shape together with their uniformity is one of the key objectives for any process [78]. Since the kinetically unstable nature of NPs, they should be stabilised against the aggregation into larger particles [132]. Micelles, polymers and coordinative ligands are frequently used as stabilisers in order to control the growth of NPs [71]. Furthermore, the development of innovative characterisation techniques could bring about some new features in the nanoscale range. For example, scanning electron microscopy (SEM), transmission electron microscopy (TEM), X-ray photoelectron spectroscopy (XPS), extended X-ray absorption fine structure (EXAFS) techniques, etc., can be applied to pinpoint active sites at the nano-size level. As a result, it is feasible greatly to attain pioneering nanocatalysts with superior activity, high selectivity, and prolonged catalytic lifetime [84]. Above all, from an industrial point of view, economical mass production of nanocrystals is indispensable to meet the demand for imminent applications [32].

In this chapter, numerous production techniques for the fabrication of inorganic NPs are addressed in detail. A careful study of their usefulness and drawbacks with special emphasis on laboratory scale synthesis has been also conferred. Compared with gas- and solid-phase strategies, wet chemical routes can be easily tailored to a great extent and can produce NPs with specific properties and fascinating designs. By utilising straightforward reactions, appropriate systems and adjusting the reaction parameters, we can prepare abundant nanopowders. Some processes can even afford hundreds of grams of high quality monodispersed NPs in the laboratory [32].

3.2   Precipitation Methods

A lot of promising synthetic techniques have been devoted to obtain a wide variety of nanoparticles. Among them, precipitation or coprecipitation is the broadly used method in both the laboratory and the industry. Notably, these processes afford ultrafine, high purity, and better dispersion of nanopowders [3]. They are simple and cost effective and reduce the use of harmful organic solvents/reagents. The synthesis procedure starts with supersaturation of metal precursor's solution by the addition of precipitating agent at around the desired pH. Afterwards, two important reactions are involved: A short burst of nucleation and subsequent growing of nuclei. These two strategies will govern the size and shape of the products. The fast nucleation and slow growth are favourable to achieve small particle size and narrow size distribution. Moreover, the size-tunable properties can be tailored easily by adjusting the pH, temperature, ionic strength, and nature of the salts [28]. Thus, a great number of nanomaterials, especially nanoscale metal oxides have been prepared using precipitation routes.

Metal oxides represent one of the successfully employed classes of nanocatalysts, either as active phases or as supports [178]. Among the various metal oxides, ceria-based nano-oxides are extensively studied materials in recent years. Principally, ceria is a key component in three-way catalytic converter (TWC) owing to its ability to store and release oxygen. This inherent oxidation and reduction tendency has been utilised to reduce hazardous gases, namely, CO, NO_x, and hydrocarbons [131]. It is worth outlining the effort of research group headed by Reddy *et al.* towards the productive synthesis of ceria-based materials [122, 123,125,129]. They developed a novel and versatile coprecipitation route using ultra-high dilute aqueous solutions at room temperature. A family of doped ceria solid solutions with transition metals (Ti, Zr or Hf) and inner transition metals (Pr, La or Tb) were prepared with aq. NH_3 solution as precipitating agent. The detailed synthesis procedure of CeO_2–ZrO_2 (1:1 mole ratio) mixed oxide is methodically illustrated in Fig. 3.2. The same procedure can be applied to the aforesaid metals. The meticulous analysis of various characterisation techniques revealed that doped ceria samples exhibit small particle size, high specific surface area, better thermal stability, and superior OSC (oxygen storage/release capacity).

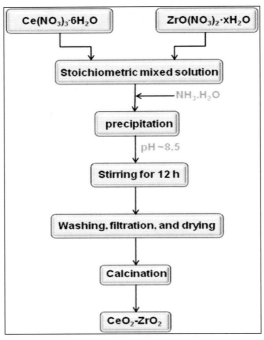

Figure 3.2 Systematic synthesis procedure of nano-sized CeO_2–ZrO_2 mixed oxide using coprecipitation method.

A very thoughtful experiment was performed to investigate the affect of reaction temperatures on coprecipitated $CoFe_2O_4$ NPs. It has been shown that the particle size of as-synthesised $CoFe_2O_4$ is effectively controlled at lower temperatures and is increased remarkably with temperature. The particle sizes ranged from 2.4 and 47.4 nm at 313 and 373 K, respectively [116]. Hydroxyapatite ($Ca_{10}(PO_4)_6(OH)_2$ or HA) nanopowders were fabricated by using calcium nitrate and phosphoric acid as the starting materials. The obtained samples are highly dispersed with an average grain size of about 60 nm and a BET surface area of 62 m^2/g. The microstructure evolution of HA ceramics was examined by varying the sintering temperature [76]. A rapid synthesis of manganites AMn_2O_4 (A = Co, Ni or Zn) was carried out using metal acetate precursors and oxalic acid in water.

The produced mixed oxides were carefully analysed by SEM, TEM, and XRD techniques and were evaluated as anode materials for Li-ion batteries. The effect of calcination temperatures (from 673 to 1273 K) on particle size was also scrutinised [26].

Recently, a cost-effective process was proposed to synthesise indium tin oxide (ITO) NPs by means of pure metals instead of metal chlorides. ITO NPs with crystallite size of 12.6 nm and a specific surface area of 45.7 m^2/g were achieved by calcining the sample at 873 K for 3 h. The prevention of chlorine ions in the reaction brought several advantages, including the short reaction time and an improvement in the particle purity as well as the dispersity [170]. At the same time, Simentsova *et al.* have explored the effect of Ni^{2+}/Cr^{3+} ratio on the formation of nanocatalyst structure [149]. An aqueous solution of Ni and Cr nitrates was precipitated with sodium or ammonium carbonate at 338–343 K. The precursors with high nickel contents led to a homogeneous system of the stichtite-type structure and showed excellent activity in hydrogenation reactions. At low nickel contents, similar structure is not perceived and hence, the catalyst manifestly no high activity. Behrens and co-authors reported a scalable nitrate-free precipitation route for the production of nano-particulate CuO/ZnO materials [9]. They used a mixed basic carbonate precursor, zincian malachite ($(Cu_{1-x}Zn_x)_2(OH)_2CO_3$) and aged in the form of needles. These needles were decomposed into nanostructured CuO/ZnO material by thermal treatment. An innovative in-line dispersion-precipitation (ILDP) technique was developed to attain nano-sized Mg–Al layered double hydroxides (LDHs). LDHs, also known as hydrotalcite-like compounds, are a class of synthetic two-dimensional nanostructured inorganic materials. By adjusting the reaction parameters, very small crystallites ranging from 3.9 to 38 nm can be achievable in ILDP method. Continuous function, tunable residence time, and proper mixing are the key benefits of this process [2].

Unfortunately, a few metal ions are unable to precipitate in aqueous medium. Thus, precipitation in non-aqueous solvents has been received great interest to undertake the above challenge. A simple non-aqueous precipitation route was conducted to prepare nanoscale $LiCoO_2$ material. In a typical reaction, adequate amounts of Li and Co precursors were dissolved in ethanol, followed by addition of 3 M KOH ethanol solution. Calcination of the resultant hydroxide product affords pure nanopowders with diameters of around 12–41 nm [51]. A highly efficient nanostructured Cu/ZnO/Al_2O_3 material was fabricated by a gel-coprecipitation process. A mixed alcoholic solution of metal nitrate reagents was taken in an alcoholic solution of 20% excess of oxalic acid at room temperature. The synthesised products are featured by high specific surface area with enhanced component dispersion [198]. Khaleel *et al.* reported that pure maghemite (γ-Fe_2O_3) as ultrafine particles can be prepared by employing a forced precipitation technique [68]. The precipitation of Fe(III) ions in 2-propanol led selectively to highly dispersed particles of ferrihydrite and subsequent calcination under dynamic vacuum yielded to nanoscale γ-Fe_2O_3.

In 2008, Taniguchi *et al.* used a Ce-oleate complex as chemically modified precursor to obtain nano-sized CeO_2 with 25 wt.% aqueous ammonia solution [161]. An additional hydrothermal treatment was performed at 423 and 473 K for 6 h and then dried at same conditions. The average sizes of nanocrystals at 423 and 473 K are measured to be 3.3 and 6.0 nm, respectively as shown in Fig. 3.3.

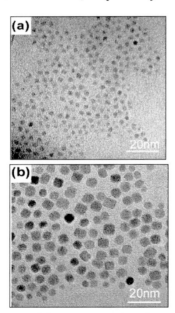

Figure 3.3 TEM images of the CeO_2 nanoparticles prepared by an oleate-modified precipitation route at (a) 423 K and (b) 473 K for 6 h. Reprinted with permission from ref. [161]. Copyright 2008 American Chemical Society.

Tang *et al.* have produced ZnO NPs using urea and sodium dodecyl sulphonate (SDS) as precipitant and surfactant, respectively [158]. A certain amount of urea and SDS was added into reaction system to inhibit the ZnO crystal growth and agglomeration. All the samples exhibited nearly spherical shape with 10–40 nm particle size. This approach can offer monodispersed particles compared with direct precipitation because of no concentration gradient of precipitates in the solution. Lv *et al.* examined the controlled growth of magnesium hydroxide NPs by XRD, TEM, and field emission SEM (FESEM) techniques [89]. The alkaline solution concentration plays a crucial role: The lower concentration of aq. NH_3 (5 wt.%) promotes the formation of needle or rod morphologies, whereas nuclei could agglomerate and form lamellar-like particles at higher concentration of aq. NH_3 (25 wt.%) solution.

3.3　Deposition-Precipitation (DP) Techniques

The preparation of supported metal nanoparticles (SMNPs) with high metal loading and better dispersion is a fruitful research area in heterogeneous catalysis. SMNPs display novel physico-chemical properties and, thereby, exhibit high activity and excellent selectivity in many catalytic formulations [101]. Deposition-precipitation (DP) is one of the notable wet chemical routes for the fabrication of SMNPs. DP strategies are usually based on precipitation of metal ions combined with deposition of precipitates on supports. They provide an easy route, which takes all the advantages of the precipitation reactions related to control of the size and size distribution of particles [69]. Numerous supports, mainly metal oxides, such as CeO_2, TiO_2, Fe_2O_3, Co_3O_4, ZrO_2, Cr_2O_3, MnO_2, NiO, CuO, ZnO, Y_2O_3, La_2O_3, U_3O_8, SnO_2, MgO, Al_2O_3, SiO_2, WO_3 and SiO_2–Al_2O_3 have been reported in the literature. Supports play a vital role for enhancing the properties of metal phase through what is usually called "metal-support interaction". They must have large specific surface areas and highly populated surface defects that could offer the strong interactions between metals and supports [88,112]. As a result, high thermal stability of NPs with inhibition of particle size could be greatly achieved. Moreover, the small crystallites are frequently anchored into support oxides, hence avoid particle migration and coalescence of particles. Therefore, the selection of an appropriate support is a decisive factor in the DP reactions [184].

Since the discovery of catalytic activity of gold (Au) in the 1985, many DP studies have been paid immense attention to synthesise nano-sized Au materials [50,87]. The pioneering work of Haruta *et al.* revealed that gold exhibits outstanding activity when it is dispersed as small particles on a selected metal oxide [59]. They developed a straightforward process with $HAuCl_4 \cdot 3H_2O$ as gold precursor and aq. NaOH solution as precipitating agent. The pH of aqueous $HAuCl_4$ solution is adjusted to the desired value (6–10) by addition of precipitant and subsequent dispersion of TiO_2 into the solution. The calcined products at 573–673 K yielded to nanoscale Au particles in the range of 2–10 nm with narrow size distribution [59]. A recent study by Lay's group illustrates the affect of specific surface area of ceria on the properties and activity of gold samples [79]. The presence of Au on the high surface area ceria showed more resistance to sintering, better surface reducibility, and thus, superior catalytic performance toward the oxidation of CO and benzene than those prepared similarly on the low-surface area oxide. An unfortunate drawback of DP in case of the gold is that it is not applicable to oxides with a point of zero charge (PZC) below 5, for instance SiO_2, WO_3, and SiO_2–Al_2O_3 [199].

Owing to its great practical simplicity, several research groups have also carried out DP method to obtain other SMNPs [106]. Iridium (Ir) NPs dispersed over different metal oxides were produced using an aqueous solution of $IrCl_4$ and 0.1 N NaOH. The influence of pH and calcination treatments were studied to evaluate the structural properties of materials. It is a noteworthy example that correlates the structural characteristics of metal with pH and thermal treatments. Figure 3.4 shows the TEM images of the Ir/TiO_2 catalysts [111]. In 2001, Kapoor *et al.* studied the deposition of palladium (Pd) particles on CeO_2, ZrO_2 and CeO_2–ZrO_2 oxides by varying the metal loadings (3–10 wt.%) [67]. The ample investigation by characterisation techniques unveiled that Pd NPs are strongly interacted with the mesoporous oxides. The specific surface areas of samples decreased with increasing metal content. A facile synthesis of nano-sized rhodium (Rh) particles on Mg–Al hydrotalcite was developed by means of Na_2CO_3 solution. Most of the particles are spherical with diameters 5, 6 and 11 nm for 1.7, 2.1 and 3.0 wt.% of Rh, respectively. The reduction of specific surface area of hydrotalcites was noticed after deposition of Rh due to incorporation of Rh particles in the bulk phase of hydrotalcite [66].

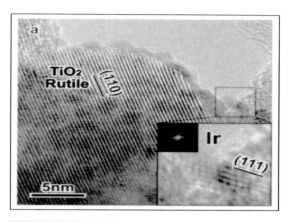

Figure 3.4 High magnification TEM pictures of the Ir/TiO_2 catalysts prepared by the DP method at pH 8: (a) calcined in air and (b) calcined in the hydrogen stream. Reprinted from ref. [111], with permission from Elsevier.

Deposition-precipitation using urea, more popularly known as homogeneous deposition-precipitation (HDP), could be an alternative technique to the conventional DP process. The key advantage of this method is that higher metal loadings can be greatly achievable, which is ascribed to the deposition of total metal present in the solution [192]. Use of urea (NH_2CONH_2) instead of typical bases, like NaOH, KOH or Na_2CO_3 is quite desirable, because solution mixing and basification can be executed separately. Moreover, it enables the steady and homogeneous dispersion of hydroxide ions throughout the solution and, hence, avoids local high pH values [13].

Wang *et al.* have fabricated monodispersed Au nanocrystals over MnO_2 oxide of different morphologies using a simple HDP approach [172]. Various characterisation results disclosed that gold exhibits strong interaction with MnO_2 nanorods when compared with MnO_2 NPs. As shown in Fig. 3.5, Au NPs are uniformly dispersed on the MnO_2 nanorod surface with semispherical shape. Bitter *et al.* reported that HDP method can even useful for deposition of metal particles on inert supports, such as graphitic carbon nanofibre [10]. Highly loaded (45 wt.%) nickel (Ni) NPs with small particles size of 9 nm were prepared. It was speculated that the oxygen-containing groups on the surface of the carbon nanofibre serve as nucleation and anchoring centres in the deposition of Ni particles. Chytil and co-authors have scrutinised the functionalisation of mesoporous silica (SBA-15) with platinum (Pt) by means of $[Pt(NH_3)_4](OH)_2$ as metal precursor [24]. The interactions between the support and the precipitating species were investigated using pH profiles. The analysis of TEM images revealed that the majority of Pt NPs on SBA-15 are in the range of 2–4 nm. However, particles in the diameter of 15 nm are also identified, which indicates that the precipitation does not occur completely within the channels of SBA-15.

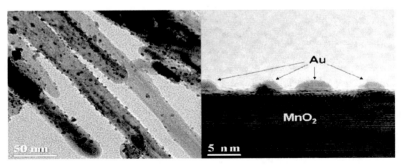

Figure 3.5 TEM (left) and HRTEM (right) images of gold nanoparticles dispersed on MnO_2 nanorod. Reprinted with permission from ref. [172]. Copyright 2008 American Chemical Society.

Deposition coprecipitation is mostly analogous to the DP method, except the nature of produced materials. This methodology affords monodispersed nanoscale metal oxides instead of pure metals over various supports. During the past decade, Reddy *et al.* have been designing supported ceria-based nanomaterials with aq. NH_3 solution as precipitating agent [121]. Major research work has been focused on deposition of CeO_2–ZrO_2 over SiO_2, TiO_2 and Al_2O_3. It is a well-known fact that SiO_2, TiO_2 and Al_2O_3 are the widely employed supports in diverse catalytic transformations. These oxides exhibit a strong influence on the redox behaviour as well as on the thermal stability of solid solutions. $Ce(NO_3)_3 \cdot 6H_2O$ and $ZrO(NO_3)_2 \cdot xH_2O$, colloidal SiO_2, anatase TiO_2 and γ-Al_2O_3 are used as source reagents. The obtained products comprise small particle size, higher surface area, superior OSC and, thereby, enhanced catalytic activity for CO oxidation.

3.4 Sol-Gel Methods

Sol-gel method is one of the well-established approaches that show great capability in the controlling of textural and surface properties of nanoscale materials. Basically, the sol-gel reactions involve the following steps: Hydrolysis, condensation, and drying process. In a few words, it entails the hydrolysis of metal precursors to give corresponding hydroxides and subsequent condensation lead to the formation of viscous gel. Gels are usually defined as a polymer of three-dimensional skeleton surrounding interconnected pores. The metal precursors, solvent systems, and electronegativity of the metals play an energetic role in the hydrolysis and condensation reactions. The resulting gel can be converted to either xerogel or aerogel that depends on the drying process. The entire reaction pathway is systematically illustrated in Fig. 3.6. This subject can be categorised into two parts, namely "aqueous" and "non-aqueous" route, where water and organic solvents are used as reaction medium, respectively. These two strategies differ in many functions because of distinctive properties of the solvents. Here, solvents not only serve as reaction media but also as an efficient controller of the reaction parameters. Their unique properties, such as surface tension, dielectric constant, and dipole moment can prominently affect the rate of gel formation, gel structure, and drying behaviour [28,77,115,124].

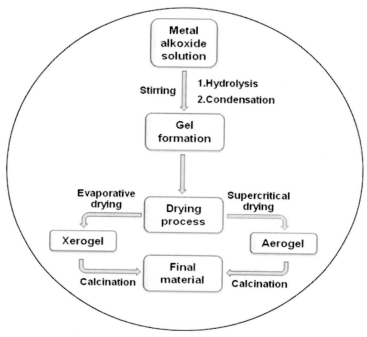

Figure 3.6   Methodical reaction pathway in sol-gel method.

3.4.1   Aqueous Sol-Gel Method

The aqueous sol-gel process was reported initially in the late 1950s for the production of $(Th,U)O_2$ oxides [7]. Metal alkoxides, acetates, chlorides, nitrates and sulphates are generally used as precursors and their chemistry in water is elaborately discoursed. During hydrolysis, metal reagents hydrolyse in water which eventually lead to the formation of analogous metal hydroxides. These metal hydroxides involve in condensation reactions either by water condensation or by an alcohol condensation, affording M–O–M bonds. Afterwards, the solvent must be removed and higher temperatures are often required to decompose the organic precursors [14]. Although this route mainly signifies the use of water, in some cases utilisation of organic solvents in addition to water is necessary to enhance the size-tunable properties.

Rajesh *et al.* have prepared cerium phosphate NPs by means of cerium nitrate and orthophosphoric acid as starting materials in aqueous media [118]. The calcination of obtained gel exhibited an average particle size of 11 and 22 nm at 673 and 1073 K, respectively. Moreover, these nanostructures showed enhanced thermal stability up to 1973 K. Around the same time, Han *et al.* studied an aqueous sol-gel process for the synthesis of hydroxyapatite (HAP) nanopowders using citric acid [57]. Here, citric acid acts as chelating agent as well as reducing agent. Consequently, strong oxidation-reduction reactions are generated between citric acid and nitrate precursors that ultimately led to formation of nanoscale HAP materials. According to Goutailler *et al.* TiO_2 nanoparticles with a mesoporous structure can be easily synthesised using ammonium bromide salts as catalysts [52]. Bromide salts are capable of inducing the crystallisation of TiO_2 and inhibit the aggregation of particles. By varying the reaction parameters, for instance reflux conditions and concentration of ammonium salts, the physico-chemical properties of TiO_2 can be tailored.

Fuentes *et al.* have fabricated nanocrystalline $Ce_xZr_{1-x}O_2$ (CZ) sold solutions with a wide range of compositions (x = 0.1, 0.25, 0.5, 0.75 and 0.9) [44]. Among the materials, composition containing 50 equivalent mol % of CeO_2 (CZ50) displayed the smallest crystallite size of 4.8 nm and the largest value of specific surface area of 45.8 m^2g^{-1}. Figure 3.7 represents the HRTEM images of CZ50 sample that aligned in the [011], [001], and [112] zone axes with the corresponding digital diffraction patterns (DDPs). Chervin *et al.* have examined both aerogel and xerogel drying for a sol-gel synthesis of yttria-stabilised zirconia (YSZ) nanomaterials in both aqueous and ethanol–water solutions [23]. YSZ is a high-temperature oxide conductor that finds extensive use as solid-state electrolyte in solid-oxide fuel cells and in oxygen sensors. Aerogel drying resulted in fine and nanoparticulate networks, whereas drying under xerogel conditions led to heterogeneous micrometre-sized hard agglomerates. Nevertheless, very little attention has been paid in the aqueous sol-gel route, probably due to the high reactivity of metal precursors in water. Moreover, both hydrolysis and condensation reactions are difficult to control individually. Also, the preparation of ternary and multi-metal oxides is often challenging issue because of different reactivities of the individual precursors in aqueous media [17,105].

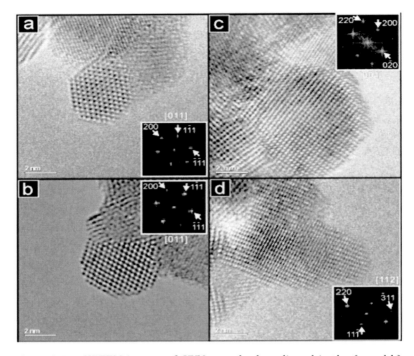

Figure 3.7   HRTEM images of CZ50 sample that aligned in the (a and b) [011], (c) [001] and (d) [112] zone axes. DDPs are inset and the diffraction spots are indexed to the fluorite structure. Reprinted with permission from ref. [44]. Copyright 2009 American Chemical Society.

3.4.2   Non-Aqueous (or Non-Hydrolytic) Sol-Gel Method

Nowadays, sol-gel synthesis of nanomaterials in organic solvents is a versatile methodology to overcome many specific problems in aqueous systems. The main advantage of non-aqueous route lies in the accessibility of highly crystalline materials at moderate temperatures. It provides high energetic reaction parameters to tune the crystal growth of final products. Another important benefit is the convenience of fabrication of the ternary and multi-metal oxide NPs. These advantages are primarily due to the manifold roles of organic solvents in the reaction mixture. They profoundly influence the particle size, shape, surface and assembly properties, and even, in some cases, composition and crystal structure. Numerous metal precursors are used to synthesise various nanoscale metal oxides, such as CeO_2, ZrO_2, TiO_2, WO_x, Sm_2O_3, MnO, Mn_3O_4, FeO_x, CuO_x, Cr_2O_3, Co_3O_4, In_2O_3, SnO_2, V_2O_5, ZnO, CoO, NiO, ReO_3, Y_2O_3:Eu, $HfO_2/Hf_xZr_{1-x}O_2$ and so on [103–105, 182].

The most explored reaction pathway is the utilisation of alcohols as solvents. Especially, benzyl alcohol serves both as a solvent and as a reducing agent. It plays an outstanding role, reacting with many precursors, thus giving access to more than 35 different nano-oxides. Figure 3.8 presents a small collection of nano-sized metal oxides prepared in benzyl alcohol [113]. A rapid synthesis of $CdIn_2O_4$ nanocrystals was performed involving the reaction of cadmium acetate and indium isopropoxide in benzyl alcohol. The crystal growth is controlled by the solvent and/or by the organic reaction products without aid of any stabilising agents. Furthermore, the molar ratio of reactants has a gigantic effect on the purity of final material. $CdIn_2O_4$ NPs with crystallite sizes of about 10 nm were obtained [17]. Jia *et al.* reported that Ni nanomaterials with different morphologies are effectively feasible in benzyl alcohol [61]. Because of the weak reducing ability of benzyl alcohol, pure metal was attained instead of metal oxide. They introduced a magnetic force into the reaction system, resulting in different shapes, such as nanospheres, nanowires, and nanoflowers. Xiao *et al.* have explored a facile non-aqueous process for the production of nanostructured rare earth oxides (Sm_2O_3, Gd_2O_3 and Dy_2O_3) [183]. The stacks of ultrathin nanodiscs for Sm_2O_3 and hierarchical nanosheet microspheres for Gd_2O_3 and Dy_2O_3 were observed. The affect of source materials in the non-hydrolytic synthesis of manganese oxides (MnO and Mn_3O_4) was investigated. The reaction of manganese (II) acetylacetonate with benzyl alcohol led to 80 and 20 wt.% of MnO (manganosite) and Mn_3O_4 (hausmannite) NPs, respectively, whereas 13 wt.% of MnO and 87 wt.% of Mn_3O_4 nanoparticulate materials were produced by $KMnO_4$ as metal precursor [34].

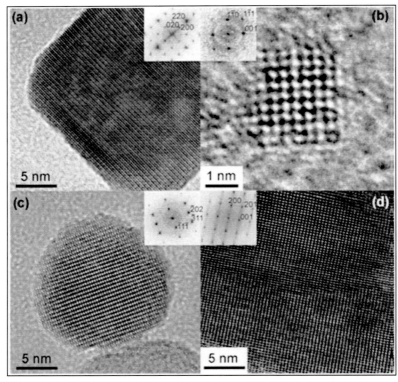

Figure 3.8 HRTEM pictures of In_2O_3 (a), $Sn_{0.95}In_{0.05}O_x$ (b), Fe_3O_4 (c), and Nb_2O_5 (d). Insets: Respective power spectra. Reprinted from ref. [113], with permission from Wiley-VCH.

Bonnetot *et al.* have performed an effective fabrication of ceria-based nanopowders Me_2O_3–CeO_2 (Me = B, Al, Ga, and In) in isopropanol [11]. The structural, textural, and surface properties of these materials were examined meticulously by means of a variety of techniques. The resultant oxides displayed more oxygen vacancies, better reduction ability and enhanced acid-base properties. The preparation of lanthanum hydroxide $(La(OH)_3)$ NPs was studied using $La(OiPr)_3$ and $KMnO_4$ in benzyl alcohol, 2-butanone and their equivolume mixture. Although $KMnO_4$ is not directly involved in the reaction process, it plays a crucial role in determining the final morphology of $La(OH)_3$ materials. Representative TEM images of various $La(OH)_3$ NPs are shown in Fig. 3.9 [33]. In 2007, Zhang *et al.* have synthesised nanocrystalline $InNbO_4$ semiconductor involving the reaction of indium acetylacetonate and niobium chloride reactants in benzyl alcohol [195]. The as-synthesised powder is featured by better crystallinity and high BET surface area of 54 m^2/g. A nonhydrolytic process was developed to produce HfO$_2$ and Hf$_x$Zr$_{1-x}$O$_2$ nano-oxides over a wide range of x values. Depending on the relative ratios of the precursors, monodispersed tetragonal nanospheres and monoclinic nanorods of Hf$_x$Zr$_{1-x}$O$_2$ are formed. Attempts were also made to obtain HfO$_2$-TiO$_2$, and ZrO$_2$-TiO$_2$ mixed oxides. However, only anatase TiO$_2$ with small incorporation of ZrO$_2$ or HfO$_2$ was perceived due to the high reactivity of Ti compounds than the Zr and Hf reagents [159].

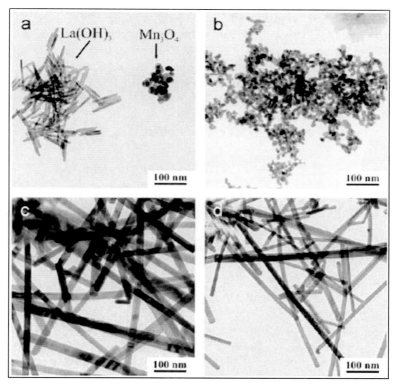

Figure 3.9 TEM profiles of La(OH)$_3$ nanoparticles synthesised under various conditions: (a) 1:1 molar ratio of La(OiPr)$_3$-to-KMnO$_4$ in benzyl alcohol, (b) in 2-butanone, (c) in their equivolume mixture, and (d) 1:0.5 molar ratio of La(OiPr)$_3$-to-KMnO$_4$ in their equivolume mixture. Reprinted from ref. [33], with permission from Elsevier.

3.5 Microemulsion Techniques

Since the early 1980s, nanomaterials production using microe-mulsions has been a hot research area in nanosynthetic chemistry [36]. Microemulsions (or micelles) are transparent and isotropic liquid systems that normally encompass a three-component assembly of water, oil, and a surfactant. Micelles provide a microenvironment to perform reactions that will eventually offer nanoscale materials with stimulated properties. Several parameters, for instance reactant concentration, the type and amount of surfactant, and calcination temperature intensely affect the size and shape of the final nanostructures [6,55]. Moreover, employ of co-surfactants (e.g. simple alcohols) can be a key factor in regulating the size-dependable properties of the crystallites [27]. This topic can be divided into two sections as oil-in-water (normal) or water-in-oil (reverse) micelles based on the several factors. In both systems, the dispersed phase comprises monodispersed droplets in the range of 100–1000 Å [96,136]. Besides the above mentioned two classes, water-in-supercritical CO_2 (W/scCO_2) has also received much attention in recent years.

3.5.1 Oil-in-Water (O/W) Microemulsion Method

An oil-in-water system is formed when oil is dispersed in the water continuous phase. These two liquids co-exist in one phase with the help of surfactant molecules. Surfactants are amphiphilic organic compounds, which possess a hydrophobic tail group as well as a hydrophilic head group. Upon the addition, surfactant molecules will migrate to the water surface, where the insoluble hydrophobic group may extend out of the bulk water phase into the oil phase and the water soluble head group remains in the water phase. This alignment and aggregation of surfactant molecules could effectively stabilise the water/oil interface. Owing to the highly soluble nature of the metal precursors in the water phase, O/W arrangement could be huge advantage for the production of inorganic NPs [48,82,138,139].

Ge *et al.* have proposed a mechanism based on the "interface-controlled reaction" to describe the formation of $BaCrO_4$ nano-structures using an O/W microemulsion [48]. Figure 3.10 depicts TEM, XRD, and ED analyses of crystallised $BaCrO_4$ nanorods and nanoparticles. A simple O/W micelle route was explored to fabricate mesoporous nanocrystalline oxides. Various nano-oxides, namely CeO_2, ZrO_2, $Ce_{0.5}Zr_{0.5}O_2$, and TiO_2 were prepared with particle size of ~2–3 nm and specific surface area of 200–370 m^2/g. These materials are found to be excellent supports for gold nanocatalysts [139]. Li and co-workers have reported the production of monodispersed Co–B (cobalt–boron) nanoalloy by chemical reduction of Co ions with borohydride in O/W system [82]. The microemulsion solution consists of cyclohexane, polyethylene glycol, and water. The size of Co–B particles is successfully controlled by modulating the cyclohexane content.

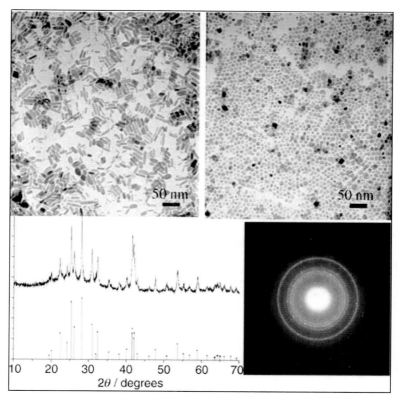

Figure 3.10   TEM, XRD, and ED analyses of crystallised $BaCrO_4$ nanorods and nanoparticles prepared ethanol- and methanol-composed microemulsions, respectively (XRD and ED results are for the nanorods). Reprinted from ref. [48], with permission from Wiley-VCH.

A novel and versatile O/W approach was used for the synthesis of Pt, Pd, Rh, and CeO_2 NPs with organometallic precursors at room temperature. Parameters, such as oil phase quantity, surfactant amount, and water content were showed remarkable effect on the size-tunable properties. Small particles of less than 7 nm with narrow size distribution were formed [138]. Jiang *et al.* have investigated the fabrication of Pd NPs and their activity for the ligand-free Heck reaction in O/W system [63]. A non-ionic surfactant Triton X10 was employed to stabilise microemulsions with enlarged interface area. Highly dispersed Pd NPs were obtained and showed excellent activity in Heck reaction. This concept of the *in situ* formation of nanoscale materials affords a wide range of applications in organic synthesis. Lead sulphide (PbS) nanocrystals were prepared in a toluene-in-water microemulsion with ultrasonic irradiation. The as-synthesised powders were prudently analysed by XRD, TEM, XPS, and energy-dispersive X-ray analysis (EDAX) techniques. It has been shown that PbS crystals are composed of uniform spherical NPs (~11 nm) with narrow size distribution and high purity [169].

3.5.2 Water-in-Oil (W/O) Microemulsion Method

Water-in-oil microemulsion is a well-documented reaction system for the production of spherical nanoscale materials. A W/O micelle is formed typically by dispersion of water in the hydrocarbon-based continuous phase. The dispersed aqueous droplets are continuously collide, come together and break apart resulting in a continuous exchange of the solute content. The exchange of reactant species by the coalescence of two droplets must happen prior to their chemical reaction between reactants. This process is totally confined within the dispersed water droplets and lends itself to small changes that can have huge effect on the particle outcome [12,150].

Bumajdad *et al.* have synthesised pure ceria nanopowders in the range of ~6–13 nm in heptane-microemulsified system [12]. The effect of surfactants was examined to understand the mechanism of micelle formation. Nanoscale ceria materials with narrow size distribution and large specific surface areas were noticed. Ferrite (Fe_3O_4) and cobalt–ferrite ($Co_xFe_{3-x}O_4$) NPs coated with silica (SiO_2) were prepared using a simple reverse micelle route. The detailed synthesis process and TEM pictures of products are illustrated in Fig. 3.11. The analysis of TEM images revealed that the Fe_3O_4 NPs are located nearly at the centre of spherical SiO_2 particles. The diameters of Fe_3O_4 and $Co_xFe_{3-x}O_4$ particles are found to be in the range of 8–12 nm and 10–14 nm, respectively [156]. In 2006, Xiong *et al.* have fabricated tungsten (W) and tungsten oxide (WO_3) samples with controllable particle size and structure [185]. Various parameters, including water-to-alkoxide ratio, water-to-surfactant ratio, reduction temperature, and drying process were scrutinised in detail. Using the same procedure, carbon-supported W and WO_3 NPs can also be produced.

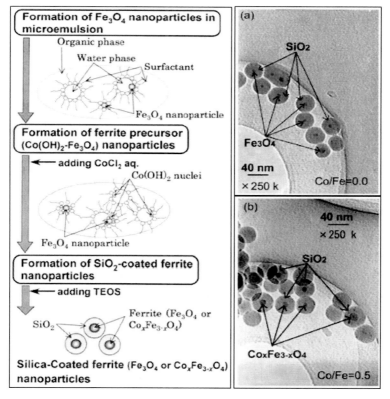

Figure 3.11   (Left) Preparation procedure for SiO_2-coated ferrite nano-particle using water-in-oil microemulsion. (Right) TEM photographs of SiO_2-coated (a) Fe_3O_4 and (b) $Co_xFe_{3-x}O_4$ nanoparticles after air calcination at 1273 K for 2 h. Reprinted from ref. [156], with permission from Wiley-VCH.

Mao *et al.* developed a hot mixing microemulsion route to attain ZnO nanostructures with stimulated morphologies [93]. The ratio of water to surfactant concentration (ω_0) was found to be a vital parameter in determining the final morphology. It has been observed that lower ω_0 values favour the formation of nanotetrahedrons, whereas nanorods are produced at higher ω_0 values. Numerous silver halides (AgCl, AgBr and AgI) in the nanoscale range were prepared in Triton X-100 microemulsions. The meticulous analysis of TEM profiles demonstrated that AgCl nanocrystals are almost spherical (20–30 nm) and AgBr particles (50–80 nm) are tadpole-like structures. The reactant molar ratio showed a massive affect on the morphology of AgI nanomaterials [187]. Recently, Oh *et al.* reported that uniform-sized tantalum oxide (TaO_x) NPs can be easily fabricated by means of a W/O emulsion route [107]. One-pot surface modification by various silane derivatives appended with functional moieties, such as polyethylene glycol (PEG) and fluorescent dye was successfully achieved. The produced TaO_x NPs exhibited extraordinary performance in the *in vivo* X-ray CT imaging and bimodal image-guided lymph node mapping.

A series of Co–Sn substituted barium ferrite particles were productively synthesised by a rapid microemulsion technique. When the Co–Sn content is rather low ($x \leq 0.5$), the products have hexagonal platelet-like structure with particle sizes about 600–800 nm in length and 80–100 nm in thickness. By increasing the Co–Sn content further led to the formation of disordered particles [47]. Perovskite $LaNiO_3$ NPs were fabricated by a microemulsion system consisted of cetyltrimethylammonium bromide, 1-butanol, cyclohexane, and the metal salt solution. A pure rhombohedral $LaNiO_3$ structure was readily formed at a calcination temperature of 1023 K for 4 h. The solar photocatalytic activity of the $LaNiO_3$ nano-oxides was found to be influenced by the size of particles [5]. Nanoplatelet Co–Al LDHs were synthesised by combining the reverse microemulsion process with homogeneous precipitation. All the samples possess a layered structure with long dodecyl chains located between the LDH platelets. The size of uniformly dispersed LDH platelets are estimated to be around 100–200 nm in width, while a few platelets are overlapped to generate large aggregates [168]. Zhang *et al.* have produced Pt–Ru bimetallic NPs using water/Triton X-100/propanol-2/cyclohexane microemulsion [197]. The composition of Pt–Ru NPs can be easily tuned by adjusting the initial metal precursor solution. Particle sizes of about 2.5 and 4.5 nm were achieved at low and high precursor concentration, respectively.

3.5.3   Water-in-scCO$_2$ (W/scCO$_2$) Microemulsion Method

There has been a special emphasis on the topic of green nano-synthesis of materials over the past decades. Many endeavours aim at the total elimination or at least the minimisation of produced waste and the implementation of sustainable methods [120]. Interestingly, supercritical carbon dioxide (scCO$_2$) has drawn substantial curiosity in green chemical synthesis due to its noteworthy benefits over the conventional organic solvents. It has many advantageous, such as non-toxic nature, non-flammability, inexpensive, chemically inert, and low viscosity. This last property furnishes a high diffusion coefficient of reactants in scCO$_2$, which enhance the reaction rates significantly. Finally, the properties of scCO$_2$ can be stimulated by temperature and pressure, providing a tunable medium to attain functional nanomaterials [80]. Due to non-polar and weak van der Waals forces, scCO$_2$ is not suitable for dissolving polar substances that limits its applications in chemical processes. One of the probable routes for enhancing the solubility of polar compounds in scCO$_2$ is to form reversed microemulsions (W/scCO$_2$). Consequently, many research works have been under-taken using supercritical CO$_2$ as a hydrophobic continuous phase in W/scCO$_2$ systems [110,137,147].

Lim *et al.* have synthesised TiO$_2$ NPs by a controlled hydrolysis of titanium tetraisopropoxide in W/scCO$_2$ microemulsion [86]. The particle size and stability of dispersions in scCO$_2$ are affected by the water to surfactant ratio, precursor concentration, and precursor injection rate. The crystallite size of particles is increased by increasing the molar ratio of water to surfactant. Nanoscale Ag and Cu metal particles can be prepared by chemical reduction of Ag$^+$ and Cu^{2+} ions in W/scCO$_2$ arrangement. Diffusion and distribution of the oxidised form of the reducing agent between micellar core and supercritical CO$_2$ could be the rate-determining step in the preparation of the silver NPs [109]. Fernandez and Wai have fabricated semiconductor NPs of CdS and ZnS involving the reaction of nitrate precursors with sodium sulphide in W/scCO$_2$

reverse micelle. A fluorinated-thiol stabiliser is introduced to inhibit the particle growth [41]. The size and size dispersion of CdS and ZnS particles can be continuously controlled over a wide range of values by tailoring the density of the fluid phase in the microemulsion system. Ohde *et al.* reported a rapid synthesis of nanoscale silver halide (AgI, AgBr and AgCl) particles by means of W/scCO$_2$ micro-emulsion [108]. AgNO$_3$ and NaI, NaBr and NaCl were taken as silver and halide precursors, respectively. The formation of products involves collision, inter-micellar exchange and chemical reaction between silver ion and halide ions. The inter-micellar exchange process is found to be the rate-determining step for the formation of nanoparticles.

3.6 Hydrothermal and Solvothermal Methods

Hydrothermal and/or solvothermal synthesis of nanostructures have drawn enormous scientific attentiveness in the last couple of years. The preparation of nanomaterials using non-aqueous solvents at temperatures above their boiling points is referred to solvothermal method or, in the case of the aqueous media known as hydrothermal process [28,181]. The use of special apparatus, for example autoclave, is necessary to withstand high pressures involved in these reactions. The materials chosen for the manu-facture of autoclaves play an imperative role in hydrothermal syntheses. Teflon lined autoclave is the ideal container in the perspective of corrosion. It sustains in alkaline media and exhibits a strong resistance to hydrofluoric acid compared with glass and quartz autoclaves. A wide variety of autoclaves and their functions are elaborately discussed in the literature [56,117]. To understand the selective role of reaction medium as well as mechanism of the reaction process, hydrothermal and solvothermal processes are addressed separately in detail.

3.6.1 Hydrothermal Method

According to Rabenau, hydrothermal method can be defined as the synthesis of nanoscale particles in aqueous media above 373 K and 1 bar [117]. However, the ionic product (Kw) of water is around 523–573 K. Thus, it is normally carried out below 573 K without the necessity of post-calcination [60]. By tuning the various reaction parameters, for example pH values, reaction temperatures, and reactant concentrations, highly crystalline nano-powders with different compositions and morphologies can be prepared [163,179,183].

Subramanian *et al.* have synthesised MnO_2 nanoarchitectures by adjusting the hydrothermal reaction time at 413 K [152]. The mixture of amorphous and nanocrystalline particles are formed at shorter dwell time and increasing the time led to better crystallinity with high specific surface area. Furthermore, diverse structures ranging from plate-like morphology to nanorods are generated depending upon the reaction time. Kaneko *et al.* have explored the structural and morphological characteristics of CeO_2 nanocrystals by means of hydrothermal reaction of $Ce(NO_3)_3$ solution in 0.3 M NaOH [65]. Decanoic acid was used additionally to modify the surface and anisotropic growth of the nanocrystals. The cautious study of TEM images revealed that the edge length of the nanocrystals is between 5 and 8 nm with average edge length about 6.7 nm. Large-scale synthesis of uniform 1D tellurium (Te) nanostructures, including nanotubes, nanowires and nanorods was performed by a hydrothermal reduction of K_2TeO_3. The affect of pH value, reactant concentration, reducing agent, and reaction temperature on the size and morphology of the Te nanostructures was examined. The obtained Te nanostructures are ideal templates to produce other Te-related nanomaterials [200].

A one-pot hydrothermal reaction was carried out to fabricate 3D layered self-assembled β-FeOOH nanorods based on a pH-induced strategy. In this process, urea plays a critical role that could continuously change the pH of the reaction system and adjust the inter-particle interaction. Also, it favours the formation of self-assembled nanomaterials due to its slow decomposition. Figure 3.12 demonstrates the morphological evaluation of β-FeOOH nanostructures at different time intervals [39]. A novel hydrothermal route was developed to attain cobalt-based nanomaterials. The Co_3O_4 nanocubes possess an average diameter of 350 nm with a perfect cubic shape, whereas $Co(OH)_2$ nanodiscs are uniform hexagonal platelets and $Co(OH)_2$ nanoflowers are assembled from large sheet-like subunits. All these materials were converted into spinel Co_3O_4 without significant variation in morphology after thermal annealing in air at moderate tempera-tures [20]. A two-step hydrothermal process was studied for the production of nanocrystalline $(ZrO_2)_{1-x}(Sc_2O_3)_x$ $(x = 0.02-0.16)$ powders. The crystallite size, surface area, microstrain, and growth activation energy of the products showed a strong dependence on the scandium doping concentration. Their complex phase evolution upon calcination over 673–1673 K was also explored [186]. Xu *et al.* reported a multi-step synthesis of MgAl LDHs from physically mixed MgO and Al_2O_3 oxides under hydrothermal conditions [188]. It was suggested that MgO and Al_2O_3 can be hydrated into $Mg(OH)_2$ and $Al(OH)_3$, when these oxides are placed in an autoclave with 40 mL of deionised water, respectively. The formation of LDHs is strongly affected by the initial pH of the suspension. On the basis of the experimental analyses, a general dissociation-deposition diffusion mechanism that includes two possible pathways was proposed.

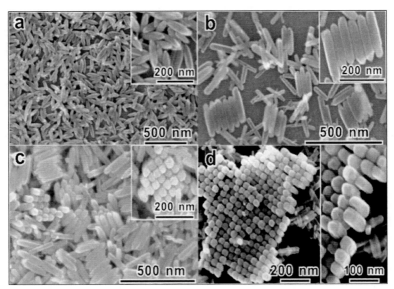

Figure 3.12 SEM pictures of the morphological evolution of β-FeOOH nanorods in the time-dependent reactions: (a) 1, (b) 3, (c) 5, and (d) 7 h (the insets: typical high-magnification images). Reprinted with permission from ref. [39]. Copyright 2010 American Chemical Society.

Interestingly, hydrothermal synthesis under supercritical water conditions ($T = 647$ K and $P = 22.1$ MPa) provides an excellent reaction environment for the nanofabrication of materials. At about supercritical circumstances, solvent properties, such as density, solubility, ionic product, and dielectric constant vary dramatically. For example, dielectric constant of water is radically changes from 78 to below 10 at supercritical conditions, which is a key driving force to induce the both supersaturation and precipitation. As a result, supercritical water enhances the reaction rate more than 10^3 times in comparison with the conventional hydrothermal route [60]. Moreover, this process has capability to produce homogeneous phases with oxygen and hydrogen, and hence, redox reactions can be easily controlled [153].

Highly crystallised NPs of CeO_2–ZrO_2 were prepared in supercritical water and the diverse characteristics of CeO_2–ZrO_2 samples were compared with coprecipitated products. It was found that the supercritical synthesis could afford highly crystalline particles, better thermal stability and superior OSC. Figure 3.13 presents the experimental diagram of the continuous hydrothermal synthesis in supercritical water [70]. According to Sue *et al.*, different nano-oxides (e.g. $AlOOH/Al_2O_3$, CuO, Fe_2O_3, NiO, and ZrO_2) could be synthesised at supercritical conditions (T = 673 K and P = 25–37.5 MPa) [154]. Among them, ZrO_2 exhibits low solubility, fast supersaturation, resulting in the small particle size. Recently, Sasakia *et al.* have produced magnesium ferrite ($MgFe_2O_4$) NPs

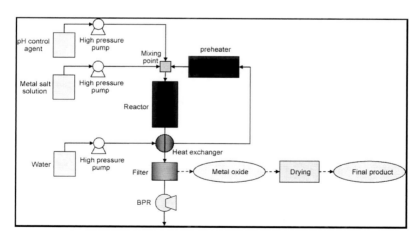

Figure 3.13 Experimental diagram of continuous hydrothermal synthesis in supercritical water. Reprinted from ref. [70], with permission from Elsevier.

involving the supercritical hydrothermal reaction of $Mg(NO_3)_2 \cdot 6H_2O$ and $Fe(NO_3)_3 \cdot 9H_2O$ [140]. The stoichiometric molar ratio of metal nitrate solutions was mixed with equivalent amount of KOH to yield $Fe(OH)_3$ and $Mg(OH)_2$ precipitates. The reaction mixture was then pressurised to 30 MPa, fed into a tubular reactor, and rapidly heated to desired temperature. The size of $MgFe_2O_4$ particles is about 20 nm and showed outstanding superpara-magnetic behaviour. A simple supercritical hydrothermal reaction was developed to synthesise potassium hexatitanate (KTO) materials. The obtained XRD, SEM, TEM, and thermal analysis results unveiled that long and felt-like fibres of KTO were produced and these fibres are thermally stable up to 1273 K [134].

3.6.2 Solvothermal Method

The solvothermal fabrication of nanomaterials using non-aqueous solvents instead of water has been gained extensive devotion due to some salient features. Since the higher tempera-tures of the solvents used, pressure can be autogenously generated in the autoclaves. The specific physico-chemical properties of solvents can substantially enhance the diffusion of reactants and, thereby, can improve size distribution and degree of crystallisation. As well, the morphology of products obtained by solvothermal process is robustly dependent on the solvents used, the ions involved, and other environmental aspects. The commonly used solvents are NH_3, HF, HBr, HCl, Br_2, Cl_2, CO_2, SO_2, H_2S, CS_2, CCl_4, S_2Cl_2, S_2Br_2, $SeBr_2$, C_6H_6, C_2H_5OH, CH_3NH_2, CH_3OH, HCOOH and so on [31,180,190].

A great number of schemes were reported for the preparation of nanoscale materials under solvothermal conditions. Yang *et al.* reported a one-pot solvothermal synthesis of silver nanocrystals to obtain different shapes, namely rods, cubes, spheres, triangular plates, hexagonal plates, and quasi-spheres [190]. The required amounts of $AgNO_3$ and poly(vinyl pyrrolidone) (PVP) were taken in dimethylformamide or ethanol. It has been observed that the dimension and shape of the Ag particles depend on the molar ratio of $PVP/AgNO_3$, $AgNO_3$ concentration, reaction temperature, and reaction time. Chaianansutcharit *et al.* used a solvothermal reaction for the production of Fe_2O_3 NPs at 523–573 K and explored the affect of solvents (e.g. 1,4-butanediol and toluene) over the stability of materials. Iron oxides prepared in 1,4-butanediol displayed superior thermal stability than those synthesised in toluene, which is attributed to phase transformation from α- to β-iron oxide in toluene [19]. At the same time, Shen's group has prepared various ZnO architectures using $Zn(CH_3COO)_2 \cdot 2H_2O$ and ethylenediamine in the presence of $HAuCl_4 \cdot 4H_2O$. The size, morphology, and structure of ZnO oxides were examined by varying the amounts and the sequence addition of the metal reactants. Here, the $Au(ethylenediamine)_2^{3+}$ complex plays a key role in controlling the morphology and structure of these nano-structures [146].

Nguyen *et al.* have developed a solvo-hydrothermal process for the synthesis of monoclinic ErOOH and cubic Er_2O_3 nano-oxides in the water/ethanol medium [102]. By simply tuning the experimental parameters, a variety of shapes, including cores/dots to spheres, wires, rods, flowers, dog bonds, bundles, brooms, straw sheaves, and wrinkle-surfaced spheres can be achieved. Furthermore, when anhydrous ethanol used instead of water/ethanol medium, the particle size markedly decreased from 18 nm (nanospheres) to 3 nm (nanocores). Figure 3.14 depicts the

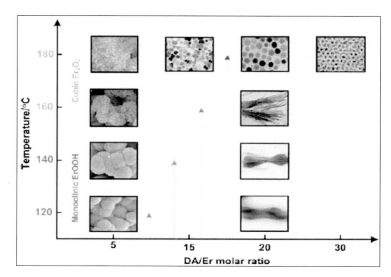

Figure 3.14 Morphologic evolution of ErOOH and Er_2O_3 nanostructures as a function of reaction temperature and decanoic acid/erbium (DA/Er) molar ratio. Reprinted with permission from ref. [102]. Copyright 2010 American Chemical Society.

formation of ErOOH and Er_2O_3 micro- and nano-structures as a function of reaction temperature and decanoic acid/erbium (DA/Er) molar ratio. Kominami and coworkers have synthesised tantalum(V)oxide (Ta_2O_5) nanopowders involving a solvo-hydro-thermal reaction of tantalum pentabutoxide (TPB) in water–toluene system [75]. X-ray diffraction patterns are evidenced for crystalline β-phase of Ta_2O_5 obtained at 573 K. Grocholl *et al.* have performed a supercritical solvothermal reaction to attain nanoscale gallium nitride (GaN) structures [54]. The reaction between $GaCl_3$ and NaN_3 produces an insoluble azide precursor that solvothermally decomposes to GaN in toluene and/or THF at their critical points. Nanoparticles with spherical (ca. 50 nm) and rod like (ca. 300 nm lengths) morphologies were formed at or below 533 K.

3.7   Microwave-Assisted Techniques

It is a known fact that the solution-based processes generally utilise the conventional thermal heating due to the necessity of highly drastic reaction environments. However, in case of the conventional heating, heat energy initially transfers to the solvents and later will have an impact on the reactants. Consequently, high thermal gradient effects, non-consistent and fruitless reaction conditions will generate throughout the bulk solution. These are the most problematic aspects in nanomaterials production, where uniform nucleation and growth rates are very crucial to materials quality. A promising route to undertake the above challenges is microwave-assisted method because of undesirable high temp-erature or pressure conditions. Since 1986, this technique has been progressed extensively in many academic and industrial labora-tories. Microwave-assisted methodologies are unique in their ability to scaled-up and providing an industrially novel process to obtain nanostructured materials. Fast and uniform heating of the reaction mixture is greatly achievable in a short period. As a result, an instantaneous internal temperature increases throughout the solution. This allows the rapid decomposition of the reactants, resulting in highly supersaturated solutions. After-wards, nucleation and subsequent growth takes place to obtain the desired materials [49,97,202].

Figure 3.15 illustrates the generation of heat energy by microwave irradiation (MWI) of water [165]. In the microwave frequency range (900 MHz to 2.45 GHz), water molecules try to orientate with the electric field. If the water molecules make an effort to re-orientate regarding to an alternating electric field, they lose heat energy spontaneously by molecular friction. Water, ethylene glycol (EG), dimethyl formamide (DMF) and simple alcohols have high dielectric losing capacity and great reduction ability. Therefore, they can be considered ideal solvents for microwave-assisted reactions [119,165].

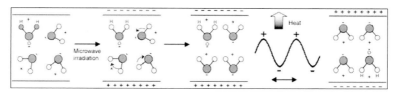

Figure 3.15 Generation of heat energy by microwave irradiation of water molecules. Reprinted from ref. [165], with permission from Wiley-VCH.

A facile one-pot MWI method has been used to synthesise various pure metallic and bimetallic NPs with stimulated properties. All the reactions were performed in DMF with an equimolar mixture of oleic acid and oleylamine ligands. A microwave power of 1000 W was operated for a microwaving time that varied from 1 to 15 min. Moreover, the same process can be used to attain bimetallic nanoalloys (CuPd, CuRh and AuPd) dispersed on ceria NPs [1]. Uniformly dispersed and highly luminescent $NaYF_4$-based upconverting NPs were prepared under MWI conditions. Various doped nanocrystals with mean diameter of 10–11 nm were obtained by MWI for 5 min. Although the same reactants and solvents are used in classical conductive heating, this route affords a better reaction control, resulting in particles of different sizes and shapes [171]. A very attentive experiment was conducted to correlate the MWI and conventional thermal heating for the production of TiO_2 NPs. A significant reduction in the processing time from 24 h to few minutes was observed in the microwave approach. The convenience of using the microwave heating is due to the function of organic cations present in the precursor [38]. A variety of metal tungstates with interesting morphologies was fabricated in a microwave-assisted oven at 373 and 423 K. The crystals formed are of submicrometre size and possess equidimensional and needle-like structures. By increasing the reaction time and temperature, disappearance of the needles and the growth of the equidimensional crystals were noticed [74].

More recently, we have synthesised various ceria-based nano-oxides by adopting a microwave-induced combustion process. For example, to synthesise $Ce_{0.5}Zr_{0.5}O_2$ (1:1 molar ratio based on oxides), the desired quantity of solid urea was added to the aqueous solution of stoichiometric mole ratios of cerium (III) nitrate and zirconyl (IV) nitrate precursors. The resulting mixture was introduced into a modified domestic microwave oven. The entire synthesis process in the microwave oven takes place approximately 40 min. The produced materials were analysed by means of diverse techniques. The obtained results showed the existence of more oxygen vacancies and lattice defects in the MW samples than those prepared by coprecipitation method. The same course of action was used to obtain numerous nanoscale mixed oxides namely, CeO_2–SiO_2, CeO_2–TiO_2, CeO_2–Al_2O_3, TiO_2–Al_2O_3, TiO_2–SiO_2, and TiO_2–ZrO_2 (Fig. 3.16) [126–128,130].

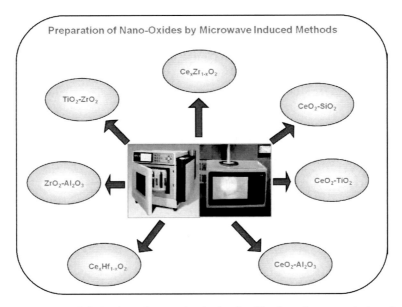

Figure 3.16 Fabrication of nano-sized mixed oxides by microwave-induced combustion method.

Zhu *et al.* reported that nano-sized CdSe, PbSe and $Cu_{2-x}Se$ materials can be prepared by MWI of metal acetates or sulphates [202]. The reaction rate and the particle growth are effectively controlled by complexating agents. Among them, PbSe sample exhibited lower monodispersity and larger particles size due to the simultaneous and fast nucleation as well as growth rates. Interestingly, Polshettiwar *et al.* synthesised nanocrystalline metal oxides under microwave irradiation in pure water without using any reducing or capping reagent. The resultant nano-oxides are self-assembled into three-dimensional morphologies with particles size in the range of 100–500 nm. Pine structured iron oxides were explored as a novel support for various catalytic organic transformations [114]. In a recent work, Yamauchi *et al.* examined a facile one-pot MWI process for the production of Ni–Co core–shell NPs [189]. The characterisation results unveiled that the nanostructured materials are composed of a Co-rich shell and a Ni-rich core. The shape of the Ni core played a key role in determining the final morphology of Ni–Co nanocrystals as shown in Fig. 3.17.

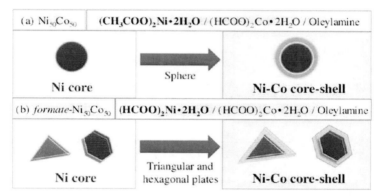

Figure 3.17 Formation of Ni core and Ni–Co core–shell nanoparticles with different morphologies. Reprinted with permission from ref. [189]. Copyright 2011 American Chemical Society.

3.8 Polyol Methods

The polyol method is one of the recently emerged techniques that provide a versatile route to obtain nanostructures with uniform sizes and shapes. This process was first developed by Fievet's group for the finely divided metal powders in 1980. Fundamentally, it signifies the utilisation of polyalcohol (polyol), which plays a two-fold role as a solvent and as a reducing agent. Also, it acts as a stabiliser to control the particle growth. The choice of polyol mainly depends on its boiling point and reduction potential. For instance, simply reducible metals (Pt, Pd or Cu) do not require high temperatures and can be easily reduced in propylene glycol (bp ~461 K), whereas mildly reducible metals (Co, Fe or Ni) need elevated temperatures, for those, trimethylene glycol (bp ~600 K) could be an appropriate polyol. Furthermore, the redox potential of polyol must possess high negative values than that of the metal species. A wide range of polyols, such as ethylene glycol (EG), propylene glycol (PG), butylene glycol (BG), diethylene glycol (DEG), tetraethylene glycol (TEG) and so forth have been reported in the literature [18,53,135].

Carroll *et al.* have studied the relationship between reaction conditions, crystal morphology, and theoretical modelling of copper and nickel NPs prepared by polyol process [18]. Numerous polyols, namely EG, PG, BG, DEG, and TEG, including a base (NaOH) and methanol were used. TEM results shown in Fig. 3.18 revealed that

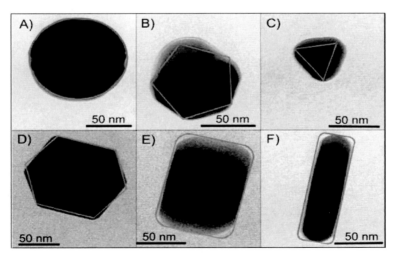

Figure 3.18   Variation in Cu morphologies by simply varying the reaction medium with Cu hydroxide as a precursor salt in polyol method: (A) EG under refluxing, (B) EG under distillation, (C) PG under refluxing, (D) DEG under refluxing, (E) DEG under distillation, and (F) TEG under distillation. Reprinted with permission from ref. [18]. Copyright 2011 American Chemical Society.

the morphologies of Cu NPs changes from spheres to rods, which is attributed to the nature of the polyol that alters the nucleation and growing steps. The synthesis of nanocrystalline MS particles (*M* = Zn, Cd and Hg) was examined using a polyol reaction of metal acetates and thiourea in DEG. The thorough analysis of various characterisation techniques disclosed that the produced materials are almost non-agglomerated with the sphalerite type of crystal structure. The particles are spherical with average diameters of between 30 and 250 nm [40].

Monodispersed silver nanocubes were synthesised in large quantities by means of a modified polyol process. The presence of PVP and its molar ratio relative to silver nitrate were displayed a vast effect in determining the geometric shape and size of the product. The obtained silver cubes are single crystals and could serve as sacrificial templates to generate single-crystalline nanoboxes of gold. Figure 3.19 illustrates the (A) low- (B) high-magnification SEM

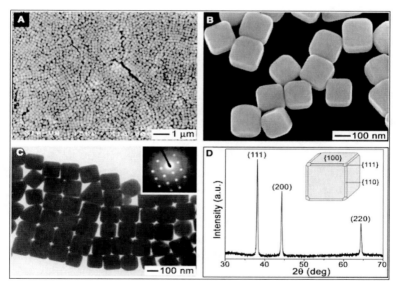

Figure 3.19 (A) Low- and (B) high-magnification SEM images of slightly truncated silver nanocubes prepared by a polyol process. (C) TEM image of the same batch of silver nanocubes. The inset shows the diffraction pattern recorded by aligning the electron beam perpendicular to one of the square faces of an individual cube. (D) XRD pattern of the same batch of sample. Reprinted from ref. [155], with permission from Science.

images, (C) TEM image, and (D) XRD pattern of silver anocubes [155]. Joo *et al.* prepared Ru NPs by a polyol reduction of Ru(acac)$_3$ in EG and butanediol [64]. Uniform Ru particles with tunable sizes from 2 to 6 nm were obtained. The smaller Ru NPs are mostly composed of spherical particles, whereas the bigger NPs contain a portion of well-faceted particles. A low-temperature polyol reaction was developed for the synthesis of nanocrystalline ternary Au–Cu–Sn system. The analysis of TEM micrographs indicates the average crystallite size of AuCuSn$_2$ is about 27 ± 14 nm. Using the same method, ordered AuNiSn$_2$ nanocrystalline alloy can also be prepared [81].

3.9 Liquid Feed Flame Spray Pyrolysis (LF-FSP) Methods

Flame spray pyrolysis (FSP) is broadly used gas phase technique for the large-scale manufacture of nanomaterials. As per the literature, there are two types of approaches, namely vapour feed- and liquid feed-flame spray pyrolysis (LF-FSP). However, it is often difficult to produce multi-component materials with homogeneous composition in the vapour feed process. This was accredited to the differences in the vapour pressure of the gaseous species and/or in the surface energies of the condensed oxides. One component nucleates initially and the other component(s) may nucleate subsequently on the surface of existing particles or as new single-component particles. Therefore, non-stoichiometric and inhomogeneous powders are normally formed. In contrast, LF-FSP provides a potential route by means of alcohol solutions of organometallic precursors, which can be aerosolised with oxygen and ignited to acquire pure crystalline nanomaterials [8,73,98]. The flame-made materials are usually characterised by high specific surface areas, non-porous structures, and show improved resistance against sintering at elevated temperatures [176]. Figure 3.20 represents the reaction setup of the FSP process [160]. **Caution!** Reaction of large quantities of NP precursors in ethanol in an oxygen atmosphere can be extremely hazardous. Good exhaust systems and a detailed standard operating process should be used prior to attempting the experiment [72].

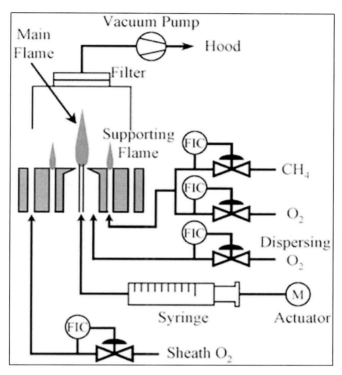

Figure 3.20 Schematic reaction setup of the FSP process for the synthesis of nanoparticles. Reprinted from ref. [160], with permission from Springer.

Mueller *et al.* reported a dynamic LF-FSP synthesis of ZrO_2 NPs using zirconium n-propoxide as Zr precursor [99]. The influence of zirconium n-propoxide concentration, ZrO_2 powder production rate, and oxidant dispersion gas flow rate were examined. It was noticed that the primary particle size of pure ZrO_2 was controlled from 6 to 35 nm by varying the aforementioned factors (Fig. 3.21). The crystal structure mostly consisted of tetragonal phase (80–95 wt.%) and balance the monoclinic phase at all reaction conditions. A LF-FSP fabrication of tellurium dioxide (TeO_2) NPs using an inexpensive and eco-friendly precursor (telluric acid in water) has been demonstrated. Precursor concentration, flow velocity and reaction pressure were studied on particle size, composition, and production rate. This laboratory scale route is capable of producing up to 80 mg/h amorphous TeO_2 NPs with particle diameters of between 10 and 40 nm [194]. Nanostructured bismuth oxide (Bi_2O_3) particles were prepared by LF-FSP reaction of nitrate precursor in a solution of ethanol/nitric acid or in acetic acid. The use of ethanol/nitric acid solutions led to a mixture of hollow, shell-like, and solid nanograined particles. The homogeneity of materials was improved as the content of acetic acid in the precursor solution increased [91]. Nanoscale particles of ZnO, MgO and NiO were produced from nitrate and acetate salt precursor solution in a modified LF-FSP reactor. An electrical furnace reactor with low pressure setup (~60 torr) and propane-oxygen diffusion

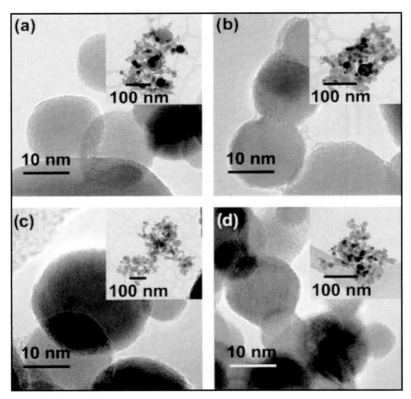

Figure 3.21 TEM micrographs of FSP-made ZrO_2 nanoparticles at a production rate of 200 g/h using dispersion gas flow rates of 25 and 50 L/min and 0.5 M and 1 M zirconium n-propoxide in EtOH at (a) 25 L/min and 0.5 M, (b) 50 L/min and 0.5 M, (c) 25 L/min and 1 M, and (d) 50 L/min and 1 M. Reprinted from ref. [99], with permission from Wiley-VCH.

flame is used in order to attain high decomposition temperatures. The effect of flame temperatures was scrutinised on the formation of nanoparticles. The obtained products were fully crystallised and the particle size measured from TEM pictures is ~30 nm [142].

A versatile LF-FSP process has been investigated for the synthesis of single component NPs by means of Ag, Pd, and Fe nitrate precursors. The systematic study of various reaction parameters suggested that the particle size is greatly controlled by the mass flow rate of metal precursor. Ag and Pd particles are consisting of pure metals, while three different Fe structures are observed, including pure Fe, hematite (α-Fe_2O_3) and magnetite (Fe_3O_4) [92].

V_2O_5/TiO_2 particles with a specific surface area up to 195 m^2/g and V_2O_5 content up to 40% (w/w) were synthesised by a novel FSP reaction. TiO_2 exhibited predominately spherical particles with a low degree of aggregation. Addition of V_2O_5 has no affect on the particle shape and aggregation state. Monomeric, polymeric and crystalline vanadia species are formed on TiO_2 particles, which depend on vanadia content [141]. Ru-doped Co–Zr (20 wt.% Co) nanocomposites were made by rapid and scalable flame spray pyrolysis. The analysis of products revealed that cobalt clusters are highly dispersed within the zirconia matrix. More importantly, the well-dispersed Ru promoter within the composite particles enhances the reducibility of cobalt [162].

3.10   Template-Directed Synthetic Techniques

The template-directed syntheses have been extensively investigated because of their effectiveness in the controlling of inner structural characteristics. Primarily, the pore size and overall porosity along with the outer shape and size of nanomaterials can be easily controlled by using different templates. Template preformation (or *in situ* formation), generation of desired nanostructures within (or around) templates, and removal of templates by appropriate routes (e.g. chemical etching and/or calcination) are key steps in template-assisted reactions. Templates can be categorised as "soft" and "hard" based on their structure and the confinement ability. Commonly used hard templates are porous silica, porous carbon, anodised alumina, polystyrene (PS) beads and nanoparticles of carbon, metals and metal oxides. Soft templates represent the structure-directing molecules, for example emulsion droplets, surfactant micelles, polymer vesicles, bubbles, and so on. The template removal is the major drawback that severely affects the purity of final nanopowders. Strong chemical treatments (e.g. hydrofluoric acid) are required to remove the templates [58,85,90,145].

Triangular Ag nanoplates were prepared by reduction of Ag ions with ascorbic acid in the presence of a soft template (CTAB). The resulting particles have an average edge size of 68 nm, thickness of 24 nm, and the degree of truncation of 0.35. Figure 3.22 represents TEM image of the purified Ag NPs and abundance of different shapes. Particles with different morphologies, for example rods,

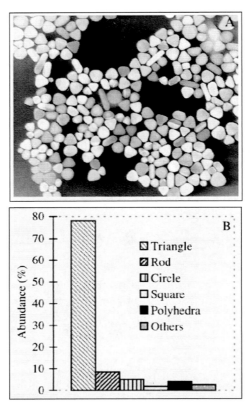

Figure 3.22 (A) TEM image of the flat-lay purified truncated silver triangular particles. (B) Histogram showing the abundance of different shapes. Reprinted with permission from ref. [21]. Copyright 2002 American Chemical Society.

discs, spheres, cubes, tetrahedra and squares were observed [21]. Interestingly, Fischer *et al.* developed a "reactive hard templating" approach with mesoporous carbon nitride to obtain ternary metal nitride NPs (Al–Ga–N and Ti–V–N) [42]. Carbon nitride plays the role of both template and reactant. In other words, mesopores of carbon nitride first serve as nanoconfinement to generate amorphous mixed oxides. During heating and decomposition, carbon nitride acts as a nitrogen source that eventually results into the corresponding nitride NPs. Self-assembled mesoporous ZrO_2 NPs were fabricated with an average diameter of 7 nm using a block copolymer (pluoronic F127) as template. N_2 sorption and HRTEM studies revealed that mesopores of ZrO_2 were formed by the regular arrangement of nanoparticles. The impregnation of ZrO_2 nanopowder by dilute sulphuric acid affords strong acidic nature and showed high catalytic activity in Friedel–Crafts benzylation of aromatics [29].

Recently, Chen *et al.* effectively synthesised $ZnAl_2O_4:Eu^{3+}$ hollow nanophosphors using carbon nanospheres [22]. The N_2 adsorption and desorption studies are evidenced for the porous nature of $ZnAl_2O_4:Eu^{3+}$. The schematic illustration of the reaction pathway is presented in Fig. 3.23. A "soft-hard dual template process" was proposed for the production of Group IIB selenides using a celloidin membrane as template. Celloidin membrane is composed of both properties of hard and soft template characteristics. The obtained materials are cubic ZnSe, hexagonal CdSe and cubic HgSe with average diameters of 3.1, 2.0 and 44 nm, respectively. Several uses of Group IIB selenides are anticipated, including as non-linear optic devices, photocatalysts, and light emitting diodes [173]. Maye *et al.* studied a mediator-template strategy for the size-controllable assembly of gold NPs [94]. Multi-dentate thioether ligands and tetraalkylammonium-capped gold NPs (5 nm) were used as molecular mediators and as template, respectively. The combination of mediation force of the ligand and the hydrophobic force of the template result into the monodispersed spherical assemblies of diameters 20–300 nm. The theoretical and experimental correlations of the morphological and optical properties support the fundamental basis for the mediator-template strategy as a versatile assembly technique.

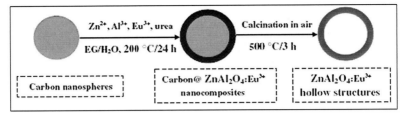

Figure 3.23 Preparation of $ZnAl_2O_4$:Eu^{3+} hollow nanophosphors using carbon nanospheres as hard templates. Reprinted from ref. [22], with permission from Elsevier.

3.11 Mechanochemical Processing

A broad range of innovative techniques spring to mind when one thinks of mixing and combining the materials in the solid state down to atomic scale [45]. Principally, solid-state reactions involve the unification of fine powders without the assistance of any solvent. In such condition, the resulting samples can be different from those of the similarly prepared materials by wet chemical routes. This could open innovative directions for modern nanosynthetic research, where the possibility of obtaining novel materials [16]. Mechanochemical processing is one of vividly described solid-phase methods that define acceleration of the chemical reactions by the mechanical activation of reactants in ball milling tools. An employed mechanical power will impel an excess energy into the reactants that can spontaneously reduce the activation energy of the reaction. Moreover, it induces the interfaces needed to the chemical reactions, affording the mild reaction conditions. This method is initially used for the fabrication of alloys and inter-metallic compounds in 1966. By choosing proper experimental parameters, such as reaction paths, precursor molar ratios, and milling conditions, nanoscale particles with substantial properties can be produced. The obtained nanopowders show lattice imperfections, phase transformations and, thereby, exhibit promising applications in many interesting fields [133,151,167]. Figure 3.24 shows the exploded view of the improved ball mill [4].

In view of those interesting features, the mechanochemical technique has been realised impressively in the synthesis of nanoscale particles. CdSe nanocrystals with zinc blende structure were prepared by mechanochemical activation of Cd and Se elemental powders for 40 h. The ball milling was carried out using ball-to-powder mass ratio of 10:1 with different diameters of balls. Subsequent capping of the surface of CdSe particles with organic-inorganic composite ligands has showed similar optical properties to those of CdSe nanocrystals obtained by the wet chemical routes [157]. Nanoscale $MgFe_2O_4$ particles with an average crystallite size of about 8.5 nm were synthesised. Stoichiometric mixtures of α-Fe_2O_3 and MgO oxides were milled for different uninterrupted times (up to 12 h) in air. An unusual magnetisation enhancement was noticed in $MgFe_2O_4$, which is attributed to the nearly random distribution of magnetic cations in the surface regions of NPs [143]. A fruitful synthesis of CeO_2 NPs was conducted by means of a solid-state reaction in steady state manner. Calcination of as-milled samples at 773 K resulted in the formation of ultrafine powder in the NaCl matrix. This matrix was gently removed by washing with deionised water and then, oven drying at 333 K led to CeO_2 particles of diameter ~10 nm [166].

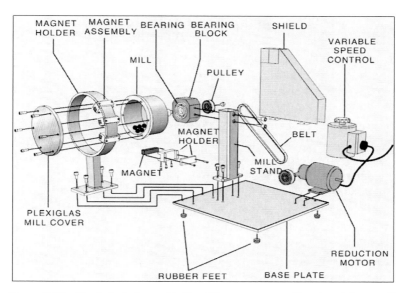

Figure 3.24 Exploded view of ball milling apparatus. Reprinted from ref. [4], with permission from Canadian Inst of Mining, Metallurgy & Petroleum.

Da Silva *et al.* used α-Fe$_2$O$_3$ and Bi$_2$O$_3$ as starting reagents to produce BiFeO$_3$ nano-oxides at room temperature [30]. An investigation of *in situ* HRTEM analysis of synthesis process reveals the inherent instability and high reactivity of oxide reactants. The particle sizes of BiFeO$_3$ are ranging from 5 to 40 nm and exhibited a partial superparamagnetism. A single-step mechanochemical reaction was conducted for the synthesis of calcium stannate (Ca$_2$SnO$_4$) NPs. A mixture of CaO and SnO$_2$ oxides were taken in a Pulverisette 6 planetary ball mill and milled up to 4 h. All the experiments were performed in air at 600 rpm. HRTEM studies revealed that the calcium stannate powders consist of an ordered core surrounded by a disordered surface shell region [144]. Zhang *et al.* have studied a mechanochemical reaction of La$_2$O$_3$ with different phases of Al$_2$O$_3$ to attain lanthanum aluminate (LaAlO$_3$) oxides [196]. Grinding for 120 min with transition Al$_2$O$_3$ results to a single-phase of LaAlO$_3$, whereas in case of α-Al$_2$O$_3$, no formation of LaAlO$_3$ was achieved. The same process was applied to produce other rare-earth (RE) aluminates (REAlO$_3$).

3.12 Ionic Liquid-Assisted Methods

In recent years, ionic liquids (ILs) have received a great deal of attention throughout both industry and academia. Ionic liquids are prevalently known as low temperature molten salts that consist of organic cations and inorganic or organic anions. In 1914, the first attempt was made to synthesise ethylammonium nitrate ([C$_2$H$_5$NH$_3$] NO$_3$) IL, which has a melting point of 285 K. Some attractive properties of ILs that make them effective "green solvents" alternative to the traditional organic solvents are the following: (a) they have ability to stand for wide range of melting points and, thereby, can be used extensively for the preparation of various novel materials. (b) The ILs display limited miscibility with both water and organic solvents, which facilitate easy extraction of products from the reaction mixture. (c) They are usually less toxic, low volatile, non-flammable, and exhibit low interface energies, high ionic conductivities, and broad electrochemical windows. (d) Also, ILs can be retrieved completely after reaction, hence they can afford an eco-friendly and economical route to obtain inorganic NPs

[35,37,64,83,177,201]. Figure 3.25 represents the categorisation of ionic liquids [164].

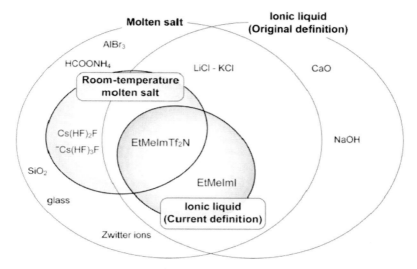

Figure 3.25   Systematic categorisation of ionic liquids. Reprinted from ref. [164], with permission from Wiley-VCH.

Among various ILs used, imidazolium-based ionic liquids are of particular interest because of their potential use in well-controllable synthesis of NPs. Moreover, the resulting products can be automatically separated out through a settling procedure without employing any antisolvent reagents [175]. Wang *et al.* studied a facile synthetic tactic to prepare Ag and Pt NPs in 1-butyl-3-methylimidazolium bis(trifluoromethylsulphonyl)imide [BMIM][Tf$_2$N] IL [174]. The reaction was carried out at desired temperatures by using oleic acid as capping agent. The utilisation of oleic acid with [BMIM][Tf$_2$N] leads to an automatic separation of powders from IL solutions. The produced Ag and Pt NPs were observed to be in the nano-size range with uniform size distribution. Recently, a sustainable easy route was reported for the production of EuF$_3$ nanospheres without use of any surfactants and templates at room temperature. Eu(NO$_3$)$_3$ and 1-butyl-3-methylimidazolium terafluoroborate [BMIM]BF$_4$ IL are used as starting reagents. The obtained EuF$_3$ nanospheres showed better photoluminescence properties and could possess the diverse applications, such as phosphors, field emitters, and optoelectronic devices [193].

The diameter and size-distribution of nickel NPs were investigated by varying the alkyl side-chain of the imidazolium cation. It was found that the particle size and size-distribution were decreased slightly from 5.9 ± 1.4 nm ($[C_1C_4Im]\cdot NTf_2$) to 5.1 ± 0.9 nm ($[C_1C_{14}Im]\cdot NTf_2$) and then an increase of 5.5 ± 1.1 nm was observed for $[C_1C_{16}Im]\cdot NTf_2$ IL [95]. Fonseca's group has shown that 1-n-butyl-3-methylimidazolium hexafluorophosphate (BMI. PF_6) IL is an excellent medium for the synthesis of Ir and Rh NPs [43]. They used dry IL conditions because a small quantity of water causes the partial decomposition of IL with the formation of HF, phosphates, and transition-metal fluorides. The careful study by various characterisation results revealed the formation of Rh^0 and Ir^0 metal phases with an average particle size of 2.0–2.5 nm. Jiang and coworkers have explored the synthesis of Bi_2S_3 flowers that are composed of uniform nanowires (60–80 nm) by an ionic liquid-assisted reaction [62]. The shape evolution and phase transformation of Bi_2S_3 strongly depended on the pH values, reaction temperatures, and reaction times. The template effect of the IL and the formation mechanism of the flowers have been discussed elaborately in their publication.

3.13 Conclusions

A wide variety of fabrication methods were studied extensively to synthesise various inorganic nanoparticles. One could choose different synthetic routes as mentioned in this chapter depending on the application of nanoparticles as well as the quantity of material required. Mainly, wet chemical approaches offer potential reaction pathways to obtain better materials in terms of chemical homogeneity and size-tunable properties. Nevertheless, it is difficult to separately control the crystallisation, surface structure, and agglomeration of particles. Gas- and solid-phase processes have also been found versatile utilisation in the nanosynthetic chemistry. Tedious multi-step processing, unfavourable reaction parameters, and expensive synthesis devices are some of the drawbacks in these techniques. Interest in this research area is expected to grow further in the coming years in view of numerous application benefits of nanomaterials.

Acknowledgements

We wish to specially acknowledge all the researchers whose work is described in this chapter for their valuable contributions. Financial support was received from the Department of Science and Technology, New Delhi, under the SERC Scheme (SR/S1/PC-63/2008).

References

1. Abdelsayed, V., Aljarasah, A., El-Shall, M. S., Al-Othman, Z. A., and Alghamdi, A. H. (2009). Microwave synthesis of bimetallic nanoalloys and CO oxidation on ceria-supported nanoalloys, *Chem. Mater.*, **21**, pp. 2825–2834.

2. Abelló, S., and Pérez-Ramírez, J. (2006). Characteristics by a miniaturized in-line dispersion–precipitation method: application to hydrotalcite synthesis, *Adv. Mater.*, **18**, pp. 2436–2439.

3. Abelló, S., Mitchell, S., Santiago, M., Stoica, G., and Pérez-Ramírez, J. (2010). Perturbing the properties of layered double hydroxides by continuous coprecipitation with short residence time, *J. Mater. Chem.*, **20**, pp. 5878–5887.

4. Ajaal, T., Smith, R. W., and Yen, W. T. (2002). The development and characterization of a ball mill for mechanical alloying, *Can. Metallurgical Q.*, **41**, pp. 7–14.

5. Aman, D., Zaki, T., Mikhail, S., and Selim, S. A. (2011). Synthesis of a perovskite $LaNiO_3$ nanocatalyst at a low temperature using single reverse microemulsion, *Catal. Today*, **164**, pp. 209–213.

6. Andersson, M., Pedersen, J., and Palmqvist, A. (2005). Silver nanoparticle formation in microemulsions acting both as template and reducing agent, *Langmuir*, **21**, pp. 11387–11396.

7. Atkinso, A., and Sega, D. L. (1998). Some recent developments in aqueous sol-gel processing, *J. Sol-Gel Sci. Technol.*, **13**, pp. 133–139.

8. Baranwal, R., Villar, M. P., Garcia, R., and Laine, R. M. (2001). Flame spray pyrolysis of precursors as a route to nano-mullite powder: powder characterization and sintering behavior, *J. Am. Ceram. Soc.*, **84**, pp. 951–961.

9. Behrens, M., Kissner, S., Girsgdies, F., Kasatkin, I., Hermerschmidt, F., Mette, K., Ruland, H., Muhler, M., and Schlögl, R. (2011). Knowledge-based development of a nitrate-free synthesis route for Cu/ZnO methanol synthesis catalysts via formate precursors, *Chem. Commun.*, **47**, pp. 1701–1703.

10. Bitter, J. H., van der Lee, M. K., Slotboom, A. G. T., van Dillen, A. J., and de Jong, K. P. (2003). Synthesis of highly loaded highly dispersed nickel on carbon nanofibers by homogeneous deposition-precipitation, *Catal. Lett.*, **89**, pp. 139–142.

11. Bonnetot, B., Rakić, V., Yuzhakova, T., Guimon, C., and Auroux, A. (2008). Preparation and characterization of Me_2O_3–CeO_2 (Me = B, Al, Ga, In) mixed oxide catalysts. 2. Preparation by sol-gel method, *Chem. Mater.*, **20**, pp. 1585–1596.

12. Bumajdad, A., Zaki, M. I., Eastoe, J., and Pasupulety, L. (2004). Microemulsion-based synthesis of CeO_2 powders with high surface area and high-temperature stabilities, *Langmuir*, **20**, pp. 11223–11233.

13. Burattin, P., Che, M., and Louis, C. (1997). Characterization of the Ni(II) phase formed on silica upon deposition-precipitation, *J. Phys. Chem. B*, **101**, pp. 7060–7074.

14. Burda, C., Chen, X., Narayanan, R., and El-Sayed, M. A. (2005). Chemistry and properties of nanocrystals of different shapes, *Chem. Rev.*, **105**, pp. 1025–1102.

15. Cabanas, A., and Poliakoff, M. (2001). The continuous hydrothermal synthesis of nano-particulate ferrites in near critical and supercritical water, *J. Mater. Chem.*, **11**, pp. 1408–1416.

16. Calandra, P., Longo, A., and Liveri, V. T. (2003). Synthesis of ultrasmall ZnS nanoparticles by solid–solid reaction in the confined space of AOT reversed micelles, *J. Phys. Chem. B*, **107**, pp. 25–30.

17. Cao, M. H., Wang, Y. D., Chen, T., Antonietti, M., and Niederberger, M. (2008). A highly sensitive and fast-responding ethanol sensor based on $CdIn_2O_4$ nanocrystals synthesized by a nonaqueous sol-gel route, *Chem. Mater.*, **20**, pp. 5781–578.

18. Carroll, K. J., Reveles, J. U., Shultz, M. D., Khanna, S. N., and Carpenter, E. E. (2011). Preparation of elemental Cu and Ni nanoparticles by the polyol method: an experimental and theoretical approach, *J. Phys. Chem. C*, **115**, pp. 2656–2664.

19. Chaianansutcharit, S., Mekasuwandumrong, O., and Praserthdam. P. (2006). Effect of organic solvents on iron oxide nanoparticles by the solvothermal method, *Cryst. Growth Des.*, **6**, pp. 40–45.

20. Chen, J. S., Zhu, T., Hu, Q. H., Gao, J. J., Su, F. B., Qiao, S. Z., and Lou, X. W. (2010). Shape-controlled synthesis of cobalt-based nanocubes, nanodiscs, and nanoflowers and their comparative lithium-storage properties, *ACS Appl. Mater. Interfaces*, **2**, pp. 3628–3635.

21. Chen, S., and Carroll, D. L. (2002). Synthesis and characterization of truncated triangular silver nanoplates, *Nano Lett.*, **2**, pp. 1003–1007.

22. Chen, X. Y., Ma, C., Bao, S. P., and Li, Z. (2010). Synthesis and photoluminescence of $ZnAl_2O_4:Eu^{3+}$ hollow nanophosphors using carbon nanospheres as hard templates, *J. Colloid Interf. Sci.*, **346**, pp. 8–11.

23. Chervin, C. N., Clapsaddle, B. J., Chiu, H. W., Gash, A. E., Satcher, J. H., and Kauzlarich, S. M. (2006). Role of cyclic ether and solvent in a non-alkoxide sol-gel synthesis of Yttria-stabilized zirconia nanoparticles, *Chem. Mater.*, **18**, pp. 4865–4874.

24. Chytil, S., Glomm, W. R., Kvande, I., Zhao, T., Walmsley, J. C., and Blekkan, E. A. (2007). Platinum incorporated into the SBA-15 mesostructure via depositionprecipitation method: Pt nanoparticle size estimation and catalytic testing, *Top. Catal.*, **45**, pp. 93–99.

25. Cortie, M. B., and van der Lingen, E. (2002). Catalytic gold nanoparticles, *Mater. Forum*, **26**, pp. 1–14.

26. Courtel, F. M., Duncan, H., Abu-Lebdeh, Y., and Davidson, I. J. (2011). High capacity anode materials for Li-ion batteries based on spinel metal oxides AMn_2O_4 (A = Co, Ni, and Zn), *J. Mater. Chem.*, **21**, pp. 10206–10218.

27. Curri, M. L, Agostiano, A., Manna, L., Della, M. M., Catalano, M., Chiavarone, L., Spagnolo, V., and Lugara, M. (2000). Synthesis and characterization of CdS nanoclusters in a quaternary microemulsion: The role of the cosurfactant, *J. Phys. Chem. B*, **104**, pp. 8391–8397.

28. Cushing, B. L., Kolesnichenko, V. L., and O'Connor, C. J. (2004). Recent advances in the liquid-phase syntheses of inorganic nanoparticles, *Chem. Rev.*, **104**, pp. 3893–3946.

29. Das, S. K., Bhunia, M. K., Sinha, A. K., and Bhaumik, A. (2009). Self-assembled mesoporous zirconia and sulfated zirconia nanoparticles synthesized by triblock copolymer as template, *J. Phys. Chem. C*, **113**, pp. 8918–8923.

30. Da Silva, K. L., Menzel, D., Feldhoff, A., Kübel, C., Bruns, M., Paesano, A., Düvel, A., Wilkening, M., Ghafari, M., Hahn, H., Litterst, F. J., Heitjans, P., Becker, K. D., and Sepelak, V. (2011). Mechanosynthesized $BiFeO_3$ nanoparticles with highly reactive surface and enhanced magnetization, *J. Phys. Chem. C*, **115**, pp. 7209–7217.

31. Deng, D., and Lee, J. Y. (2008). Hollow core–shell mesospheres of crystalline SnO_2 nanoparticle aggregates for high capacity Li^+ ion storage, *Chem. Mater.*, **20**, pp. 1841–1846.

32. Dingsheng, W., Ting, X., and Yadong, L. (2009). Nanocrystals: solution-based synthesis and applications as nanocatalysts, *Nano Res.*, **2**, pp. 30–46.

33. Djerdj, I., Garnweitner, G., Su, D. S., and Niederberger, M. (2007). Morphology-controlled nonaqueous synthesis of anisotropic lanthanum hydroxide nanoparticles, *J. Solid State Chem.*, **180**, pp. 2154–2165.

34. Djerdj, I., Arcon, D., Jaglicic, Z., and Niederberger, M. (2008). Nonaqueous synthesis of metal oxide nanoparticles: Short review and doped titanium dioxide as case study for the preparation of transition metal-doped oxide nanoparticles, *J. Solid State Chem.*, **181**, pp. 1571–1581.

35. Duan, X., Lian, J., Ma, J., Kim, T., and Zheng, W. (2010). Shape-controlled synthesis of metal carbonate nanostructure via ionic liquid-assisted hydrothermal route: the case of manganese carbonate, *Cryst. Growth Des.*, **10**, 4449–4455.

36. Eastoe, J., Hollamby, M. J., and Hudson, L (2006). Recent advances in nanoparticle synthesis with reversed micelles, *Adv. Colloid. Interface. Sci.*, **128–130**, pp. 5–15.

37. Endres, F., and Abedin, S. Z. E. (2006). Air and water stable ionic liquids in physical chemistry, *Phys. Chem. Chem. Phys.*, **8**, pp. 2101–2116.

38. Estruga, M., Domingo, C., and Ayllón, J. A. (2010). Microwave radiation as heating method in the synthesis of titanium dioxide nano-particles from hexafluorotitanate-organic salts, *Mater. Res. Bull.*, **45**, pp. 1224–1229.

39. Fang, X. L., Li, Y., Chen, C., Kuang, Q., Gao, X. Z., Xie, Z. X., Xie, S. Y., Huang, R. B., and Zheng, L. (2010). pH-Induced simultaneous synthesis and self-assembly of 3D layered β-FeOOH nanorods, *Langmuir*, **26**, pp. 2745–2750.

40. Feldmann, C., and Metzmacher, C. (2001). Polyol mediated synthesis of nanoscale MS particles (M~Zn, Cd, Hg), *J. Mater. Chem.*, **11**, pp. 2603–2606.

41. Fernandez, C. A., and Wai, C. M. (2007). Continuous tuning of cadmium sulfide and zinc sulfide nanoparticle size in a water-in supercritical carbon dioxide microemulsion, *Chem. Eur. J.*, **13**, pp. 5838–5844.

42. Fischer, A., Müller, J. O., Antonietti, M., and Thomas, A. (2008). Synthesis of ternary metal nitride nanoparticles using mesoporous carbon nitride as reactive template, *ACS Nano*, **2**, pp. 2489–2496.

43. Fonseca, G. S., Umpierre, A. P., Fichtner, P. F. P., Teixeira, S. R., and Dupont, J. (2003). The use of imidazolium ionic liquids for the formation

and stabilization of Ir^0 and Rh^0 nanoparticles: efficient catalysts for the hydrogenation of arenas, *Chem. Eur. J.*, **9**, pp. 3263–3269.

44. Fuentes, R. O., and Baker, R. T. (2009). Synthesis of nanocrystalline CeO_2–ZrO_2 solid solutions by a citrate complexation route: a thermochemical and structural study, *J. Phys. Chem. C*, **113**, pp. 914–924.

45. Gaffet, E., Bernard, F., Niepce, J. C., Charlot, F., Gras, C., Le Caer, G., Guichard, J. L., Delcroix, P., Mocellin, A., and Tillement, O. (1999). Some recent developments in mechanical activation and mechanosynthesis, *J. Mater. Chem.*, **9**, pp. 305–314.

47. Gao, X., Du, Y., Liu, X., and Xu, P. (2011). Synthesis and characterization of Co–Sn substituted barium ferrite particles by a reverse microemulsion technique, *Mater. Res. Bull.*, **46**, pp. 643–648.

48. Ge, J. P., Xu, S., Liu, L. P., and Li, Y. D. (2006). Formation of disperse nanoparticles at the oil/water Interface in normal microemulsions, *Chem. Eur. J.*, **12**, pp. 6552–6558.

49. Gerbec, J. A., Magana, D., Washington, A., and Strouse, G. F. (2005). Microwave-enhanced reaction rates for nanoparticle synthesis, *J. Am. Chem. Soc.*, **127**, pp. 15791–15800.

50. Gómez-Cortés, A., Díaz, G., Zanella, R., Ramírez, H., Santiago, P., and Saniger, J. M. (2009). Au-Ir/TiO_2 prepared by deposition precipitation with urea: improved activity and stability in CO oxidation, *J. Phys. Chem. C*, **113**, pp. 9710–9720.

51. Gonsalves, K. E., Rangarajan, S. P., and Wang, J. (2002). *Nanostructured Materials and Nanotechnology*, ed. *Hari, S. N.*, Chapter 1 "Chemical synthesis of nanostructured metal, metal alloys, and semicoductors", (Academic Press, San Diego) pp. 1–56.

52. Goutailler, G., Guillard, C., Daniele, S., and Hubert-Pfalzgraf, L. G. (2003). Low temperature and aqueous sol-gel deposit of photo-catalytic active nanoparticulate TiO_2, *J. Mater. Chem.*, **13**, pp. 342–346.

53. Grisaru, H., Palchik, O., Gedanken, A., Palchik, V., Slifkin, M. A., and Weiss, A. M. (2003). Microwave-assisted polyol synthesis of $CuInTe_2$ and $CuInSe_2$ nanoparticles, *Inorg. Chem.*, **42**, pp. 7148–7155.

54. Grocholl, L., Wang, J., and Gillan, E. G. (2001). Solvothermal azide decomposition route to GaN nanoparticles, nanorods, and faceted crystallites, *Chem. Mater.*, **13**, pp. 4290–4296.

55. Hadi, A., and Yaacob, I. I., (2007). Novel synthesis of nanocrystalline CeO_2 by mechanochemical and water-in-oil microemulsion methods, *Mater. Lett.*, **61**, pp. 93–96.

56. Hakuta, Y., Ura, H., Hayashi, H., and Arai, K. (2005). Continuous production of $BaTiO_3$ nanoparticles by hydrothermal synthesis, *Ind. Eng. Chem. Res.*, **44**, pp. 840–846.

57. Han, Y. C., Li, S. P., Wang, X. Y., and Chen, X. M. (2004). Synthesis and sintering of nanocrystalline hydroxyapatite powders by citric acid sol-gel combustion method, *Mater. Res. Bull.*, **39**, pp. 25–32.

58. Han, Y. S., Fuj, M., Shchukin, M., Mohwald, H., and Takahashi, M. (2009). A new model for the synthesis of hollow particles via the bubble templating method, *Cryst. Growth Des.*, **9**, pp. 3771–3775.

59. Haruta, M. (1997). Novel catalysis of gold deposited on metal oxides, *Catal. Surv. Jpn*, **1**, pp. 161–173.

60. Hayashi, H., and Hakuta, Y. (2010). Hydrothermal synthesis of metal oxide nanoparticles in supercritical water, *Materials*, **3**, pp. 3794–3817.

61. Jia, F. L., Zhang, L. Z., Shang, X. Y., and Yang, Y. (2008). Non-aqueous sol–gel approach towards the controllable synthesis of nickel nanospheres, nanowires, and nanoflowers, *Adv. Mater.*, **20**, pp. 1050–1054.

62. Jiang, J., Yu, S. H., Yao, W. T., Ge, H., and Zhang, G. Z. (2005). Morphogenesis and crystallization of Bi_2S_3 nanostructures by an ionic liquid-assisted templating route: Synthesis, formation mechanism, and properties, *Chem. Mater.*, **17**, pp. 6094–6100.

63. Jiang, J. Z., and Cai, C. (2006). *In situ* formation of dispersed palladium nanoparticles in microemulsion: efficient reaction system for ligand-free Heck reaction, *J. Colloid Interf. Sci.*, **299**, pp. 938–943.

64. Joo, S. H., Park, J. Y., Renzas, J. R., Butcher, D. R., Huang, W., and Somorjai, G. A. (2010). Size effect of ruthenium nanoparticles in catalytic carbon monoxide oxidation, *Nano Lett.*, **10**, pp. 2709–2713.

65. Kaneko, K., Inoke, K., Freitag, B., Hungria, A. B., Midgley, P. A., Hansen, T. W., Zhang, J., Ohara, S., and Adschiri, T. (2007). Structural and morphological characterization of cerium oxide nanocrystals prepared by hydrothermal synthesis, *Nano Lett.*, **7**, pp. 421–425.

66. Kapoor, M. P., and Matsumura, Y. (2000). Liquid phase methanol carbonylation catalysed over rhodium supported on hydrotalcite, *Chem. Commun.*, pp. 95–96.

67. Kapoor, M. P., Raj, A., and Matsumura, Y. (2001). Methanol decomposition over palladium supported mesoporous CeO_2-ZrO_2 mixed oxides, *Micropor. Mesopor. Mater.*, **44–45**, pp. 565–572.

68. Khaleel, A. A. (2004). Nanostructured pure gamma-Fe_2O_3 via forced precipitation in an organic solvent, *Chem. Eur. J.*, **10**, pp. 925–932.

69. Khodakov, A. Y., Chu, W., and Fongarland, P. (2007). Advances in the development of novel cobalt fischer-tropsch catalysts for synthesis of long-chain hydrocarbons and clean fuels, *Chem. Rev.*, **107**, pp. 1692–1744.

70. Kim, J. R., Myeong, W. J., and Ihm, S.K. (2007). Characteristics in oxygen storage capacity of ceria-zirconia mixed oxides prepared by continuous hydrothermal synthesis in supercritical water, *Appl. Catal. B*, **71**, pp. 57–63.

71. Kim, K. S., Demberelnyamba, D., and Lee, H. (2004). Size-selective synthesis of gold and platinum nanoparticles using novel thiol-functionalized ionic liquids, *Langmuir*, **20**, pp. 556–560.

72. Kim, M., Hinklin, T. R., and Laine, R. M. (2008). Core–shell nanostructured nanopowders along $(CeO_x)_x(Al_2O_3)_{1-x}$ tie-line by liquid-feed flame spray pyrolysis (LF-FSP), *Chem. Mater.*, **20**, pp. 5154–5162.

73. Kim, M., and Laine, R. M. (2009). One-step synthesis of core–shell $(Ce_{0.7}Zr_{0.3}O_2)_x(Al_2O_3)_{1-x}$ $[(Ce_{0.7}Zr_{0.3}O_2)@Al_2O_3]$ nanopowders via liquid-feed flame spray pyrolysis (LF-FSP), *J. Am. Chem. Soc.*, **131**, pp. 9220–9229.

74. Kloprogge, J. T., Weier, M. L., Duong, L. V., and Frost, R. L. (2004). Microwave-assisted synthesis and characterisation of divalent metal tungstate nanocrystalline minerals: ferberite, hübnerite, sanmartinite, scheelite and stolzite, *Mater. Chem. Phys.*, **88**, pp. 438–443.

75. Kominami, H., Miyakawa, M., Murakami, S., Yasuda, T., Kohno, M., Onoue, S., Kera, Y., and Ohtani, B. (2001). Solvothermal synthesis of tantalum(V) oxide nanoparticles and their photocatalytic activities in aqueous suspension systems, *Phys. Chem. Chem. Phys.*, **3**, pp. 2697–2703.

76. Kong, L. B., Ma, J., and Boey, F. (2002). Nanosized hydroxyapatite powders derived from coprecipitation process, *J. Mater. Sci.*, **37**, pp. 1131–1134.

77. Krumm, M., Pueyo, C. L., and Polarz, S. (2010). Monolithic zinc oxide aerogels from organometallic sol-gel precursors, *Chem. Mater.*, **22**, pp. 5129–5136.

78. Kwon, S. G., and Hyeon, T. (2008). Colloidal chemical synthesis and formation kinetics of uniformly sized nanocrystals of metals, oxides, and chalcogenides, *Acc. Chem. Res.*, **41**, pp. 1696–1709.

79. Lay, S., Qiu, Y., and Wang, S. (2006). Effects of the structure of ceria on the activity of gold/ceria catalysts for the oxidation of carbon monoxide and benzene, *J. Catal.*, **237**, pp. 303–313.

80. Lee, M. H., Lin H. Y., *and* Thomas, J. L. (2006). A novel supercritical CO_2 synthesis of amorphous hydrous zirconia nanoparticles, and their calcination to zirconia, *J. Am. Ceram. Soc.*, **89**, pp. 3624–3630.

81. Leonard, B. M., Bhuvanesh, N. S. P., and Schaak, R. E. (2005). Low-temperature polyol synthesis of $AuCuSn_2$ and $AuNiSn_2$: using solution chemistry to access ternary intermetallic compounds as nanocrystals, *J. Am. Chem Soc.*, **127**, pp. 7326–7327.

82. Li, H., Liu, J., Xie, S., Qiao, M., Dai, W., and Li, H. (2008). Highly active Co–B amorphous alloy catalyst with uniform nanoparticles prepared in oil-in-water microemulsion, *J. Catal.*, **259**, pp. 104–110.

83. Li, L. L., Zhang, W. M., Yuan, Q., Li, Z. X., Fang, C. J., Sun, L. D., Wan, L. J., and Yan, C. H. (2008). Room temperature ionic liquids assisted green synthesis of nanocrystalline porous SnO_2 and their gas sensor behaviors, *Cryst. Growth Des.*, **8**, pp. 4165–4172.

84. Li, Y., and Somorjai, G. A. (2010). Nanoscale advances in catalysis and energy applications, *Nano Lett.*, **10**, pp. 2289–2295.

85. Liang, H. W., Liu, S., and Yu, S. H. (2010). The controlled synthesis of one-dimensional inorganic nanostructures using pre-existing one-dimensional nanostructures as templates, *Adv. Mater.*, **22**, pp. 3925–3937.

86. Lim, K. T., Hwang, H. S., Ryoo, W., and Johnston, K. P. (2004). Synthesis of TiO_2 nanoparticles utilizing hydrated reverse micelles in CO_2, *Langmuir*, **20**, pp. 2466–2471.

87. Lim, S. H., Phonthammachai, N., Zhong, Z. Y., Teo, J., and White, T. J. (2009). Robust gold-decorated silica-titania pebbles for low-temperature CO catalytic oxidation, *Langmuir*, **25**, pp. 9480–9486.

88. Longo, A., Liotta, L. F., Carlo, G. D., Giannici, F., Venezia, A. M., and Martorana, M. A. (2010). Structure and the metal support interaction of the Au/Mn oxide catalysts, *Chem. Mater.*, **22**, pp. 3952–3960.

89. Lv, J. P., Qu, L. Z., and Qu, B. J. (2004). Controlled growth of three morphological structures of magnesium hydroxide nanoparticles by wet precipitation method, *J. Cryst. Growth*, **267**, pp. 676–684.

90. Ma, Y., and Qi, L. (2009). Solution-phase synthesis of inorganic hollow structures by templating strategies, *J. Colloid Interf. Sci.*, **335**, pp. 1–10.

91. Maedler, L., and Pratsinis, S. E. (2002). Bismuth oxide nanoparticles by flame spray pyrolysis, *J. Am. Ceram. Soc.*, **85**, pp. 1713–1718.

92. Mäkelä, J. M., Keskinen, H., Forsblom, T., and Keskinen, J. (2004). Generation of metal and metal oxide nanoparticles by liquid flame spray process, *J. Mater. Sci.*, **39**, pp. 2783–2788.

93. Mao, J., Li, X. L., Qin, W. J., Niu, K. Y., Yang, J., Ling, T., and Du, X. W. (2010). Control of the morphology and optical properties of ZnO nanostructures via hot mixing of reverse micelles, *Langmuir*, **26**, pp. 13755–13759.

94. Maye, M. M., Lim, I. I. S., Luo, J., Rab, Z., Rabinovich, D., Liu, T., and Zhong, C. J. (2005). Mediator-template assembly of nanoparticles, *J. Am. Chem. Soc.*, **127**, pp. 1519–1529.

95. Migowski, P., Machado, G., Texeira, S. R., Alves, M. C. M., Morais, J., Traversec, A., and Dupont, J. (2007). Synthesis and characterization of nickel nanoparticles dispersed in imidazolium ionic liquids, *Phys. Chem. Chem. Phys.*, **9**, pp. 4814–4821.

96. Misra, N., Roy, M., Mohanta, D., Baruah, K. K., and Choudhury, A. (2008). Photochromism and magneto-optic response of ZnO:Mn semiconductor quantum dots fabricated by microemulsion route, *Cent. Eur. J. Phys.*, **6**, pp. 109–115.

97. Mohamed, M. B., AbouZeid, K. M., Abdelsayed, V., Aljarash, A., and El-Shall, M. S. (2010). Growth mechanism of anisotropic gold nanocrystals via microwave synthesis: Formation of dioleamide by gold nanocatalysis, *ACS Nano*, **4**, pp. 2766–2772.

98. Mueller, R., Maedler, L., and Pratsinis, S. E. (2003). Nanoparticle synthesis at high production rates by fame spray pyrolysis, *Chem. Eng. Sci.*, **58**, pp. 1969–1976.

99. Mueller, R., Jossen, R., Pratsinis, E., Watson, M., and Akhtar, K. (2004). Zirconia nanoparticles made in spray flames at high production rates, *J. Am. Ceram. Soc.*, **87**, pp. 197–202.

100. Narayanan, R., and El-Sayed, M. A. (2005). Catalysis with transition metal nanoparticles in colloidal solution: Nanoparticle shape dependence and stability, *J. Phys. Chem. B*, **109**, pp. 12663–12676.

101. Nares, R., Ramírez, J., Gutiérrez-Alejandre, A., Louis, C., and Klimova, T. (2002). Ni/Hβ-Zeolite catalysts prepared by deposition-precipitation, *J. Phys. Chem. B*, **106**, pp. 13287–13293.

102. Nguyen, T. D., Dinh, C. T., and Do, T. O. (2010). Shape- and size-controlled synthesis of monoclinic ErOOH and Cubic Er_2O_3 from micro- to nanostructures and their upconversion luminescence, *ACS Nano*, **4**, pp. 2263–2273.

103. Niederberger, M., and Garnweitner, G. (2006). Organic reaction pathways in the nonaqueous synthesis of metal oxide nanoparticles, *Chem. Eur. J.*, **12**, pp. 7282–7302.

104. Niederberger, M., Garnweitner, G., Buha, J., Polleux, J., Ba, J., and Pinna, N. (2006). Nonaqueous synthesis of metal oxide nanoparticles: review and indium oxide as case study for the dependence of particle morphology on precursors and solvents, *J. Sol-Gel Sci. Technol.*, **40**, pp. 259–266.

105. Niederberger, M. (2007). Sol-Gel routes to metal oxide nanoparticles, *Acc. Chem. Res.*, **40**, pp. 793–800.

106. Nunez, F., Angel, G. D., Tzompantzi, F., and Navarrete, J. (2011). Catalytic wet-air oxidation of p-Cresol on Ag/Al_2O_3–ZrO_2 catalysts, *Ind. Eng. Chem. Res.*, **50**, pp. 2495–2500.

107. Oh, M. H., Lee, N., Kim, H., Park, S. P., Piao, Y., Lee, J., Jun, S. W., Moon, W. K., Choi, S. H., *and* Hyeon, T. (2011). Large-scale synthesis of bioinert tantalum oxide nanoparticles for X-ray computed tomography imaging and bimodal image-guided sentinel lymph node mapping, *J. Am. Chem. Soc.*, **133**, pp. 5508–5515.

108. Ohde, H., Rodriguez, J. M., Ye, X. R., and Wai, C. M. (2000). Synthesizing silver halide nanoparticles in supercritical carbon dioxide utilizing a water-in-CO_2 microemulsion, *Chem. Commun.*, pp. 2353–2354.

109. Ohde, H., Hunt, F., and Wai, C. M. (2001). Synthesis of silver and copper nanoparticles in a water-in-supercritical-carbon dioxide microemulsion, *Chem. Mater.*, **13**, pp. 4130–4135.

110. Ohde, H., Wai, C., Kim, H., Kim, J., and Ohde, M. (2002). Hydrogenation of olefins in supercritical CO_2 catalyzed by palladium nanoparticles in a water-in-CO_2 microemulsion, *J. Am. Chem. Soc.*, **124**, pp. 4540–4541.

111. Okumura, M., Masuyama, N., Konishi, E., Ichikawa, S., and Akita, T. (2002). CO oxidation below room temperature over Ir/TiO_2 catalyst prepared by deposition-precipitation method, *J. Catal.*, **208**, pp. 485–489.

112. Patil, N. S., Uphade, B. S., McCulloh, D. G., Bhargava, S. K., and Choudhary, V. R. (2004). Styrene epoxidation over gold supported on different transition metal oxides prepared by homogeneous deposition-precipitation, *Catal. Commun.*, **5**, pp. 681–685.

113. Pinna, N., and Niederberger, M. (2008). Surfactant-free nonaqueous synthesis of metal oxide nanostructures, *Angew. Chem. Int. Ed.*, **47**, pp. 5292–5304.

114. Polshettiwar, V., Baruwati, B., and Varma, R. S. (2009). Self-assembly of metal oxides into three-dimensional nanostructures: Synthesis and application in catalysis, *ACS Nano*, **3**, pp. 728–736.

115. Popa, A. F., Courtheoux, L., Gautron, E., Rossignol, S., and Kappenstein, C. (2005). Aerogel and xerogel catalysts based on θ-alumina doped with silicon for high temperature reactions, *Eur. J. Inorg. Chem.*, pp. 543–554.

116. Qu, Y., Yang, H., Yang, N., Fan, Y., Zhu, H., and Zou, G. (2006). The effect of reaction temperature on the particle size, structure and magnetic properties of coprecipitated $CoFe_2O_4$ nanoparticles, *Mater. Lett.*, **60**, pp. 3548–3552.

117. Rabenau, A. (1985). The role of hydrothermal synthesis in preparative chemistry, *Angew. Chem. Int. Ed. Engl.*, **24**, pp. 1026–1040.

118. Rajesh, K., Mukundan, P., Pillai P. K., Nair, V. R., and Warrier, K. G. K. (2004). High surface area nanocrystalline cerium phosphate through aqueous sol-gel route, *Chem. Mater.*, **16**, pp. 2700–2705.

119. Rao, K. J., Vaidhyanathan, B., Ganguli, M., and Ramakrishnan, P. A. (1999). Synthesis of inorganic solids using microwaves, *Chem. Mater.*, **11**, pp. 882–895.

120. Raveendran, P., Fu, J., and Wallen, S. L. (2003). Completely "green" synthesis and stabilization of metal nanoparticles, *J. Am. Chem. Soc.*, **125**, pp. 13940–13941.

121. Reddy, B. M., Bharali, P., Saikia, P., Khan, A., Loridant, S., Muhler, M., and Grunert, W. (2007). Influence of alumina, silica, and titania supports on the structure and CO oxidation activity of $Ce_xZr_{1-x}O_2$ nanocomposite oxides, *J. Phys. Chem. C*, **111**, pp. 10478–10483.

122. Reddy, B. M., Bharali, P., Saikia, P., Park, S. E., Muhler, M., and Grünert, W. (2008). Structural characterization and catalytic activity of nanosized $Ce_xM_{1-x}O_2$ (M = Zr & Hf) mixed oxides, *J. Phys. Chem. C*, **112**, pp. 11729–11737.

123. Reddy, B. M., Saikia, P., Bharali, P., Yamada, Y., Kobayashi, T., Muhler, M., and Grünert, W. (2008). Structural characterization and catalytic activity of nanosized ceria-terbia solid solutions, *J. Phys. Chem. C*, **112**, pp. 16393–16399.

124. Reddy, B. M., Bharali, P., and Saikia, P. (2009). *New Nanotechniques*, eds. Malik, A., and Rawat, R. J. Chapter 6 "A Comprehensive Overview Synthesis Techniques of Nanostructured Oxides," (Nova Science Publishers, New York), pp. 243–276.

125. Reddy, B. M., Gode, T., Lakshmi, K., Yamada, Y., and Park, S. E. (2009). Structural characterization and catalytic activity of nanocrystalline ceria-praseodymia solid solutions, *J. Phys. Chem. C*, **113**, pp. 15882–15890.

126. Reddy, B. M., Reddy, G. K., Ganesh, I., and Ferreira, J. M. F. (2009). Microwave-assisted synthesis and structural characterization of nanosized $Ce_{0.5}Zr_{0.5}O_2$ for CO oxidation, *Catal. Lett.*, **130**, pp. 227–234.

127. Reddy, B. M., Reddy, G. K., Ganesh, I., and Ferreira, J. M. F. (2009). Single step synthesis of nanosized CeO_2–M_xO_y mixed oxides ($M_xO_y = SiO_2$, TiO_2, ZrO_2, and Al_2O_3) by microwave-induced solution combustion synthesis-characterization and CO oxidation, *J. Mater. Sci.*, **44**, pp. 2743–2751.

128. Reddy, B. M., Reddy, G. K., Rao, K. N., Ganesh, I., and Ferreira, J. M. F. (2009). Characterization and photocatalytic activity of TiO_2-M_2O_3 ($M_2O_3 = SiO_2$, Al_2O_3 and ZrO_2) mixed oxides synthesized by microwave-induced solution combustion technique, *J. Mater. Sci.*, **44**, pp. 4874–4882.

129. Reddy, B. M., Katta, L., and Thrimurthulu, G. (2010). Novel nanocrystalline $Ce_{1-x}La_xO_{2-\delta}$ ($x = 0.2$) solid solutions: Structural characteristics and catalytic performance, *Chem. Mater.*, **22,** pp. 467–475.

130. Reddy, G. K., Katta, L., and Reddy, B. M. (2010). Structural characterization and dehydration activity of CeO_2–SiO_2 and CeO_2–ZrO_2 mixed oxides prepared by a rapid microwave-assisted combustion synthesis method, *J. Mol. Catal. A: Chem.*, **319**, pp. 52–57.

131. Reddy, B. M., Thrimurthulu, G., and Katta, L. (2011). Design of efficient $Ce_xM_{1-x}O_{2-\delta}$ (M = Zr, Hf, Tb, and Pr) nanosized model solid solutions for CO oxidation, *Catal. Lett.*, **141**, pp. 572–581.

132. Richter, K., Birkner, A., and Mudring, A. V. (2010). Stabilizer-free metal nanoparticles and metal–metal oxide nanocomposites with long-term stability prepared by physical vapor deposition into ionic liquids, *Angew. Chem. Int. Ed.*, **49**, pp. 2431–2435.

133. Rojac, T., Kosec, M., Malic, B., and Holc, J. (2005). Mechanochemical synthesis of $NaNbO_3$, *Mater. Res. Bull.*, **40**, pp. 341–345.

134. Rosiyah, B. Y., Hiromichi, H., Takako, N., Takeo, E., Yoshio, O., and Norio, S. (2001). Hydrothermal synthesis of potassium hexatitanates under subcritical and supercritical water conditions and its application in photocatalysis, *Chem. Mater.*, **13**, pp. 842–847.

135. Roychowdhury, C., Matsumoto, F., Mutolo, P. F., Abruna, H. D., and DiSalvo, F. J. (2005). Synthesis, characterization, and electrocatalytic

activity of PtBi nanoparticles prepared by the polyol process, *Chem. Mater.*, **17**, pp. 5871–5876.

136. Ruckenstein, E. (1996). Microemulsions, macroemulsions, and the bancroft rule, *Langmuir*, **12**, pp. 6351–6353.

137. Sagisaka, M., Iwama, S., Hasegawa, S., Yoshizawa, A., Mohamed, A., Cummings, S., Rogers, S. E., Heenan, R. K., and Eastoe, J. (2011). Super-efficient surfactant for stabilizing water-in-carbon dioxide microemulsions, *Langmuir*, **27**, pp. 5772–5780.

138. Sanchez-Dominguez, M., Boutonnet, M., and Solans, C. (2009). A novel approach to metal and metal oxide nanoparticle synthesis: the oil-in-water microemulsion reaction method, *J. Nanopart. Res.*, **11**, pp. 1823–1829.

139. Sanchez-Dominguez, M., Liotta, L. F., Di Carlo, G., Pantaleo, G., Venezia, A. M., Solans, C., and Boutonnet, M. B. (2010). Synthesis of CeO_2, ZrO_2, $Ce_{0.5}Zr_{0.5}O_2$, and TiO_2 nanoparticles by a novel oil-in-water microemulsion reaction method and their use as catalyst support for CO oxidation, *Catal. Today*, **158**, pp. 35–43.

140. Sasaki, T., Ohara, S., Naka, T., Vejpravova, J., Sechovsky, V., Umetsu, M., Takami, S., and Jeyadevan, T. (2010). Continuous synthesis of fine $MgFe_2O_4$ nanoparticles by supercritical hydrothermal reaction, *J. Supercritical Fluids*, **53**, pp. 92–94.

141. Schimmoeller, B., Schulz, H., Ritter, A., Reitzmann, A., Kraushaar, C. B., Baiker, A., and Pratsinis, S. E. (2008). Structure of flame-made vanadia/titania and catalytic behavior in the partial oxidation of o-xylene, *J. Catal.*, **256**, pp. 74–83.

142. Seo, D. J., Park, S. B., Kang, Y. C., and Choy, K. L. (2003). Formation of ZnO, MgO and NiO nanoparticles from aqueous droplets in flame reactor, *J. Nanopart. Res.*, **5**, pp. 199–210.

143. Sepelak, V., Feldhoff, A., Heitjans, P., Krumeich, F., Menzel, D., Litterst, F. J., Bergmann, I., and Becker, K. D. (2006). Nonequilibrium cation distribution, canted spin arrangement, and enhanced magnetization in nanosized $MgFe_2O_4$ prepared by a one-step mechanochemical route, *Chem. Mater.*, **18**, pp. 3057–3067.

144. Sepelak, V., Becker, K. D., Bergmann, I., Suzuki, S., Indris, S., Feldhoff, A., Heitjans, P., and Grey, C. P. (2009). A one-step mechanochemical route to core–shell Ca_2SnO_4 nanoparticles followed by ^{119}Sn MAS NMR and [119]Sn Mössbauer spectroscopy, *Chem. Mater.*, **21**, pp. 2518–2524.

145. Shchukin, D. G., and Caruso, R. A. (2004). Template synthesis and photocatalytic properties of porous metal oxide spheres formed by nanoparticle infiltration, *Chem. Mater.*, **16**, pp. 2287–2292.

146. Shen, L., Bao, N., Yanagisawa, K., Domen, K., Grimes, C. A., and Gupta, A. (2007). Controlled synthesis and assembly of nanostructured ZnO architectures by a solvothermal soft chemistry process, *Cryst. Growth Des.*, **7**, pp. 2742–2748.

147. Shimizu, K., Cheng, I. F., Wang, J. S., Yen, C. H., Yoon, B., and Wai, C. M. (2008). Water-in-supercritical CO_2 microemulsion for synthesis of carbon-nanotube-supported Pt electrocatalyst for the oxygen reduction reaction, *Energy Fuels*, **22**, pp. 2543–2549.

148. Shylesh, S., Schünemann, V., and Thiel, W. R. (2010). Magnetically separable nanocatalysts: Bridges between homogeneous and heterogeneous catalysis, *Angew. Chem. Int. Ed.*, **49**, pp. 3428–3459.

149. Simentsova, I. I., Minyukova, T. P., Khassin, A. A., Dokuchits, E. V., Davydova, L. P., Yu, M. I., Plyasova, L. M., Kustova, G. N., and Yurieva, T. M. (2010). The effect of the precursor structure on the catalytic properties of the nickel-chromium catalysts of hydrogenation reactions, *Russ. Chem. Bull.*, **59**, pp. 2055–2060.

150. Smetana, A. B., Wang, J. S., Boeckl, J., Brown, G. J., and Wai, C. M. (2007). Fine-tuning size of gold nanoparticles by cooling during reverse micelle synthesis, *Langmuir*, **23**, pp. 10429–10432.

151. Smolyakov, V. K., Lapshin, O. V., and Boldyrev, V. V. (2008). Mathematical simulation of mechanochemical synthesis in a macroscopic approximation, *Theor. Found. Chem. Eng.*, **42**, pp. 54–59.

152. Subramanian, V., Zhu, H., Vajtai, R., Ajayan, P. M., and Wei, B. (2005). Hydrothermal synthesis and pseudocapacitance properties of MnO_2 manostructures, *J. Phys. Chem. B*, **109**, pp. 20207–20214.

153. Sue, K., Murata, K., Kimura, K., and Arai, K. (2003). Continuous synthesis of zinc oxide nanoparticles in supercritical water, *Green Chem.*, **5**, pp. 659–662.

154. Sue, K., Suzuki, M., Arai, K., Ohashi, T., Ura, H., Matsui, K., Hakuta, Y., Hayashi, H., Watanabe, M., and Hiaki, T. (2006). Size-controlled synthesis of metal oxide nanoparticles with a flow-through super-critical water method, *Green Chem.*, **8**, pp. 634–638.

155. Sun, Y., and Xia, Y. (2002). Shape-controlled synthesis of gold and silver nanoparticles, *Science*, **298**, pp. 2176–2179.

156. Tago, T., Hatsuta, T., Miyajima, K., Kishida, M., Tashiro, S., and Wakabayashi, K. (2002). Novel synthesis of silica-coated ferrite nanoparticles prepared using water-in-oil microemulsion, *J. Am. Ceram. Soc.*, **85**, pp. 2188–2194.

157. Tan, G. L., and Yu, X. F. (2009). Capping the ball-milled CdSe nanocrystals for light excitation, *J. Phys. Chem. C*, **113**, pp. 8724–8729.

158. Tang, E., Tian, B., Zheng, E., Fu, C., and Cheng, G. (2008). Preparation of zinc oxide nanoparticle via uniform precipitation method and its surface modification by methacryloxypropyltrimethoxysilane, *Chem. Eng. Comm.*, **195**, pp. 479–491.

159. Tang, J., Fabbri, J., Robinson, R. D., Zhu, Y., Herman, I. P., Steigerwald, M. L., and Brus, L. E. (2004). Solid-solution nanoparticles: Use of a nonhydrolytic sol-gel synthesis to prepare HfO_2 and $Hf_xZr_{1-x}O_2$ nanocrystals, *Chem. Mater.*, **16**, pp. 1336–1342.

160. Tani, T., Mädler, L., and Pratsinis, S. E. (2002). Homogeneous ZnO nanoparticles by flame spray pyrolysis, *J. Nanopart. Res.*, **4**, pp. 337–343.

161. Taniguchi, T., Watanabe, T., Sakamoto, N., Matsushita, N., and Yoshimura, M. (2008). Aqueous route to size-controlled and doped organophilic ceria nanocrystals, *Cryst. Growth Des.*, **8**, pp. 3725–3730.

162. Teoh, W. Y., Setiawan, R., Mädler, L., Grunwaldt, J. D., Amal, R., and Pratsinis, S. E. (2008). Ru-doped cobalt-zirconia nanocomposites by flame synthesis: Physicochemical and catalytic properties, *Chem. Mater.*, **20**, pp. 4069–4079.

163. Tian, X., Li, J., Chen, K., Han, J., Pan, S., Wang, Y., Fan, X., Li, F., and Zhou, Z. (2010). Nearly monodisperse ferroelectric $BaTiO_3$ hollow nanoparticles: Size-related solid evacuation in Ostwald-ripening-induced hollowing process, *Cryst. Growth Des.*, **10**, pp. 3990–3995.

164. Torimoto, T., Tsuda, T., Okazaki, K., and Kuwabata, S. (2010). New frontiers in materials science opened by ionic liquids, *Adv. Mater.*, **22**, pp. 1196–1221.

165. Tsuji, M., Hashimoto, M., Nishizawa, Y., Kubokawa, M., and Tsuji, T. (2005). Microwave-assisted synthesis of metallic nanostructures in solution, *Chem. Eur. J.*, **11**, pp. 440–452.

166. Tsuzuki, T., and McCormick, P. G. (2001). Synthesis of ultrafine ceria powders by mechanochemical processing, *J. Am. Ceram. Soc.*, **84**, pp. 1453–1458.

167. Tsuzuki, T., and McCormick, P. G. (2004). Mechanochemical synthesis of nanoparticles, *J. Mater. Sci.*, **39**, pp. 5143–5146.

168. Wang, C. J., Wu, Y. A., Jacobs, R. M. J., Warner, J. H., Williams, G. R., and O'Hare, D. (2011). Reverse micelle synthesis of Co–Al LDHs: Control of particle size and magnetic properties, *Chem. Mater.*, **23**, pp. 171–180.

169. Wang, H., Zhang, J. R., and Zhu, J. J. (2002). Sonochemical preparation of lead sulfide nanocrystals in an oil-in-water microemulsion, *J. Cryst. Growth*, **246**, pp. 161–168.

170. Wang, H., Xu, X., Zhang, J., and Li, C. (2010). A Cost-effective co-precipitation method for synthesizing indium tin oxide nano-particles without chlorine contamination, *J. Mater. Sci. Technol.*, **26**, pp. 1037–1040.

171. Wang, H. Q., and Nann, T. (2009). Monodisperse upconverting nanocrystals by microwave-assisted synthesis, *ACS Nano*, **3**, pp. 3804–3808.

172. Wang, L. C., Liu, Y. M., Chen, M., Cao, Y., He, H. Y., and Fan, K. N. (2008). MnO_2 nanorod supported gold nanoparticles with enhanced activity for solvent-free aerobic alcohol oxidation, *J. Phys. Chem. C*, **112**, pp. 6981–6986.

173. Wang, Y., Wu, Q. S., and Ding, Y. P. (2004). Preparation of Group IIB selenide nanoparticles using soft-hard dual template method, *J. Nanopart. Res.*, **6**, pp. 253–257.

174. Wang, Y., and Yang, H. (2006). Oleic acid as the capping agent in the synthesis of noble metal nanoparticles in imidazolium-based ionic liquids, *Chem. Commun.*, pp. 2545–2547.

175. Wang, Y., Maksimuk, S., Shen, R., and Yang, H. (2007). Synthesis of iron oxide nanoparticles using a freshly-made or recycled imidazolium-based ionic liquid, *Green Chem.*, **9**, pp. 1051–1056.

176. Weidenhof, B., Reiser, M., Stöwe, K., Maier, W. F., Kim, M., Azurdia, J., Gulari, E., Seker, E., Barks, A., and Laine, R. M. (2009). High-throughput screening of nanoparticle catalysts made by flame spray pyrolysis as hydrocarbon/NO oxidation catalysts, *J. Am. Chem. Soc.*, **131**, pp. 9207–9219.

177. Welton, T. (1999). Room temperature ionic liquids. Solvents for synthesis and catalysis, *Chem. Rev.*, **99**, pp. 2071–2083.

178. White, R. J., Luque, R., Budarin, V. L., Clark, J. H., and Macquarrie, D. J. (2009). Supported metal nanoparticles on porous materials. Methods and applications, *Chem. Soc. Rev.*, **38**, pp. 481–494.

179. Wu, H., Xu, H., Su, Q., Chen, T., and Wu, M. (2003). Size- and shape-tailored hydrothermal synthesis of YVO_4 crystals in ultra-wide pH range conditions, *J. Mater. Chem.*, **13**, pp. 1223–1228.

180. Wu, M., Long, J., Huang, A., Luo, Y., Feng, S., and Xu, R. (1999). Microemulsion-mediated hydrothermal synthesis and characteri-zation of nanosize rutile and anatase particles, *Langmuir*, **15**, pp. 8822–8825.

181. Wu, M., Lin, G., Chen, D., Wang, G., He, D., Feng, S., and Xu, R. (2002). Sol-hydrothermal synthesis and hydrothermally structural evolution of nanocrystal titanium dioxide, *Chem. Mater.*, **14**, pp. 1974–1980.

182. Xiao, H. Y., Ai, Z. H., and Zhang, L. Z. (2009). Nonaqueous sol-gel synthesized hierarchical CeO_2 nanocrystal microspheres as novel adsorbents for wastewater treatment, *J. Phys. Chem. C*, **113**, pp. 16625–16630.

183. Xiao, H. Y., Li, P. N., Jia, F. L., and Zhang, L. Z. (2009). General nonaqueous sol-gel synthesis of nanostructured Sm_2O_3, Gd_2O_3, Dy_2O_3, and Gd_2O_3: Eu^{3+} phosphor, *J. Phys. Chem. C*, **113**, pp. 21034–21041.

184. Xing, Y., Liu, Z., and Suib, S. L. (2007). Inorganic synthesis for the stabilization of nanoparticles: Application to Cu/Al_2O_3 nanocomposite materials, *Chem. Mater.*, **19**, pp. 4820–4826.

185. Xiong, L., and He, T. (2006). Synthesis and characterization of ultrafine tungsten and tungsten oxide nanoparticles by a reverse microemulsion-mediated method, *Chem. Mater.*, **18**, pp. 2211–2218.

186. Xu, G., Zhang, Y. W., Liao, C. S., and Yan. C. H. (2004). Doping and grain size effects in nanocrystalline ZrO_2–Sc_2O_3 system with complex phase transitions: XRD and Raman studies, *Phys. Chem. Chem. Phys.*, **6**, pp. 5410–5418.

187. Xu, S., and Li, Y. (2003). Different morphology at different reactant molar ratios: synthesis of silver halide low-dimensional nanomaterials in microemulsions, *J. Mater. Chem.*, **13**, pp. 163–165.

188. Xu, Z. P., and Lu, G. Q. (2005). Hydrothermal synthesis of layered double hydroxides (LDHs) from mixed MgO and Al_2O_3: LDH formation mechanism, *Chem. Mater.*, **17**, pp. 1055–1062.

189. Yamauchi, T., Tsukahara, Y., Yamada, K., Sakata, T., and Wada, Y. (2011). Nucleation and growth of magnetic Ni–Co (core–shell) nanoparticles in a one-pot reaction under microwave irradiation, *Chem. Mater.*, **23**, pp. 75–84.

190. Yang, Y., Matsubara, S., Xiong, L., Hayakawa, T., and Nogami, M. (2007). Solvothermal synthesis of multiple shapes of silver nanoparticles and their SERS properties, *J. Phys. Chem. C*, **111**, pp. 9095–9104.

191. Yu, C. H., Oduro, W., Tam, K., and Edman, S. C. T. (2009). *Metallic Nanoparticles*, ed. John, B., Chapter 10 "Some Applications of Nanoparticles," (Elsevier, Oxford, UK) pp. 365–378.

192. Zanella, R., Giorgio, S., Shin, C. H., Henry, C. R., and Louis, C. (2004). Characterization and reactivity in CO oxidation of gold nanoparticles supported on TiO_2 prepared by deposition-precipitation with NaOH and urea, *J. Catal.*, **222**, pp. 357–367.

193. Zhang, D., Yan, T., Li, H., and Shia, L. (2011). Ionic liquid-assisted synthesis and photoluminescence property of mesoporous EuF$_3$ nanospheres, *Micropor. Mesopor. Mater.*, **141**, pp. 110–118.

194. Zhang, H. W., and Swihart, M. T. (2007). Synthesis of tellurium dioxide nanoparticles by spray pyrolysis, *Chem. Mater.*, **19**, pp. 1290–1301.

195. Zhang, L. Z., Djerdj, I., Cao, M. H., Antonietti, M., and Niederberger, M. (2007). Nonaqueous sol-gel synthesis of a nanocrystalline InNbO$_4$ visible-light photocatalyst, *Adv. Mater.*, **19**, pp. 2083–2086.

196. Zhang, Q., and Saito, F. (2000). Mechanochemical synthesis of lanthanum aluminate by grinding lanthanum oxide with transition alumina, *J. Am. Ceram. Soc.*, **83**, pp. 439–441.

197. Zhang, X., and Chan, K. Y. (2003). Water-in-oil microemulsion synthesis of platinum-ruthenium nanoparticles, their characterization and electrocatalytic properties, *Chem. Mater.*, **15**, pp. 451–459.

198. Zhang, X. R., Wang, L. C., Yao, C. Z., Cao, Y., Dai, W. L., He, H. Y., and Fan, K. N. (2005). A highly efficient Cu/ZnO/Al$_2$O$_3$ catalyst via gel-coprecipitation of oxalate precursors for low-temperature steam reforming of methanol, *Catal. Lett.*, **102**, pp. 183–190.

199. Zheng, N., and Stucky, G. (2006). A general synthetic strategy for oxide-supported metal nanoparticle catalysts, *J. Am. Chem. Soc.*, **128**, pp. 14278–14280.

200. Zhu, H. T., Zhang, H., Liang, J. K., Rao, G. H., Li, J. B., Liu, G. Y., Du, Z. M., Fan, H. M., and Luo, J. (2011). Controlled synthesis of Te nano-structures from nanotubes to nanorods and nanowires and their template applications, *J. Phys. Chem. C*, **115**, pp. 6375–6380.

201. Zhu, J., Shen, Y., Xie, A., Qiu, L., Zhang, Q., and Zhang, S. (2007). Photoinduced synthesis of anisotropic gold nanoparticles in room-temperature ionic liquid, *J. Phys. Chem. C*, **111**, pp. 7629–7633.

202. Zhu, J. J., Palchik, O., Chen, S., and Gedanken, A. (2000). Microwave assisted preparation of CdSe, PbSe, and Cu$_{2-x}$Se nanoparticles, *J. Phys. Chem. B*, **104**, pp. 7344–7347.

Chapter 4

Commercial-Scale Production of Nanoparticles

Takuya Tsuzuki

Research School of Engineering, College of Engineering and Computer Science, Australian National University, Ian Ross Building 31, North Road, Canberra ACT 0200, Australia

takuya.tsuzuki@anu.edu.au

4.1 Introduction

As described in the previous chapters, a large number of nano-enabled commercial products have already appeared in a broad range of markets and industries [38,39,42,50,57]. However, most of the commercialised applications are based on only a limited number of nanoparticle materials such as silica, alumina, silver, gold, fullerenes, titanium oxide and zinc oxide. Many other nanoparticulate materials that show unique and useful properties still remain on the laboratory bench. One of the most prominent hurdles that must be overcome for the commercial application of nanoparticles is production scale-up. For example, the four major commercial markets of nanoparticles by volume are

Nanotechnology Commercialisation
Edited by Takuya Tsuzuki
Copyright © 2013 Pan Stanford Publishing Pte. Ltd.
ISBN 978-981-4303-28-6 (Hardcover), 978-981-4303-29-3 (eBook)
www.panstanford.com

estimated as (i) automotive catalysts (11,500 tonnes), (ii) chemical mechanical planarisation slurry (9,400 tonnes), (iii) magnetic recording media (3,100 tonnes) and (iv) sunscreens (1,500 tonnes) [16]. Although the required volume of nanoparticles depends on the applications, successful scale-up of nanoparticle production to a similar volume range to those is necessary to take nanotechnology innovation to the general public [50].

This chapter focuses on current trends in the commercial-scale production methods of nanoparticles. The limiting factors for the scalability of synthesis methods are explained and the relationship between commercial nanoparticle materials and production methods is discussed. Particular emphasis is placed on the fact that different synthesis techniques lead to different properties of nanoparticles even when the qualities such as particle size and crystal phase appear quite similar. The production techniques of nanoparticles need to be carefully selected based not only on the scalability and production costs but also on the properties of nanoparticles required for specific applications.

4.2 Methods Used in the Commercial-Scale Production of Nanoparticles

4.2.1 Challenges in Production Scale-Up

Since the mid-1980s, significant research efforts have been made to develop the techniques to synthesise nanoparticles [1,18,28,40]. As a result, a large number of synthesis methods are currently used on a laboratory scale (Chapter 3) [38]. However, the batch size of most of the laboratory scale synthesis is around 1 g. In some cases, 10–100 g per batch is possible and is regarded as large-scale synthesis in the laboratory [28], but such an amount is often too little for the commercial application of nanoparticles.

To date, a few methods have been demonstrated to produce a wide range of nanoparticulate materials in large quantities, and those methods are evolving day by day. The process development for the commercial-scale production of nanoparticles goes through stages similar to standard product development processes, namely, proof-of-concept through proof-of-process (pilot plant), proof-of-production (process parameter optimisation, HAZOP review, etc.) before full-scale production (Fig. 4.1). In some cases, small quantities of nanomaterials are sold by research organisations and the activity is called commercialisation. However, in general, the name "commercial product" can be used only after all the quality control protocols are in place, raw material supply is secured and the production volume (the supply of the product) is ensured. Strictly speaking, the products in the development stages should be called "engineering products" or "concept products", instead of "commercial product", even if they are sold commercially.

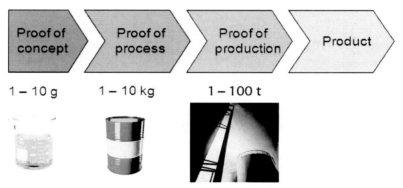

Figure 4.1   Typical development stages for the commercial scale production of nanoparticulate materials.

Although the activity to scale up the nanoparticle production is not generally considered scholarly research and is often looked down on by the academic community, a great deal of technological and scientific investigation is required to overcome the rather monumental challenge in the production scale-up of nanomaterials. Just to give an example, the control of reaction environments such as temperature, concentration, pressure and gas/liquid flow rates becomes exponentially difficult when the batch size increases from 100 mL to 100 L. Hence, by simply making the same shape and design of the processing facility and/ or equipment into a larger dimension, one will fail to obtain satisfactory control of the quality in the end products. In particular, as the physical dimension of the product becomes smaller, quality control of the products (i.e. consistency in the properties of nanoparticles) becomes more difficult. To overcome these technical challenges, it is necessary to carefully combine engineering approaches with fundamental scientific knowledge on the unique features of nanoparticles.

Where commercialisation is concerned, not only technological considerations but also other factors need to be taken into account. Those factors include costs that are related with raw materials, operation, equipment, safety and waste disposal. For example, some methods to produce nanomaterials use non-conventional techniques and hence require a high capital cost compared with other industrial products. Some techniques demand expensive raw materials and create toxic by-products. Some production processes are only possible through high energy consumption [35]. Therefore, even if a particular method is technically capable of scaling up the production, it is a necessary but insufficient condition for the commercialisation of nanotechnology innovation. It is not uncommon for an attempt to scale-up a production method to meet a dead-end before reaching the commercialisation stage, due to those technological and cost-related issues. In this regard, it is a highly risky exercise to scale up the production capacity of nanoparticulate products. Entrepreneurs also need to keep an eye on the constantly changing market and regulatory trends as well as consumer sentiment (see later chapters), as these can alter the course of process and product development.

In the following section, examples of the methods used for the commercial production of inorganic nanopowders and carbon-based nanomaterials are described and the relationship between the production methods and the types of available nanoparticle products is discussed.

4.2.2 Inorganic Nanoparticles

Commercial nanoparticles of metals, nitrides, semiconductor quantum dots and metal oxides are produced primarily using techniques that can be categorised into five areas, namely, comminution (mechanical grinding/milling), vapour phase, liquid-phase and solid-phase techniques and the combinations of those (Table 4.1). In the following subsections, general overviews of these methods are given.

Table 4.1   Techniques used for commercial scale production of inorganic nanoparticles

Approach	Synthesis environment	Methods
Top down		Mechanical milling (dry, wet), etching
Bottom up	Vapour phase	Flame pyrolysis, spray pyrolysis, laser ablation, plasma atomisation, gas condensation, electron-beam valorisation, cluster-beam, combustion, etc.
	Liquid phase	Sol-gel, hydrothermal, solvothermal, reverse-micelle, sonochemical, microwave, electro-deposition, etc.
	Solid phase	Mechanochemical, thermo-mechanical

4.2.2.1 Mechanical grinding/milling (top-down)

The top-down approach is an extension of the traditional methods used to produce large quantities of fine powders from bulk materials by grinding. The process generally involves high energy milling under a dry or wet condition, often with the addition of milling aids such as surfactants and diluents [6,25,30,44,59]. The equipment is commercially available and the number of process parameters is relatively small, so that the process is easily scalable. Much improvement has been made in recent years on the instrumental design and the quality of grinding media. This has led to significant progress in the use of this technique for the production of a wide range of nanoparticles. However, this approach suffers from difficulties in ensuring that all the particles are milled uniformly, especially for hard raw materials. The drawback typically results in a wide particle size distribution with a long "tail" on the larger particle side, representing the un-milled precursors in the final commercial products. In addition, typical milling times span from several hours up to many days and long milling times result in substantial contamination from the milling balls and mill components.

4.2.2.2 Vapour phase technique (bottom-up)

In vapour phase techniques, nanoparticles are built from atoms or molecules by the rapid solidification of a liquid or vapour in a gaseous medium [12,24]. The vaporisation of raw materials can be carried out using a variety of techniques including laser ablation, explosion, electric arc discharge, plasma, electric resistance and electron beam. Particle size, agglomeration and size distribution can be controlled by the vaporisation rate, gas pressure and the flow of the newly formed particles. Since the melting temperatures of metal oxides are normally extremely high, corresponding metals are often used as the precursors and vaporised metals form metal oxide nanoparticles by reacting with an oxidative gas.

Because of the fewer sources of contaminants and the high temperatures involved, this technique can produce nanoparticles with high purity and high crystallinity. However, the uniformity in stoichiometry, oxidation states and dopant concentration as well as particle size among the nanoparticles is often difficult to control by this technique. Owing to high-temperature operations and lack of a solid medium that hinders agglomeration, the resulting nanoparticles in commercial-scale production have characteristics of high crystallinity, geometrical shapes, a wide size distribution and evidence of agglomeration or particle sintering. The method often suffers from the trade-off between particle size and throughput. Increasing production rates will cause greater difficulty in controlling particle growth and in preventing agglomeration [50].

4.2.2.3 Liquid-phase technique (bottom-up)

Liquid-phase processes are widely used in many industries to make conventional micron-size powders as an economical manu-facturing process. Generally, the technique offers the ability to control particles sizes, shapes and stoichiometry in a precise manner. The technique also boasts flexibility in reaction paths and versatility in producing a variety of particulate materials. To date, the smallest commercial nanoparticles such as quantum dots are produced using liquid-phase techniques. Most of the recent research efforts have been dedicated to the control of particle size and shape, by the use of various surfactants, templates and microe-mulsions [10,37,53]. The growth-limiting agents may remain on the particle surface unless additional steps to remove the agents are introduced in the production process.

The difficulty in scaling up the production using the liquid-phase techniques lies in achieving a uniform reaction environment (temperature, pH, chemical concentration, etc.) in large chemical baths to ensure uniform quality of the resulting nanoparticles. Increase in production rate requires a high particle concentration that causes particle agglomeration during particle growth.

In commercial-scale production, an aqueous reaction environment, instead of organic solvents, is normally used for cost and safety reasons. For the production of metal oxide nanoparticles, the use of an aqueous environment can result in the formation of hydroxide or carbonate nanoparticles and hence additional processes to thermally decompose the salts into oxides are necessary, thus introducing the chance of severe particle agglomeration. As a result, commercial nanoparticles produced using liquid-phase techniques often have characteristics of a narrow size distribution of primary particles but high degrees of agglomeration.

4.2.2.4 Solid-phase technique; mechanochemical processing (bottom-up)

Solid-phase synthesis of nanoparticles is mainly represented by mechanochemical methods [3,29,30,43,59]. The technology uses high energy dry milling to mechanically induce chemical reactions to occur at a low temperature in a ball mill. In order to synthesise nanoscale particles, a nanoscopically uniform reaction environment is created by the milling of precursor powders to form a nanoscale composite structure of the starting materials. In such nanoscopic reaction environments, only short diffusion paths are required for the reaction to take place. Hence, the chemical reactions that normally occur only at high temperatures can be induced at room temperature during milling or during subsequent heat treatment at a relatively low temperature.

The reactants are often milled in the presence of an inert diluent phase to control the agglomeration of resulting nanoparticles. By carefully controlling the volume ratio between nanoparticle and the matrix phases, the precipitated nanoparticles can be separated from each other by the solid matrix. The nanoparticles can be further heat treated in the solid matrix that prevents temperature-induced agglomeration from occurring. Then the nanoparticles are collected simply by selective removal of the inert matrix phase, often by washing away the soluble diluent phase with a solvent [49,52]. This final process can increase production costs and the chance of contamination. The advantages of the technique include relatively simple operations and ease in creating a large-scale uniform reaction environment that leads to uniform size and shape of nanoparticles.

4.2.2.5 Which methods and why?

Although accurate data are difficult to obtain, currently there are over 100 commercial companies that produce large quantities of inorganic nanoparticles for sale [33]. The trend in the methods of commercial-scale production was studied by randomly selecting 75 companies ranging from large multinational corporations to small start-ups [50]. Figure 4.2 shows the statistics of the production methods the selected companies use. The production methods were divided into the five generic categories already mentioned, namely, top-down (comminution, mechanical milling), vapour, liquid and solid-phase techniques and the combinations of those. As shown in Fig. 4.2, over 80% of the selected companies use either vapour phase synthesis (39%) or liquid-phase synthesis (45%) techniques.

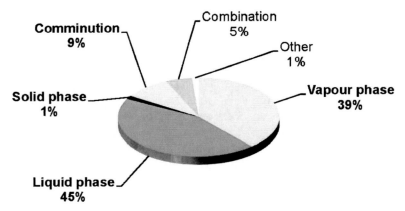

Figure 4.2 Techniques used for commercial scale production of inorganic nanoparticles. Top-down and bottom-up approaches are indicated in red and blue fonts, respectively. Updated the data presented in ref. [50].

It should be noted that the popularity of a particular method is not an indication of the superiority or versatility of the technique.

Many companies choose production techniques in order to leverage other business activities in the company or for historical reasons. The reason why there is only one company operating using a solid-phase bottom-up synthesis technique is due to the fact that the company processes a strong patent which covers most of the aspects of the production technology (mechanochemical process) and hence no other company can use the same production method.

Table 4.2 shows the relationship between the nanoparticle products and the commercial production techniques used by the selected 75 commercial companies [50].

Table 4.2 Relationship between commercial inorganic nanoparticle products and production methods

	Top-down	Bottom-up		
Products	**Mechanical grinding**	**Solid phase synthesis**	**Liquid phase synthesis**	**Vapour phase synthesis**
Metals (excluding silver and gold)			×	×
Silver, gold			×	
Carbides, nitrides				×
Semiconductor quantum dots			×	
Metal oxides	×	×	×	×

Source: Reproduced with permission from ref. [50]. Copyright 2009 Inderscience Publishers.

On a laboratory scale, all the five production techniques listed in Table 4.2 are potentially capable of producing most of the material in those product groups. For example, it is demonstrated that mechanochemical and thermo-mechanical processes can produce metals, nitrides, semiconductor quantum dots and metal oxide nanoparticles [49]. However, on a commercial scale, only a certain technique is used to produce selected materials. The reason is due mainly to the fundamental capability of the technique to produce certain materials on a large scale. For example, materials that are reactive to oxygen and moisture, such as non-precious metals, are commercially produced using only vapour phase methods due to the ease of scale-up under a tight quality control. Silver and gold are almost exclusively produced using liquid-phase precipitation methods to take advantage of the ease of particle size control. Carbides and nitrides often require high crystallinity and hence demand the use of high-temperature synthesis, i.e. vapour phase techniques. Commercial semiconductor quantum dots are frequently produced using liquid-phase synthesis methods, as liquid-phase techniques allow for the precise control of the stoichiometry and crystalline phases to obtain desired quantum effects.

Owing to these technical and commercial restrictions on large-scale production, only a few types of nanoparticulate materials are used in the commercialised applications.

4.2.3 Carbon-Based Nanoparticles

The commercial products of carbon-based nanoparticulate materials such as carbon nanotubes (CNTs), carbon nanofibres (CNFs), fullerenes and diamond nanoparticles have been available since the early 1990s. Recently, graphene-containing composites entered the commercial market as a heat sink material in mobile devices to replace heavier copper. In order to meet the requirement for high-temperature and low-impurity reaction conditions, the synthesis of the carbon-based products is generally based on vapour/vacuum phase techniques.

4.2.3.1 Carbon nanotubes

There are a number of excellent review articles about CNTs [9,41], and hence, in this section, only a brief overview is given. Carbon nanotubes are the strongest and stiffest materials in terms of tensile strength and elastic modulus and about 300 times stronger than high-carbon steels. The metallic form of single-walled CNTs (SWCNTs) can carry 1,000 times greater current than copper [17]. Because of these unique properties, along with their light weight and thermal stability, CNTs are expected to find a wide range of applications including electronics and structural materials.

Although the discovery of CNTs is later than that of fullerene, the commercial development of CNTs progressed rapidly [23]. In 2011, there were 75 CNT manufacturers in 19 countries posted on the Internet and the majority of them were located in the United States (Fig. 4.3). The unit sales sizes of fullerenes are typically as small as 1–100 g. However, production scale has been increasing. In 2009, 500 tonnes per year for CNTs was achieved by CNano Technology [36]. Because of the excellent mechanical properties of CNTs, polymer nanocomposites that contain CNTs have already appeared in consumer products, mostly in the form of sporting gear such as tennis racquets, golf clubs and bicycle frames [57].

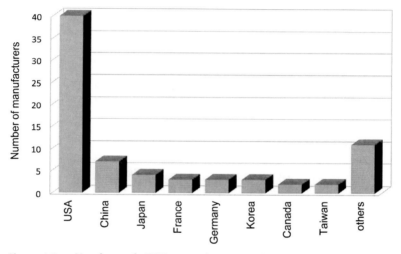

Figure 4.3   Number of CNT manufacturers that are posted on the Internet.

In laboratories, CNTs are produced using the reaction that occurs in a vacuum or in inert gas atmospheres. The most commonly used methods are (1) vaporisation of carbon from raw materials such as graphite blocks by laser ablation and arc discharge and (2) chemical vapour deposition (CVD) by decomposing organic gas molecules such as acetylene and methane [11,23,27]. The same techniques are used for the commercial production of CNTs.

Arc-discharge vaporisation of graphite was used when CNTs were first scientifically recognised [20]. This technique uses two graphite rods placed end to end, on which a large current (~100 A) is applied to create a high-temperature arc discharge between the two electrodes. The discharge vaporises the surface of one of the carbon electrodes, and forms a small rod-shaped deposit on the other electrode [23].

Laser ablation methods were originally developed by Smalley's group for the investigation of fullerenes [14]. A laser beam is introduced through the window and focused onto the graphite target and the vapour condenses into CNTs in an Ar-gas flow. The CNTs are carried by the gas and collected at a cold trap. Because of the high-power laser vaporisation and homogeneous annealing conditions, this process results in higher crystallinity of the CNTs [23]. The morphology of CNTs can be controlled by changing the gas flow rate, laser pulse duration, gas temperature and catalyst metals. This technique has disadvantages as a commercial process because of its low yield and high operation costs.

CVD involves passing a vaporised hydrocarbon through a tube furnace at a sufficiently high temperature (600–1,200°C) and added catalyst materials decompose the hydrocarbon into carbon. CNTs grow over the catalyst and are collected upon cooling the system to room temperature [23]. CVD has been used for the production of carbon fibres since the 1950s [54]. Compared with arc-discharge and laser ablation methods, CVD is a simple, economic technique and easy to scale up for synthesising CNTs [23]. In fact, many commercial CNT manufacturers use this technique. However, since the temperature used is lower than for the arc-discharge and laser ablation methods, the crystallinity of CNTs is compromised.

The most challenging issue for the large-scale production of CNT is to achieve high purity in terms of structure and catalyst residuals. CNTs can take various forms which have different properties. For example, SWCNTs can have three different structures, namely, armchair, zigzag and chiral structures, depending on the orientation of the C–C bond to the direction of the tube length (Fig. 4.4) [4]. These different structures exhibit different electrical properties even if the physical dimension of CNTs is the same [9,41]. For example, the bandgap energy of SWCNTs can vary from zero to ~2 eV and their electrical conductivity can show metallic or semiconducting behaviour, depending on the structure. Multi-walled CNTs (MWCNTs) can have different internal diameters and numbers of layers. In addition, length, entangled states and straightness all depend on the synthesis conditions, gas environment and often production batch size. Although most of the schematic drawings of CNTs in nanotechnology textbooks show straight tubular structures, it is very rare to actually have long and straight CNTs in commercial products. In reality, commercial CNTs have an appearance similar to small black cotton balls if they are supplied as dry, and on close examination under a scanning electron microscope, they more or less resemble entangled spaghetti.

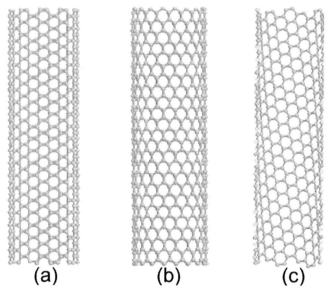

(a) **(b)** **(c)**

Figure 4.4 Structure of CNT; (a) zigzag, (b) armchair and (c) chiral structure. Reproduced with permission from ref. [5]. Copyright 2002 American Chemical Society.

Impurities such as carbon nanoparticulates and amorphous carbon can be removed by infrared irradiation or heating in air at 420–500°C. Amorphous carbon has many dangling bonds that are highly reactive to oxygen, whereas CNTs have few defects and are more resistant to oxidation [23]. This fact allows selective removal of amorphous carbon.

Unlike the production of fullerenes, SWCNTs normally require catalyst nanoparticles during the growth stage to improve the production rate and purity. The catalysts normally used are metal nanoparticles from which CNTs grow. The activities of various kinds of metal catalysts (Ni, Co, Fe, etc.) have been investigated [58]. Those catalyst nanoparticles have to be removed at the final stage of production by washing the CNT products with acid. Oxidation treatments also help remove the catalysts by attacking the protective layer of catalytic metal particles [5]. It is reported that the agglomeration states and remaining catalyst nano-particles play a significant role in the toxicity of CNTs and other properties [19,56].

As discussed above, not only the synthesis techniques but also the detailed process parameters affect the characteristics of end CNT products. Hence, it is not surprising that the properties, purity, morphology and functionality vary among CNT commercial products from different manufacturers. This variation in commercial CNTs should be taken into consideration in academic research as well as in the commercial development of CNT-containing devices and products.

4.2.3.2 Fullerenes

Fullerenes or Bucky balls are spherical molecules of ~1 nm in diameter and possess unique properties such as super con-ductivity and anti-oxygen activity [34]. A large array of potential applications has been identified, including IT devices, diagnostics, pharmaceuticals, environmental and energy industries [32]. Most commonly produced commercial fullerene is the so-called C60, a Bucky ball that consists of 60 carbon atoms.

The laboratory-based synthesis methods for fullerenes have been extensively reviewed by Lamb and Huffman [26]. The most common methods used are (1) the carbon arc method, where graphite rods are vaporised with electrical currents or laser in low-pressure inert gas and (2) the gas combustion method, where a continuous low flow of hydrocarbon fuel is burned at low pressure. Those methods have been successfully adapted to the commercial-scale production of fullerenes [2,32,45]. In particular, the development of gas combustion methods enabled tonnes-per-year production. The mass production of fullerenes allowed the first commercial usage of fullerenes in 2003 in 10-pin bowling balls. A fullerene-containing resin is applied on the surface of bowling balls to enhance controllability of the ball on the bowling lane as well as to reduce wear [32]. The C60-containing bowling balls captured 70% of market share in Japan [46].

The author counted 17 registered commercial manufacturers of fullerenes, of which 11 companies are based in the United States and the rest scattered in 6 other countries. The unit sales sizes of fullerenes are still very small, typically from 1–500 g. However, the production of those carbon-based nanoparticulate materials are said to be commercially viable, if the price is above $1 per gram. In most cases, the price of C60 (~99.0% purity) is around US$20 per gram. The high cost and small unit sales sizes compared to other nanoparticle products are due to the low production rates associated with the synthesis technology.

4.2.3.3 Diamond nanoparticles

Because of their excellent hardness, chemical stability, high refractive index and high thermal capacity, diamond nanoparticles find use in many applications including polishing media, optical composites, structural composites, dry lubricants for metal industries and coolant liquids [15].

There are only few manufacturers of nano-diamond operating in the market. In 2011, 10 companies appeared on the Internet, of which 7 were in China. Most of the nano-diamond products are not white but black or grey. Some products have particle sizes as large as 60 μm regardless of the claim to be nanoscale, though they appear to have nanoscale grains. Hence, one should not confuse the name "nano-diamond" with "diamond nanoparticles".

The common method used for the commercial production of diamond nanoparticles is the detonation technique. In this method, the high pressure and high temperature required for the synthesis of diamond, are created by the detonation and vaporisation of organic compounds $C_xH_yN_zO_w$ (trinitrotoluene and hexogen mixtures) under the condition of negative oxygen balance [8]. The raw product after detonation contains non-diamond carbons as well as metal impurities that originate from the detonation chamber. Those unwanted materials are removed by boiling the raw product in acid at a high temperature and under high-pressure. The technique is relatively low-cost and capable of a large-scale production of diamond nano-particles with an average grain size of ~4 nm. As such, ~1 kg unit sales size is achievable. By careful control of the process parameters, well dispersed diamond nanoparticles of <100 nm can be produced [22]. The detonation method is also used to produce ceramics that only form under high-pressure/high-temperature conditions, such as carbide and nitride powders.

4.3 Effects of Production Methods on the Properties of Commercial Nanoparticles

For applied research and product development, it is useful to use commercial nanoparticles instead of nanoparticles made in a small quantity in the laboratory. The properties of nanoparticles synthesised in the laboratory may not be replicated in the commercial nano-particle products. The reason for this is not only the availability of bulk quantity but also the fact that the properties of nanoparticles synthesised in the laboratory may not be replicated in the commercial nanoparticle products. The properties of nanoparticles depend on the synthesis methods, production scale and process parameters, even if average particle sizes are similar.

In order to elucidate the effects of synthesis methods on the properties of nanoparticle products, the following section discusses the research undertaken by the author's group on commercial ZnO and CeO$_2$ nanoparticles that were produced using different techniques.

4.3.1 ZnO

First, the properties of commercial ZnO nanoparticles that were synthesised using solid, liquid and vapour phase techniques are compared [7,21,48,51]. The particle sizes of the nanoparticle samples were measured using different sizing methods. Crystallite size was estimated from the full-width at half-maximum of the diffraction peaks using the Scherrer equation. The average particle diameter (the so-called BET particle size) was estimated from the Brunauer–Emmett–Teller (BET) specific surface area using the equation $D = 6/S\rho$, where D is the average particle diameter, S is the specific surface area, and ρ is the density 5.61 g/cm^3. Number weighted particle size was also measured by photo correlation spectroscopy (PCS).

Table 4.3 lists the average particle sizes measured by the different techniques. The variation between the sizes measured using different methods indicates the degree of agglomeration [47]. Better agreement between different sizing methods indicates less degrees of agglomeration. The results in Table 4.3 suggest that the degree of agglomeration in the mechanochemically synhesised nanoparticles was less than that for commercial nanoparticles produced using liquid and vapour methods.

Table 4.3 Particle sizes of ZnO nanoparticles synthesised using different techniques

Synthesis methods	BET surface area (m²/g)	BET particle size (nm)	Crystallite size (nm)	PCS particle size (nm)
Solid phase method	41	26	24	24
Liquid phase method	41	26	20	160
Vapour phase method	14	77	50	170

Source: Reproduced with permission from ref. [21]. Copyright 2002 BSSRL.

Note: Particle sizes were estimated from specific surface are X-ray diffraction peak width and dynamic light scattering.

For all the three samples, at least one sizing method gave average particle sizes between 20–50 nm. Hence, the average particle size of the three samples may appear similar to each other on the product specification sheet. In fact, there is no regulation or guideline about which sizing methods should be used to indicate the particle size of commercial nanoparticles in the product specification sheet. However, it is rather important to know the particle size measured using different methods, as the agglomeration largely influences the property of the nanoparticles and in turn the performance of the commercial products that contains the nanoparticles. It is highly recommended that the users of commercial nanoparticles evaluate the quality of nanoparticle products by employing more than one sizing method.

Particle shapes can also be significantly affected by synthesis methods. Figure 4.5 shows typical transmission electron microscopy (TEM) images of the commercial ZnO nanopowder samples.

It is evident that different synthesis methods lead to different morphology. Solid-phase and liquid-phase precipitation methods result in near spherical shape particles, whereas vapour phase methods lead to geometric shape particles with high crystallinity.

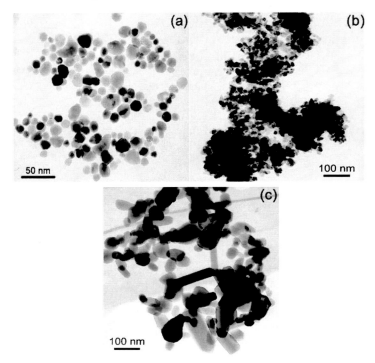

Figure 4.5 Transmission electron microscopy images of ZnO nano-powders produced by (a) a solid-phase method, (b) a liquid-phase method, (c) a vapour phase method. Reproduced with permission from ref. [21]. Copyright 2002 B5SRL.

UV-screening is one of the major applications of ZnO nano-particles in cosmetic, plastic and paint industries, where high transparency of nanoparticle suspension systems to visible light is required. Figure 4.6 shows the UV-Vis specular transmission spectra of the three commercial ZnO nanoparticles suspended in water. In the UV light region, all the samples show similar UV screening properties. However, in the visible light region, the nanoparticles synthesised using the solid-phase method exhibit the highest transmittance. This is because the light scattering power of a nanoparticle decreases proportional to the 6th power of particle diameter. The overall light scattering by a nanoparticle suspension system is smaller if the average particle size is small, the particle size distribution is narrow and the degree of agglomeration is low. The nanoparticles synthesised using the solid-phase method fit into this condition more than the nanoparticles produced using the other two methods, realising low light scattering and in turn high transparency to the visible light.

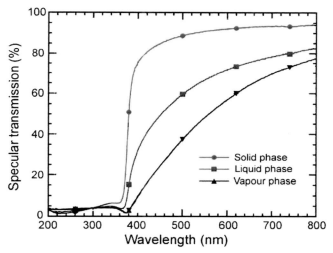

Figure 4.6   Synthesis method dependence of the UV-Vis specular transmission spectra of ZnO nanoparticle suspensions in deionised water, at the particle concentration of 0.01 wt% and the optical path length of 10 mm. Polyacrylic electrolyte was added 10 wt % relative to ZnO. Prior to measurement, the suspensions were subjected to intense ultrasonication for 15 min. Reproduced from ref. [50]. Copyright 2009 Inderscience Publishers.

Although the nanoparticles produced using a liquid-phase method had small crystallite sizes, the optical transparency was significantly lower, due to the presence of agglomeration of ~160 nm in size as measured by PCS (Table 4.3). The nanoparticles produced using a vapour phase method showed the lowest transmission in the visible wavelength region as a consequence of their significantly larger crystallite and agglomerate sizes. This application as a transparent UV screening agent is one of the examples where the quality of nanoparticles should be assessed not only by the average particle size but also by the degree of agglomeration.

Figure 4.7 shows the photoluminescence spectra of the three commercial ZnO nanoparticles suspended in water. The photoluminescence was weak for the nanoparticles synthesised using solid and liquid-phase methods. In contrast, the nanoparticles produced using a vapour phase method gave comparatively strong emission. This is due to the fact that the intensity of photoluminescence improves as crystallinity increases. In general, crystal defects act as non-radiative recombination sites for photo-generated electrons and holes, which leads to the weakening of photoluminescence. Because of the high temperature involved in the process, vapour phase techniques tend to produce nanoparticles having higher crystallinity and fewer defects than solid and liquid-phase methods. The application as phosphors is another example where the quality of nanoparticles cannot be assessed solely by their average size or even by the degree of agglomeration [50].

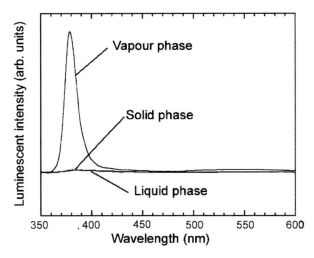

Figure 4.7   Synthesis method dependence of the photoluminescence spectra of ZnO nanoparticle suspensions in deionised water. Reproduced with permission from ref. [7]. Copyright 2008, Springer.

Photocatalysis is another important application of ZnO nanoparticles. ZnO is a semiconductor with a bandgap energy of 3.3 eV, and hence, it absorbs light having a wavelength shorter than 375 nm. On absorption of UV light, the electrons in the occupied valence band are excited by the UV light into the unoccupied conducting band. The conduction band electrons and valence band holes thus created can migrate out to the particle surface to react with electron donors or acceptors that are adsorbed on the particle surface. The photo-generated electrons and holes can be used in many applications, including solar cells and hydrogen production by water-splitting. In ZnO, the photo-excited electrons and holes have suitable redox potentials for inducing a wide range of catalytic reactions [7]. For example, when the photo-generated electrons and holes react with water or oxygen, they produce reactive superoxide ($\bullet O_2$) and hydroxyl ($\bullet OH$) free radicals. Those photo-generated free radicals are useful for applications in wastewater treatment via organic pollutant decomposition, and antibacterial and anti-fouling coatings. At the same time, the photocatalysis of ZnO may be harmful in sunscreen and skin-care applications.

Figure 4.8a shows the photocatalytic activity of the three samples characterised by measuring the hydroxyl radical concentration using a spin-trapping technique with electron paramagnetic resonance (EPR) spectroscopy [13]. The values are normalised against the specific surface area [7]. It is evident that the powders synthesised using a vapour phase method exhibit significantly stronger photocatalytic activity per unit surface area than the nanoparticles prepared by the other two methods. This is due to the fact that the ZnO nanoparticles synthesised using solid- and liquid-phase methods have lower crystallinity and higher numbers of crystal defects than the ones made using the vapour phase technique. The crystal defects act as the recombination sites for photo-generated charges to prevent the charges from reacting with water to form OH radicals and in turn to suppress the photocatalytic activity.

It is interesting to note that the photoactivity rate constant and specific surface area did not correlate well in those three commercial nanoparticles (Fig. 4.8b). Since photocatalytic reaction occurs on particle surfaces, in general, the rate of photocatalysis increases as the surface area increases. However, this argument is valid only when other conditions such as crystallinity, crystal phase, impurity levels and particle shapes are constant. The above three samples were made using different methods that led to the differences in particle shape, crystallinity and degree of agglomeration. Such differences cannot be identified by looking at the product specification sheets and yet have critical implications for the quality and performance of the nanoparticle-enabled commercial products.

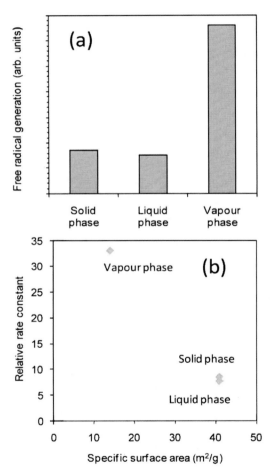

Figure 4.8 (a) Synthesis-method dependence of the surface area normalised photocatalytic activity of ZnO nanoparticle suspensions in deionised water. Reproduced with permission from ref. [7]. Copyright 2008, Springer. (b) Correlation between photoactivity rate constant and specific surface area.

4.3.2 CeO$_2$

In 2007, the Organization for Economic Cooperation and Development (OECD) Working Party on Manufactured Nanomaterials (WPMN) launched the sponsorship program for safety testing of nanomaterials [31]. As a part of this program, three commercial CeO$_2$ powders that were selected by OECD-WPNM, namely, NM–211, NM–212 and NM–213, were characterised for their physical and chemical properties by many laboratories around the world [51]. This section compares the characteristics of the three commercial CeO$_2$ powders.

The results of X-ray diffraction spectroscopy (XRD) and N$_2$-gas absorption analysis are listed in Table 4.4. The XRD study showed that all the samples had a cubic fluorite structure (JCPDS No. 43-1002). The crystallite sizes were estimated from the full-width at half-maximum of the X-ray diffraction peaks using the Scherrer equation and found that the values were close to the nominal particle sizes. However, BET Specific surface areas were largely different among the samples. Their average particle size was estimated from the specific surface area through the equation $D = 6/S\rho$, where D is the average particle diameter, S is the specific surface area, and ρ is the density 7.21 g/cm^3, assuming that the sample consisted of mono-dispersed spherical particles. In NM–213, the resulting BET particle size showed significant difference from the nominal value. As discussed earlier, the difference between the crystallite size estimated by XRD and the particle size estimated from surface area is a good indication of particle agglomeration [47]. The close values of the crystallite size estimated by XRD and the particle size estimated from the surface area suggest that the particles had a low degree of agglomeration and hence will be easy to re-disperse in liquid media.

Although nominal particle sizes of the three samples are all less than 30 nm, their morphologies were significantly different from each other, as shown in the TEM images in Fig 4.9. NM–211 has a near-spherical morphology, a typical result of the sol-gel or wet precipitation technique. NM–212 has geometrical shapes due to the vapour phase synthesis and NM–213 has various shapes where 50–200 nm sized grains are evident.

Table 4.4 Characteristics of three commercial CeO_2 powders that are selected by OECD Working Party on Manufactured Nanomaterials

	NM–211	NM–212	NM–213
Estimated synthesis method	Solid- or liquid-phase method	Vapour phase method	unknown
Nominal particle size (nm)	10	40	<5000 nm
XRD crystallite size (nm)	10	18	22
BET surface area (m^2/g)	61	23	3.5
BET particle size (nm)	14	35	241

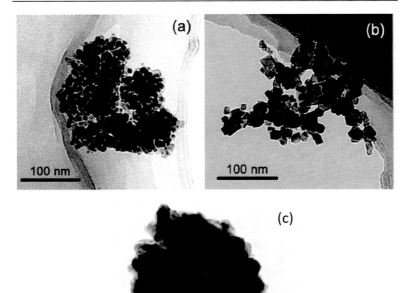

Figure 4.9 Transmission electron microscopy images of three commercial CeO_2 powders that are selected by OECD Working Party on Manufactured Nanomaterials; (a) NM–211, (b) NM–212, (c) NM–213.

It is commonly believed that CeO_2 does not have photo-catalytic properties. However, CeO_2 is still a wide bandgap semiconductor and photo-generation of electrons and holes is expected. In fact, the three OECD samples showed weak photo-activity. The degradation rates of Rhodamine-B dye under simulated sunlight in the presence of the samples are shown in Fig. 4.10a [55]. The degradation rate that was normalised with the BET specific surface area is also shown in Fig. 4.10a. It is evident that the degradation rate, which is proportional to the strength of the photocatalytic activity of CeO_2 powders, significantly varied amongst the three samples. Although it is generally believed that photocatalytic activity increases as the surface area increases, there was no correlation between the photodegradation rate and specific surface area among the three samples (Fig. 4.10b).

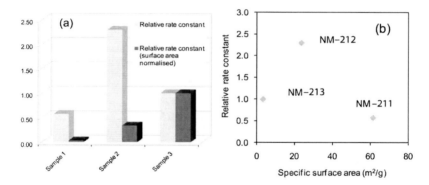

Figure 4.10 (a) Photocatalytic activity of three commercial CeO_2 powders that were selected by the OECD Working Party on Manufactured Nanomaterials. The relative intensities against the photocatalytic activity of NM–213 are shown. (b) Correlation between photoactivity rate constant and specific surface area.

High photocatalytic activity of NM–212 and NM–213 are related to the high crystallinity on the particle surface that was formed by the synthesis methods involving a high temperature. The surfaces of spherical particles in NM–211 have more physical defects than the surfaces of geometrically shaped particles in NM–212 and NM–213. Those physical defects act as trapping and recombination sites for photo-generated electrons and holes, contributing to the reduction of photocatalysis. Hence, NM–211 has lower photocatalytic activity than NM–212 and NM–213.

The results shown in this section indicate that, even if nominal particle sizes are similar, the characteristics of commercial CeO_2 nanoparticles vary significantly. In particular, the difference in specific surface area, the degree of agglomeration and particle shapes are important parameters that will influence the performance of CeO_2 nanoparticles as a catalyst or photocatalyst in many applications.

4.4 Summary

One of the key requirements for a nanotechnology innovation to be applicable commercially is the large-scale production of high quality nanoparticles. Production rates of over a tonne per year are often required to create commercially viable nano-enabled products. Although many different methods are available to synthesise nanoparticles on a laboratory scale, only a few methods are currently used for commercial-scale production. This is due mainly to the technological difficulty of the particular method to scale up the production, high production costs and intellectual property issues. Furthermore, each commercial production method has a limited range of nanoparticle materials it can produce in a commercially viable manner.

Of particular significance is the fact that each production method results in nanoparticles having a unique combination of properties; different methods lead to different properties of nano-particles. Those product-specific characteristics of nanoparticles are not necessarily shown in the materials safety data sheets (MSDS) or in the product specification sheets. Even when some qualities such as particle size and crystal phase appear quite similar on the product specification sheet of those nanoparticles, some properties that are critical to specific applications may vary widely, depending on the production techniques and parameters used. It is strongly recommended that the users of commercial nanoparticles carry out the characterisation of key properties according to the target applications and research aims. If a particular commercial nanopowder product is found to be unsuitable, the reason may not be based on materials, chemical composition or particle size but may be due to other features such as shape, degree of agglomeration or size distribution. In that case, the use of nanoparticle products from other companies may give totally different outcomes.

By the same token, the manufacturers of commercial nano-particles need to be aware that their commercial nanoparticle products may not be suitable for every possible application. For effective marketing and product development, the manufacturers should recognise the unique properties of their products compared to those of other manufacturers, based on their production techniques and process conditions.

In order to introduce a wider variety of nanoparticle materials to the market and so expand the practical application of nano-particles, it is necessary to further improve the existing methods and also to develop new synthesis techniques with scalability and commercial viability in mind.

Acknowledgements

The author gratefully acknowledges Dr Jinfeng Wang and Dr Rongliang He for their assistance in acquiring the data for CeO_2 nanoparticles. The author also thanks Dr Maxine McCall and Dr Victoria Coleman for providing the OECD samples.

References

1. Baraton, M. I. (2003). *Synthesis, Functionalization and Surface Treatment of Nanoparticles* (American Scientific Publishers. USA).

2. Bogdanov, A. A., Deininger, D., and Dyuzhev, G. A. (2000). Development prospects of the commercial production of fullerenes, *Tech. Phys.*, **45**, pp. 521–527.

3. Boldyrev, V. V. (1996). Mechanochemistry and mechanical activation, *Mater. Sci. Forum*, **225–227**, pp. 511–520.

4. Charlier, J. C., Blase, X., and Roche, S. (2007). Electronic and transport properties of nanotubes, *Rev. Mod. Phys.*, **79**, pp. 677–732.

5. Dai, H. (2002). Carbon nanotubes: synthesis, integration and properties, *Acc. Chem. Res.*, **35**, pp. 1035–1044.

6. de Castro, C. L., and Mitchell, B. S. (2001). *Synthesis, Functionalization and Surface Treatment of Nanoparticles*, ed. Baraton, M. I., "Nanoparticles from Mechanical Attrition" (American Scientific Publishers, USA) pp. 1–15.

7. Dodd, A., McKinley, A., Tsuzuki, T., and Saunders, M. (2008). A comparative evaluation of the photocatalytic and optical properties of nanoparticulate ZnO synthesised by mechanochemical processing, *J. Nanopart. Res.*, **10**, pp. 243–248.

8. Dolmatov, V. Y. (2001). Detonation synthesis ultradispersed diamonds: properties and applications, *Russ. Chem. Rev.*, **70**, pp. 607–626.

9. Dresselhaus, M. S., Dresselhaus, G., and Avouris, P. (2001). *Carbon Nanotubes; Synthesis, Structure, Properties, and Applications* (Springer-Verlag, Germany).

10. Eastoe, J., Hollamby, M. J., and Hudson, L. (2006). Recent advances in nanoparticle synthesis with reversed micelles, *Adv. Colloid Interf. Sci.*, **128–130**, pp. 5–15.

11. Endo, M., Hayashi, T., and Kim, Y. A. (2006). Large scale production of carbon nanotubes and their applications, *Pure Appl. Chem.*, **78**(9), pp. 1703–1713.

12. Granqvist, C. G., Kish, L. B., and Marlow, W. H. (2004). *Gas Phase Nanoparticle Synthesis* (Springer, Netherland).

13. Grela, M., Coronel, M., and Colussi, A. J. (1996). Quantitative spin-trapping studies of weakly illuminated titanium dioxide sols. Implications of the mechanism of photocatalysis, *J. Phys. Chem.*, **100**, pp. 16940–16946.

14. Guo, T., Diener, M. D, Chai, Y., Alford M. J., Haufler, R. E., McClure, S. M., Ohno, T., Weaver, J. H., Scuseria, G. E., and Smalley, R. E. (1992). Uranium stabilization of C28: a tetravalent fullerene, *Science*, **257**, pp. 1661–1664.

15. Ho, D. (2010). Nanodiamonds: applications in biology and nanoscale medicine (Springer, New York).

16. Holister, P., and Harper, T. E. (2002). *The Nanotechnology Opportunity Report* (CMP Cientificia), No. 1 and 2, March.

17. Hong, S., and Myung, S. (2007). Nanotube electronics: a flexible approach to mobility. *Nat. Nanotechnol.*, **2**(4), pp. 207–208.

18. Hosokawa, M., Nogi, K., Naito, M., and Yokoyama, T. (2007). *Nanoparticle Technology Handbook* (Elsevier, UK).

19. Hurt, R. H., Monthioux, M., and Kane, A. (2006). Toxicology of carbon nanomaterials: status, trends and perspectives on the special issue, *Carbon*, **44**, pp. 1028–1033.

20. Iijima, S. (1991). Helical microtubules of graphitic carbon, *Nature*, **354**, pp. 56–58.

21. Innes, B., Tsuzuki, T., Dawkins, H., Dunlop, J., Trotter, G., Nearn, M. R., and McCormick, P. G. (2002). Nanotechnology and the cosmetic chemist, *Nutracos,* Sept/Oct 2002, pp. 7–12.

22. International Technology Center, Nanodiamond, http://www.itc-inc. org/psd.html, last accessed: 30 August 2011.

23. Karthikeyan, S., Mahalingam, P., and Karthik, M. (2009). Large scale synthesis of carbon nanotubes, *E-J. Chem.*, **6**, pp. 1–12.

24. Kruis, F. E., Fissan, H., and Peled, A. (1998). Synthesis of nanoparticles in the gas phase for electronic, optical and magnetic applications—a review, *J. Aerosol Sci.*, **29**, pp. 511–535.

25. Koch, C. C. (2003). Top down synthesis of nanostructured materials; mechanical and thermal processing methods, *Rev. Adv. Mater. Sci.*, **5**, pp. 91–99.

26. Lamb, L. D., and Huffman, D. R. (1993). Fullerene production, *J. Phys. Chem. Solids*, **54**, pp. 1635–1643.

27. MacKenzie, K. J., Dunens, O. M., See, C. H., and Harris, A. T. (2008). Large scale carbon nanotube synthesis, *Recent Pat. Nanotechnol.*, **2**, pp. 25–40.

28. Masala, O., and Seshadri, R. (2004). Synthesis routes for large volumes of nanoparticles, *Annu. Rev. Mater. Res.*, **34**, pp. 41–81.

29. McCormick, P. G. (1995). Application of mechanical alloying to chemical refining, *Mater. Trans. Jpn. Inst. Met.*, **36**, pp. 161–169.

30. Miami, F., and Maurigh, F. (2004). *Dekker Encyclopedia of Nanoscience and Nanotechnology*, Vol. 2, ed. Schwarz, J. A., Contescu, C. I., and Putyera, K., "Mechanosynthesis of Nanophase Powders" (Taylor & Francis, UK) pp. 1787–1796.

31. Murashov, V., Engel, S., Savolainen, K., Fullam, B., Lee, M., and Kearns, P. (2009). Occupational safety and health in nanotechnology and organisation for economic cooperation and development, *J. Nanopart. Res.*, **11**, pp. 1587–1591.

32. Murayama, H., Tomonoh, S., Michael-Alford, J., and Karpuk M. E. (2004). Fullerene production in tons and more: from science to industry, *Fuller. Nanotub. Car. N.*, **12**, pp. 1–9.

33. Nanoparticle.com, Particle suppliers and standards, http://nano-particles.org/standards/#Suppliers, last accessed: 30 June 2011.

34. Osawa, E. (2002). *Perspectives of Fullerene Nanotechnology* (Kluwer academic, UK).

35. Osterwalder, N., Capello, C., Hungerbuhler K., and Stark, W. J. (2006). Energy consumption during nanoparticle production: how economic is dry synthesis?, *J. Nanopart. Res.*, **8**, pp. 1–9.

36. Pangaea Ventures Ltd, News on Monday, 22 June 2009, CNano technology commissions world's largest carbon nanotube manufacturing plant with a capacity of 500 tons per year, http://www.pangaeaventures.com/news/41-cnano-technology-commissions-worlds-largest-carbon-nanotube-manufacturing-plant-with-a-capacity-of-500-tons-per-year/, last accessed: 30 August 2011.

37. Park, J., Joo, J, Kwon, S. G., Jang, Y., and Hyeon, T. (2007). Synthesis of monodisperse pherical nanocrystals, *Angew. Chem. Int. Ed.*, **46**, pp. 4630–4660.

38. Perez, J., Bax, L., and Escolano, C. (2005). *Roadmap Report on Nanoparticle* (Willems and van den Wildenberg, Spain).

39. Pitkethly, M. J. (2004). Nanomaterials—the driving force, *Mater. Today*, **7**(12), suppl., 1, pp. 20–29.

40. Rao, C. N. R., and Cheetham, A. K. (2006). The chemistry of nanomaterials: synthesis, properties and applications (Wiley-VCH, Weinheim).

41. Reich, S., Thomsen, C., and Maultzsch, J. (2004). *Carbon Nanotubes; Basic Concepts and Physical Propertie* (Wiley-VCH, Weinheim).

42. Roco, M. C. (1999). Nanoparticles and nanotechnology research, *J. Nanopart. Res.*, **1**, pp. 1–6.

43. Senna, M. (2001). Recent development of materials design through a mechanochemical route, *Int. J. Inorg. Mater.*, **3**, pp. 509–514.

44. Suranarayana, C. (2001). Mechanical alloying and milling, *Prog. Mater. Sci.*, **46**, pp. 1–184.

45. Takehara, H., Fujiwara, M., Arikawa, M., Diener, M. D., and Michael-Alford, J. (2005). Experimental study of industrial scale fullerene production by combustion synthesis, *Carbon*, **43**, pp. 311–319.

46. Tremblay, J. F. (2003). Fullerenes by the ton, *Chem. Eng. News*, **81**(32), pp. 13–14.

47. Tsuzuki, T., and McCormick, P. G. (2001). Synthesis of ultrafine ceria powders by mechanochemical processing, *J. Am. Ceram. Soc.*, **84**, pp. 1453–1458.

48. Tsuzuki, T., and McCormick, P. G. (2001). ZnO nanoparticles synthesised by mechanochemical processing, *Scripta Mater.*, **44**, pp. 1731–1734.

49. Tsuzuki, T., and McCormick, P. G. (2004). Mechanochemical synthesis of nanoparticles, *J. Mater. Sci.*, **39**, pp. 5143–5149.

50. Tsuzuki, T. (2009). Commercial scale production of inorganic nanoparticles, *Int. J. Nanotechnol.*, **6**, pp. 567–568.

51. Sponsorship Program for the Testing of Manufacturerd Nanomaterials, Organization for Economic Cooperation and Development, Working Party on Manufactured Nanomaterials, available at http://www.oecd.org/document/47/0,3746,en_2649_37015404_41197295_1_1_1_1,00.html/, last accessed: 30 August 2011.

52. Urakaev, F. K. H., Shevchenko, V. S., and Boldyrev, V. V. (2005). Theoretical and experimental investigation of mechanosynthesis of anoparticles by means of dilution with the final product, *Chem. Sustain. Dev.*, **13**, pp. 321–337.

53. Vayssieres, L. (2004). On the design of advanced metal oxide nanomaterials, *Int. J. Nanotechnol.*, **1**, pp. 1–41.

54. Walker Jr., P. L., Rakszawski, J. F., and Imperial, G. R. (1959). Carbon formation from carbon monoxide-hydrogen mixtures over iron catalysts. I. Properties of carbon formed, *J. Phys. Chem.*, **63**, pp. 133–140.

55. Wang, J., Tsuzuki, T., Sun, L., and Wang, X. (2009). Reducing the photocatalytic activity of zinc oxide quantum dots by surface modification, *J. Am. Ceram. Soc.*, **92**, pp. 2083–2088.

56. Wick, P., Manser, P., Limbach, L. K., Dettlaff-Weglikowska, U., Krumeich, F., Roth, S., Stark, W. J., and Bruinink, A. (2007). The degree and kind of agglomeration affect carbon nanotube cytotoxicity, *Toxicol. Lett.*, **168**, pp. 121–131.

57. Woodraw Wilson International Center for Scholars, Project on Emerging Nanotechnologies (2011), http://www.nanotechproject.org/inventories/consumer/ analysis_draft/, last accessed: 30 August 2011.

58. Yudasaka M., Kasuya Y., Kokai F., Takahashi K., Takizawa M., Bandow S., and Iijima S. (2002). Causes of different catalytic activities of metals in formation of single-wall carbon nanotubes, *Appl. Phys. A.*, **74**, 377–385.

59. Zhang, D. L. (2004). Processing of advanced materials using high-energy mechanical milling, *Prog. Mater. Sci.*, **49**, pp. 537–560.

Chapter 5

The Commercialisation of Nanotechnology: The Five Critical Success Factors to a Nanotech-Enabled Whole Product

Craig Belcher,[a,b] Richard Marshall,[a,c] Grant Edwards,[a,d] and Darren Martin[a,d,e]

[a]TenasiTech Pty Ltd, Level 7, GPS Building, St Lucia, Queensland 4072, Australia
[b]UniQuest Pty Ltd, Level 7, GPS Building, St Lucia, Queensland 4072, Australia
[c]Uniseed Management Pty Ltd, Level 7, GPS Building, St Lucia, Queensland 4072, Australia
[d]Australian Institute for Bioengineering and Nanotechnology,
The University of Queensland, St Lucia, Queensland 4072, Australia
[e]School of Chemical Engineering, The University of Queensland,
St Lucia, Queensland 4072, Australia

c.belcher@uniquest.com.au

The commercialisation of nanotechnology—the transition from research to an economically viable product—is particularly vulnerable to the commercialisation "Valley of Death" compared with other technologies for reasons relating to product focus, market engagement, scale-up and product development. For these reasons, we believe the valley of death represents more than just a funding

Nanotechnology Commercialisation
Edited by Takuya Tsuzuki
Copyright © 2013 Pan Stanford Publishing Pte. Ltd.
ISBN 978-981-4303-28-6 (Hardcover), 978-981-4303-29-3 (eBook)
www.panstanford.com

gap for nanotechnology commercialisation. In terms of funding, the lack of an absolute measure and market acceptable signal of an increase in value in the technology upon the achievement of an impartial product development milestone represents a major challenge in funding nanotechnology ventures. We review the literature on the commercialisation of nanotechnology to summarise the nanotech-specific risks and success factors, and identify gaps in the literature for nanotechnology commercialisation ventures to develop a nanotechnology-enabled whole product. We also extend the literature by proposing an end-to-end model for the commercialisation of nanotechnology to traverse the valley of death.

5.1 Introduction

Along the path to commercialisation, nanotechnology's biggest liability is its novelty. Inventions often attract attention because of their ingenuity, but a product must also be useful and compelling [23].

In this chapter, we argue that the successful commercialisation of nanotechnology needs more than the norm to cross the "Valley of Death". Before elaborating and justifying our assertion, first we explain what, in this chapter, we mean by nanotechnology, commercialisation, success and the valley of death.

The scope of nanotechnology encompassed by this chapter includes all forms and applications of nanotechnology—particles, materials, devices or other structures with at least one dimension in the 1–100 nm range—apart from the subset of nanotechnology which has a medical application.

Commercialisation is the act and process of managing the transition of research to a market application [9]. Commercialisation success is deemed to have been achieved if an end-product which is or comprises the nanotechnology achieves sustained product sales and profits [28].

The valley of death is the gap that exists between research (with all of its uncertainties) and product development (with its focus on economic returns) [10]. Regardless of the industry or the origin of the research (for simplicity in this chapter we contextualise it from a university environment), technology commercialisation must endure and traverse the valley of death if it is to become a successful product.

The objective of this chapter is to summarise the literature on the risks and success factors for nanotechnology commercialisation, identify the gaps and issues, and propose a framework for the critical success factors to develop a nanotechnology-enabled whole product. To begin with, it is worth understanding the various characteristics and perspectives on the valley of death in the innovation and technology management literature.

5.1.1 The Valley of Death

The valley of death is the transition from research activities to product development (Fig. 5.1). It is where ideas and inventions that arise from research activities must undergo a preliminary assessment of their technical feasibility, market demand and commercial case [32] and a decision made whether to proceed or terminate product development [27]. Often the decision is made to proceed, but the product fails for intrinsic or extrinsic reasons. The valley of death is so called because it is a graveyard for many technologies.

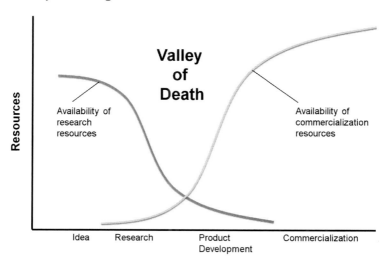

Figure 5.1 New product development, commercialisation and the "Valley of Death".

On the upstream side of the valley of death is research which is inherently uncertain and downstream is the more regimented process of product development characterised by deliverables, deadlines, budgets and their attainment against plan. Commercialisation is about the translation—from science to business—crossing these two distinct paradigms.

The valley of death is commonly understood to be attributed solely to a lack of funding. If more funding was available, the argument goes, then the technology would have undoubtedly been developed into a robust product with a sustainable competitive advantage, attained dominant market share, produced in sufficient volumes at a low cost of goods to maximise profit margins. If that were all true and the commercial case was so strong and well articulated, then it would be reasonable to assume that the commercialisation facilitator would be able to attract the necessary additional capital. Hence, there is a view that the lack of funding in the valley of death is actually a symptom of other problems and is not the root cause [9].

We categorise the root causes as either developmental or decisional.

By developmental, we mean the acts involved in translating the research outcome to a commercial product have been misdirected or mismanaged. Often the researcher follows the technology into product development but the mindset and expertise that made them a successful researcher makes them ill-suited to that required for time-tabled product development and commercial management. So, even in the scenario of optimal commercial decisions, poorly implemented and inefficient developmental activities mean they fall short of the product development milestones, resulting in a commercial failure. The technology did not make it out of the valley of death.

By decisional, we mean that the research outcome has been directed towards the development of a product for which the business case is flawed for economic, market need, cost of development, inappropriate management team or any other typical reason in business and product development.

According to Daily and Sumpter [9], the dominant cause for technology commercialisation failure is the "lack of a compelling business case leading to unsatisfied customer expectations". Christensen *et al.* [3] single out over-engineering of a product (complexity) and price point for a given market application (too expensive) as the reasons for the inability of a technology to successfully climb out of the valley of death. While it may be argued that developmental and decisional reasons are overlapping, Markham [22] fundamentally attributed the commercialisation impasse as a decisional rather than a developmental issue.

Another way at looking at the valley of death is to apply institutional theory [11,12] to understand who the "actors" involved in the process of technology commercialisation are. The actors on the research side are the purveyors of the technology (the researchers) and the facilitators of commercialisation. Their role is to "sell" the early stage technology. The product development side is populated by the companies that may ultimately convert the technology into a commercially viable product and take it to the marketplace. They may be small-to-medium sized companies but most likely will be the larger corporate entities with an existing product and market presence. On this "buy" side, financial institutions are also important actors, as they provide the capital in technology commercialisation companies for the prospect of significant future commercial returns.

Applying this point of view to nanotechnology commercialisation, McGahn [24] argues that the supply side need to do more to present the nanotechnology in a product-oriented form to demonstrate its value in a way the would-be buyer can understand. Our premise is that nanotechnology research outcomes are most often a component that improves an existing whole product—it is seldom itself the complete product sold to the end-user. This is another common causal factor to the valley of death for nanotechnology. Hence, the seller of a nanotechnology is able to increase the probability, value and velocity of the sale, if they are able to package the technology in a form that is palatable to the buyer and demonstrates its value in the context of the improvement to the buyer's existing whole product. We refer to such a product as a nanotech-enabled whole product.

On the buy side, the authors believe that many of the large existing companies that could benefit from nanotechnology have little or no experience of nanotechnology. There is little awareness and absorptive capacity [4] in those companies for nanotechnology, even when they may have a genuine culture of "Open Innovation" [2]. As a consequence, the onus is back on the supply side to do more to package their nanotechnology in a form that can be understood by the would-be buyer from the buyer's view of the world. However, there are always exceptions and the case study chapter considers such an instance with the internal venture group at Degussa (now Evonik Industries).

We will revisit the issues of the nanotech-enabled whole product, the business case, customer satisfaction and economic viability when we speak later about the commercialisation of nanotechnology.

5.2 Nanotechnology Commercialisation Critical Success Factors

From a review of the literature, combined with our theoretical knowledge and experience gained from successful and failed nanotechnology commercialisation ventures, we identified the following five critical success factors:

  (i) product orientation (and not technology admiration)
 (ii) continuous market interaction and selection of a beachhead application
(iii) application of spiral product development methodology
 (iv) attraction and retention of commercialisation partners
  (v) mitigation of nanotechnology-specific technology risks.

We now consider each of these nanotechnology commercialisation critical success factors.

5.2.1 Product Orientation (and Not Technology Admiration)

5.2.1.1 The need for focus on the single most commercially viable and attainable application

As an enabling technology, any particular single nanotechnology invention has the potential to improve or enable a multitude of products in diverse and disparate fields of use. Hence, nanotechnology with multiple fields of use is sometimes called a "platform technology" [34]. The risk is that the purveyors of the nanotechnology admire the revolutionary and platform nature of their invention but lack the necessary focus on product orientation.

Prima facie, a technology having many potential applications may equate to it being more commercially valuable than another technology with fewer or only one application. While this view is held by some nanotechnology researchers, the theory is flawed. The early indicator of potential commercial value is not the quantity of potential applications but the magnitude of the most commercially viable and attainable application. We support this proposition by two independent industry scenarios, one in the nanotechnology industry and the other, by analogy, in the biotechnology industry.

An iconic nanotechnology invention is the carbon nanotube. A carbon nanotube exhibits unique electrical, mechanical, chemical, thermal, optical and biological properties which make it a promising candidate for integration into numerous different products across a range of industries. As a company manufacturing different grades of carbon nanotubes for hundreds of customers under contract, with associated patent applications, Carbon Nanotechnologies, Inc (CNI) was a nanotechnology commercialisation company. The business lacked a focus on identifying and exploiting a single commercially viable and attainable application.

In 2007, CNI merged with the nanotubes-based electronics company Unidym to focus on a single commercially viable and attainable application, and the combined entity was immediately rewarded by the financial markets. It placed a value on the combined company greater than the sum of the constituent companies, provided that the new entity focused on a single product application rather than attempting to serve the plethora of possible carbon nanotube applications [31]. See Box 5.1 for more information about the CNI—Unidym case study.

The biotechnology industry also teaches us that the financial markets reward a focus on the single most commercially viable and attainable application and not the number of potential applications. Biotechnology intellectual property often has one dominant commercially viable application. For example, a biological marker found only on the surface of cells when they become cancerous would lend it to be the basis of an anti-cancer drug and no other product type would be as commercially attractive. The commercial value of that intellectual property can be quantitatively derived from the specifics and assumptions of the cancer market and the specific segment it can address.

Box 5.1 Carbon Nanotechnologies, Inc.–Unidym case study

This case study looks at the drivers of value in Carbon Nanotechnologies, Inc. (CNI) as a nanotechnology commercialisation company before and after its forward integration merger with Unidym in 2007 [31], and similarly in the acquisition of Unidym by Wisepower Co., Ltd in 2010.

CNI held numerous patent applications relating to carbon nanotubes and was focused on mass manufacturing and marketing different grades of carbon nanotubes for a multitude of potential applications. Carbon nanotubes exhibit unique electrical, mechanical, chemical, thermal, optical and biological properties which make them promising candidates for integration into numerous different products across a range of industries.

Prior to the 2007 merger, CNI had hundreds of customers, several substantial government grants and contracts, and at least three business relationships with multinational companies in disparate industries with non-overlapping commercial objectives. However, the business lacked focus.

On April 23, 2007, CNI announced that it was going to merge with Unidym Inc., a Silicon Valley based subsidiary of Arrowhead Research Corporation focused on the development of nanotube-based electronics. The implications of the merger are discussed in detail later, but here we wish to point out that the transaction revealed that "the company's best option was to forego the dream of single-handedly benefiting from the broad growth of carbon nanotubes. Instead stockholders were forced to settle for sharing the value that a single customer, Unidym, could derive from using these carbon nanotubes in its intended target market" [31]. The financial markets placed a value on the combined company greater than the sum of the constituent companies, provided that the new entity—Unidym, Inc.—focused on a single product application rather than attempting to serve the plethora of possible applications.

The business model of Arrowhead Research Corporation and its rationale for the Unidym–Wisepower transaction adds further evidence for our case. In Arrowhead's own words [33], it "is a nanomedicine company developing therapeutic products at the interface of biology and nanoengineering. It identifies and selectively acquires proprietary drug platforms based on novel nanotechnology intellectual property. Once acquired, [the Arrowhead business model requires that] each technology platform is housed in its own subsidiary. This structure offers a number of benefits, including flexibility in financing and portability of assets in an exit transaction, while allowing Arrowhead to guide operations and business development, as well as retain substantial upside exposure".

Wisepower Co. Ltd, is a publicly traded, Seoul, Korea-based electronics company. It is a leading supplier of Li-polymer batteries for mobile appliances in Asia, and its current customers include cellular phone manufacturers. The company's products also include wireless charging systems for electronic devices with high-quality LED packages and solid-state lighting in development in 2010. According to Unidym in its press release on 18 January 2010 about the acquisition, "Wisepower's deep relationships with many potential Unidym customers and experience growing electronics businesses fit well with Unidym's leadership in carbon nanotube technology for electronics applications. Wisepower is well positioned to help Unidym develop manufacturing scale-up and customer acquisition as it moves into the market adoption phase of its business". The same press release goes on to say.

> Completion of this transaction represents the achievement of an important strategic goal for Arrowhead. The Company has evolved from a diversified nanotechnology company addressing multiple industries to a focused nanomedicine company, and the Unidym sale is the final step in this transition. This enables Arrowhead to allocate its resources toward a single industry, operate and grow in a more cohesive fashion, and it lays the groundwork for Arrowhead to emerge as a leader in the field.

"We are able to further increase our operational efficiencies by building a team of professionals with a single industry focus and by enabling more synergies among our subsidiaries", continued Dr Christopher Anzalone, Arrowhead's CEO. "This transaction is also potentially beneficial to shareholder value from an outward-looking perspective. We now have a far simpler and more cohesive story to articulate to analysts and institutional investors, and I am confident that our value proposition will fit more neatly into investors' established frameworks".

This Carbon Nanotechnologies—Unidym—Wisepower case study exemplifies the need for facilitators of nanotechnology commercialisation to focus on identifying a single commercially viable and attainable application, and that the commercial value of the project is realised as the opportunity is translated, in a step-wise fashion, to a successful product in the marketplace.

The need for product orientation and application focus in nanotechnology commercialisation is advocated by Stewart [34] saying "Successful startup nanotechnology companies will 'pick one thing and do it really well'. Thus, if the nanotechnology intellectual property is a 'platform technology', [the commercialisation venture] should pick a lead application and focus on it to the initial exclusion of all others. The other applications can be developed later". In the words of McGahn [24], "Going after too many markets dilutes resources and reduces the ability to execute. The key to successful commercialization lies with focus".

Bearing in mind that various potential applications of any particular nanotechnology can be in "multiple fields of use", as was the case with CNI with its carbon nanotubes, the critical business question then becomes: What is the exhaustive list of potential applications? Maine and Ashby offer a methodology to systematically assess and prioritise viable applications [18], however, there is no method in widespread use.

5.2.1.2 The need to start with an exhaustive list of potential base markets and applications

The exhaustive list of potential base markets and applications of the nanotechnology invention is the critical input that will otherwise limit the quality of the analysis and the decision as to which amongst them is the single most commercially viable and attainable application. This may seem like a basic and obvious point but determining the base market and overcoming the applications knowledge gap is a significant challenge for nanotechnology commercialisation.

As described by McGahn [24], a base market "defines the greater space in which a company plays. It's usually a large and obvious market that is readily understood and addresses fundamental needs, such as healthcare, government, retail, construction or energy. The base market is often segmented into many smaller markets and offers ample opportunity for a variety of solutions to be successful". Importantly, the base market must contain potential applications for the nanotechnology that leverage its unique aspects and endows the product with a competitive edge over the incumbent products [23].

According to McGahn [24], the nanotechnology inventor and commercialisation facilitator must avoid the temptation to instinctively select the "obvious" base market and the largest market segments within them as the beachhead application. They may indeed be the most attractive applications based on economic analysis but the resources required—human and financial—may be beyond the reach of the early stage nanotechnology commercialisation venture. The reality is that the most attractive applications are not always attainable at the outset and a "swing for a home run" strategy is very risky whether or not you are in the valley of death. The apparent instinct to select these obvious and large targets is "because they have easy-to-grasp value propositions". But McGahn [24] points out that they may not be commercially viable or attainable because the "new entrants face big barriers: (1) competition is already entrenched; (2) the ability to deliver may not be commensurate with capabilities required for success, such as scale or staffing; and (3) it is difficult to compete on cost alone". The existence of the applications knowledge gap (coined by the authors) is eloquently described by Bennett [1]: "Nanotechnology participants so often lament about an unrelenting dearth of commercial applications. They say that becoming aware of solid usage opportunities represents a not insignificant barrier to commercial exploitation". The reason for the applications knowledge gap being a significant challenge, especially for nanotechnology commercialisation, is several-fold.

The first reason has been identified by Bennett [1]: "While researchers, inventors, early-stage nano-ventures and the like, are typically knowledge-rich insofar as their discoveries and/or technologies are concerned, they, more often than not, lack the critical industrial/commercial knowledge needed to effectively and efficiently innovate to fulfill unmet and/or unrecognized, commercially exploitable needs". Maine, Lubik and Garnsey find that key capabilities to value creation for nanotech ventures include technology–market matching, alliance building, and experimentation [21].

The second reason is that prospective customers do not always have the market expertise or broad background to fully evaluate the potential of a nanotechnology or the product advantages it could provide existing whole products. Bennett [1] argues that "the application of traditional market research techniques such as focus groups, questionnaires, interviews, and the like, are totally insufficient and ineffective" for nanotechnology commercialisation. Customers or users are most often unaware of what they really want until they see it, especially in the case of very innovative products. Only when they can get mock-ups or see, feel, touch and respond to a prototype, they can provide feedback about its value and identify otherwise hidden potential drivers of value. The traditional market research techniques lack the benefits that can arise from evaluating a mock-up or prototype. This is consistent with Leonard's active experimentation advice under similar conditions of uncertainty [16].

A further reason is best illustrated using the same biotechnology example described earlier, the cancer-specific biomarker as the basis of an anti-cancer therapy. There is a "clear line of sight" from the research phase to the probable eventual product. New product development in biotechnology is more analogous to a rifle shot, whereas developing a successful nano-techenabled whole product is "more like a heat-seeking missile that not only changes paths but also changes targets continuously (to avoid hitting flak) until it finally zeroes in on a real target" [33].

The extremes between nanotechnology and biotechnology on this concept of applications knowledge gap are further described by Stewart [34]: "Even at the early stage of a biotechnology invention made in a university, the intellectual property anticipates a well-defined product application, either as a therapeutic or diagnostic or both. It is known how to extrapolate from a biotech discovery to a medical product. In the case of nanotechnology IP, it may have been developed with an application in mind but, by its very nature, can find various applications in multiple industries. The applications and industries can be diverse and disparate, making the commercialization of the nanotechnology IP more challenging".

The final reason we provide is a consequence of a point made by Stewart [34], that any one nanotechnology invention can be used in diverse and disparate applications and industries. The implication is that no individual has sufficient industry and consumer experience to compare and contrast the range of applications and make a well-informed intuitive assessment as to what application is the most commercially viable and attainable application. Maine and Garnsey concur, asserting that the technologymarket matching process is particularly challenging for radical, generic technology commercialised from upstream positions in industry value chains [20].

For biotechnology start-up companies, the pool of managers with comparable broad industry experience is relatively simple to access. Indeed in California, Boston and a number of regions of the United States, Europe and the United Kingdom reside executives who have been instrumental in developing anti-cancer therapeutic start-up companies, requisite with the experience of selling the company to a pharmaceutical company. This is similarly true in the computer and IT industries.

By comparison, the commercially embryonic state of the nanotechnology industry means there is a paucity of managers available with deep experiential knowledge of a broad range of industries and product applications in them. Moreover, it will take some time for this mix of experiential knowledge to be gained by individuals and for nanotechnology to be able to attract and retain them. By way of example, a nanotechnology company developing functional proteins showing great potential to replace existing industrial chemicals may have potential uses in agricultural sprays, hair care products, oil-well drilling fluids and hospital sterilisation products. A chief executive officer with appropriate research experience almost certainly will not have worked in all of these industries, or an executive with a career in a diversified chemicals company may have sold chemical-based products in many of those industries but would probably not have had the experience of taking a university breakthrough technology to the market, let alone a nanotechnology. This example is based on an Australian nanotechnology company in which some of the authors have an involvement.

So a problem for nascent nanotechnologies is closing the applications knowledge gap given the absence of individuals who posses experience in all major industry sectors and awareness of current industry trends and nuances in unmet applications in the marketplace. A solution is the formation of an Application and Market Advisory Board to provide information, comparison, prioritisation and recommendations as to which applications to pursue. The other solution is intense and interactive market engagement. In the next two sections, we make suggestions on how to reduce this nanotechnology commercialisation risk factor.

For now, we conclude with the remarks of Bennett [1], who states that "contrary to what nanotechnology developers and facilitators would hope, nanotechnologies are chiefly technology-push technologies; that is they are technologies looking for markets... As a direct consequence, the developers and facilitators of nanotechnologies must actively go forth and aggressively (and creatively) hunt for, and even make, market opportunities".

5.2.2 Continuous Market Interaction and Selection of a Beachhead Application

Bennett believes that "the most expedient way to close the knowledge gap is to actively seek out and materially involve potential users in the innovation process [1]. Because of the platform nature of nanotechnology and its potential for disruptive innovation, prerequisite for successful nanotechnology commercialisation is early, active and ongoing participation of customers (and other stakeholders, e.g. vendors, suppliers, etc.) in the innovation process. Participation is critical because it serves as the means to expose and understand the customer's tacit needs, which is vital before effective consideration can be given to what role technology should or could play in fulfilling those needs".

Bennett advocates "experiential marketing" which goes far beyond the mere identification of explicit customer needs (i.e. traditional market research) and provides a powerful tool for unveiling not only customers' tacit needs and values, but also a body of tacit knowledge which a technology creator most frequently lacks and which is highly essential in the discontinuous innovation process [1]. Before being in a position to conduct experiential marketing it is necessary to identify a small number of lead users—individuals and companies—within the selected base market.

Once the base market is determined, the objective is to identify and engage with a small number of lead users who embrace open innovation [2] to illicit strong and often hidden drivers of value from the nanotechnology in the potential end-products that only the end-user can articulate. A "lead user" may be a prospective customer but can also include interested suppliers, distributors and other stakeholders. The hallmarks of a lead user are that they are knowledgeable, often technically trained, have considerable interest and experience with company's existing product [15] and perceives key economic benefits from an innovation or a solution to a problem, and experiences them ahead of the market [26,36,37].

Once identified, nanotechnology commercialisation facilitators should engage with the lead users in a participative learning-oriented marketing process. "The first meeting is like a 'show-and-tell' session wherein the technology originator demonstrates to the prospective customer what the technology can do. By experiencing the technology first-hand, the customer, the technology originator, and other involved stakeholders launch a collective learning process that begins to reveal otherwise latent knowledge and needs, as well as new future value for the customer" [1].

The goal at this stage of the nanotechnology commercialisation process is to make an informed business decision as to which of the exhaustive list of potential applications in the base market should become the beachhead application. If it is too early to decide, then two or more product opportunities may be explored in parallel, resources permitting, until the beachhead application—the single most commercially viable and attainable application—becomes apparent. An analysis of nanotechnology ventures by Maine, Lubik and Garnsey, for example, found that all of the successful nanotechnology ventures in their study cohort began with at least four market targets, whereas the unsuccessful ventures targeted one or two [21]. However, once the beachhead application is found, the nanotechnology commercialisation facilitators need to focus on to the exception of everything else, until it is successfully commercialised or it fails. Selection of, and focus on, the beachhead application is a major determinant of nanotechnology commercialisation success.

The concept of a "beachhead" application is that technology companies should initially focus on a single market—a beachhead—and dominating that small specific market before using it as a springboard to win adjacent extended markets [25]. It is based on the D-Day analogy (Box 5.2).

Box 5.2   The D-Day analogy—fighting your way into the mainstream

The following excerpt from Moore [25] about the Allied invasion of Normandy on D-Day, 6 June 1944, helps to conceptualise the meaning of a "beachhead" application:

> Our long-term goal is to enter and take control of a mainstream market (Eisenhower's Europe) that is currently dominated by an entrenched competitor (the Axis). For our product to wrest the mainstream market from this competitor, we must assemble an invasion force comprising other products and companies (the Allies). By way of entry into this market, our immediate goal is to transition from an early market base (England) to a strategic target market segment in the mainstream (the beaches of Normandy). Separating us from our goal is the chasm (the English Channel). We are going to cross that chasm as fast as we can with an invasion force focused directly and exclusively on the point of attack (D-Day). Once we force the competitor out of our targeted niche markets (secure the beachhead), then we will move out to take over additional market segments (districts of France) on the way toward overall market domination (liberation of Europe)....

The key to the Normandy advantage, what allows the fledgling enterprise to win over pragmatist customers in advance of broader market acceptance, is focusing an overabundance of support into a confined market niche. By simplifying the initial challenge, the enterprise can effectively develop a solid base of references, collateral, and internal procedures and documentation by virtue of a restricted set of market variables. The efficiency of the marketing process, at this point, is a function of the "boundedness" of the market segment being addressed. The more tightly bound it is, the easier it is to create and introduce messages into it, and the faster these messages travel by word of mouth.

For the beachhead application, a tactic is to identify an incremental improvement to an existing whole product and communicate the benefits to the customer without describing the science behind it. The aim is to identify existing, yet unmet, market needs and/or products that already exist that would be more valuable with extended and enhanced functionality. Maine, Lubik and Garnsey demonstrate that high-value nanotech ventures focus initially on substitution markets while securing IP to provide a "real option" on emerging markets [21]. The same tactic, in other words, is to avoid technology-push and identify an authentic market need. In our Carbon Nanotechnologies case study (Box 5.1), CNI was a technology-push to the market with no product focus, whereas Unidym was positioned as a nanotube-enabled electronics company. Unidym was improving electronics for end-users. Not surprisingly, CNI was valued more when it merged with Unidym and forfeited its hundred other product lines of carbon nanotubes for various applications.

Lastly, while the most attractive applications based on economic analysis are not always candidates to be the beachhead application, they can still feature in the overall commercialisation strategy and value proposition of a nanotechnology commercialisation plan. The ideal product strategy would pursue a beachhead and articulate follow-on applications which help to demonstrate the value of the nanotechnology in other, sometimes "blue-sky", market segments. In fact, business theory tells us that if you do not plan to grow your business like this, you may make money in the medium term, but the company will eventually stagnate. Hence, the objective is to plan a "product roadmap that aligns the functionality of each new market entry with the end game, [where] you can demonstrate evolutionary feature sets in a series of complimentary markets in such a way that, when the value is derived from the series of evolutions and is applied to the big ultimate market, the solution is truly revolutionary; the benefits are understood; and full-scale adoption is easy" [24].

Once the beachhead application has been decided, the resources at the disposal of the nanotechnology venture need to be deployed on new product development and commercialisation.

5.2.3 Application of Spiral Product Development Methodology

Of the various methods of new product development described in the literature, it is the belief of the authors that spiral product development [7] will almost always be the method most likely to increase the probability of, and reduce the time to achieving a commercially successful product.

The spiral product development process accommodates planned "build-test-feedback-and-revise" iterations that span several phases of development (Fig. 5.2). Making the right decisions in the selection of the beachhead application and setting the customer-required product specification is generally accepted in the literature as the most significant determinant of subsequent product success or failure [33]. "Simply stated, the first few plays of the game seem to decide the outcome!" [5].

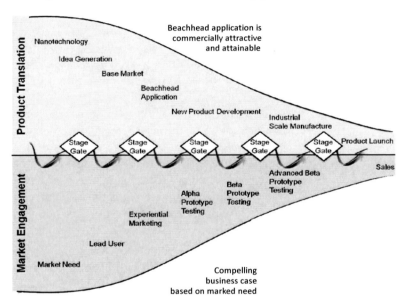

Figure 5.2   Spiral product development is a series of "build-test-feedback-and-revise" iterations. Adapted from Cooper [7].

While there may be very little to show the customer in the first iteration of experiential marketing, the second iteration of spiral development should have a representation of the proposed end-product. Moreover, it needs to represent the whole product and its nanotech-enabled improvements—by way of its form, fit and function—and not the science behind it. The representation may be a prototype, a hand-made model or mock-up, or even a computer-generated virtual prototype.

The product obviously does not have to be fully ready at this stage—in fact, a nanotech product seldom will be—but needs to give the customer a feel for what the product will be and do.

Interest, liking, preference and purchase intent are thus establi-shed even before the project is a formal development project. Feedback is sought, and the needed product revisions are made.

Regardless, the customers need the technology packaged in a form that is palatable to them and their needs. So whilst the nanotechnology may be a revolutionary rather than incremental change to the technology within the existing product, the key to end-user adoption is to present the revolutionary as the evolu-tionary: "convenient and familiar through an improved version of a product the end user already uses" [24].

Bennett [1] is a proponent of the "build-test-customer feedback-and-revise" iterative approach for nanotechnology commerciali-sation. Maine and Garnsey identify co-development with alliance partners which are prospective customers as another approach to achieving this [19].

The key elements of the experiential marketing process are experience and learning that occur in a framework where marketing and R&D functions are not conducted separately and on a one-off basis, but where they are joined together and carried out iteratively and in a highly interactive way over an interval of time.

The process begins with the learning that occurs and new knowledge that is generated through the customer's first experience with the technology, and it continues with each subsequent learning episode and knowledge-generating incident that takes place when the customer is given the opportunity to experience the technology as transformed to integrate learning and know-ledge acquired through the prior experience. A chief characteristic of this interactive, participation-style R&D process is this ability to quickly and effectively advance the innovation process in direct response to tacit knowledge and needs which are transformed into explicit knowledge and needs each time the customer uses and experiences the revised technology.

The collective learning process also leads to the creation of new knowledge which serves as critical raw material for furthering the discontinuous innovation process.

Put simply, for a new product to deliver superior benefits for the end-user, the facilitators of commercialisation and the product development managers cannot presuppose that they know the explicit and tacit needs and wants of the user, instead it must involve their lead users repeatedly starting from decision of the beachhead application and throughout the product development process. By employing the spiral development approach, they are able to identify and address product development risks early in the project when costs are still relatively low. And as the project moves through each iteration, the successive versions of the product get closer to the final product, and at the same time, closer to the customer's ideal.

Each iteration of the spiral product development process has the following five regular steps:

(i) Determine objectives, alternatives and constraints.
(ii) Identify (as much as possible) and prioritise risks.
(iii) Evaluate alternatives to key risks.
(iv) Develop the deliverables for that iteration, and verify that they are correct.
(v) Plan the next iteration (if there is one).

The spiral product development process serves to overcome difficulties presented by unclear initial product requirements, a challenge which is poorly handled by the traditional stage-gate process (see Box 5.3). While the stage-gate process is widely used in the biotechnology, aviation and aerospace industries, the spiral product develop approach is a relatively recent process design and has been adopted by many in the software and information technology industries.

The spiral process may be advantageous for increasing the probability that the resultant product meets the customer's wants and needs, but it exposes the nanotechnology venture to greater financial risk of inability to draw down investment tranches and under-capitalisation. The lack of rigid specifications (at least initially) and the dependency on lead user input in setting the milestones means that the milestones may require frequent adjustment and there may be much longer periods of time between product development iterations than budgeted. The spiral product development process requires objective, moveable but verifiable milestones to determine whether the project is ready for the next iteration of spiral development. Compared with other classes of technology commercialisation, this makes nanotechnology appear haphazard, risky and unattractive to many potential investors.

Box 5.3   Stage-gate new product development

The most widely used type of product development process is the traditional stage-gate or waterfall process. It comprises a series of discrete phases of activity with each phase being separated by a stage-gate to determine whether the objectives of the proceeding phase were successfully completed, such that the project may proceed to the next phase, otherwise it must circle back and attempt that phase again [6]. Note that the early activities of the traditional stage-gate process for new product development—idea generation, preliminary assessment and product conceptualisation—are the activities that occur in the valley of death.

There are a small number of existing product development processes, each of which manage risk differently but appropriately for the nature of its industry or type of product [35]. Fundamentally, product development is a management tool to reduce the risk and uncertainty in the process.

In terms of risk and uncertainty, there are many, but Unger and Eppinger [35] identified the four major types as technical, schedule, budget and market. Technical risk is the uncertainty regarding whether a new product is technologically feasible and will perform as expected. Schedule risk is the uncertainty regarding whether a new product can be developed in the time allowed. Budget risk is the uncertainty regarding whether a new product can be developed with the financial resources available. Lastly, market risk is the uncertainty regarding whether a new product accurately addresses changing customer needs, product positioning with respect to dynamic competition and price tolerance to ensure profitability.

5.2.4   Attraction and Retention of Commercialisation Partners

As described in Section 5.1, the main actors on the "buy" side of the valley of death are the companies with existing whole products on the market that could benefit from the nanotechnology and the financial institutions wishing to invest capital into the nanotechnology commercialisation venture to assist in its product-searching and new product development activities.

For the facilitators of nanotechnology commercialisation, we believe it is critical to attract both the large corporate companies with their whole products already incumbent in the marketplace and the financial institutions as commercialisation partners. Maine and Garnsey also argue that access to complementary assets (such as sector specific design and marketing capabilities) and access to finance are critical to demonstrating value in nanomaterials commercialisation [20].

McGahn [24] is adamant that the market-based demonstration of irrefutable value must be the immediate commercial objective of a nanotechnology commercialisation venture. He argues that, to the extent that it has the resources and capabilities to deliver prototypes, it is in its own best interest to do so because it reduces the development time (resources and risk) that the market-incumbent large corporate commercialisation partner needs to contribute. Moreover "the easier you make the sale for the partner to its own management or from the partner to their customers, the quicker you [the nanotechnology commercialisation venture] will get to revenue".

Finance permitting, McGahn [24] believes that it is advantageous for the nanotechnology venture to "develop the manufacturing process and the materials to a point where the company can sell direct to the market. It allows for direct market feedback, enabling ongoing technology development to continue with a full under standing of market needs and impact. It also better allows partners to see the technology packaged in a way that is complementary to their business. They can understand the possible versus the probable, with the probable usually being worth more to them because they can see an immediate or short path to break-even and revenue".

A point in case is the Carbon Nanotechnologies–Unidym case study (Box 5.1). Whilst CNI was manufacturing and marketing its own carbon nanotubes, it was not learning from its customers as to what its product performance needs were or what may be the single most commercially valuable and attainable application of its nanotubes. That changed immediately following the forward integration M&A transaction with Unidym being an industry leader in the use of carbon nanotubes technology in electronics applications. Interestingly, commercialisation of the nanotechnology underwent another round of forward integration with the combined entity being acquired in 2010 by Wisepower Co Ltd. Wisepower is a large multinational corporate company which has a focus on electronic products, including Li-polymer batteries for mobile appliances and wireless charging systems for electronic devices.

We believe that the CNI–Unidym–Wisepower case study illustrates the business model that will become the cornerstone for nanotechnology commercialisation: intense and interactive market engagement, beachhead application, spiral product development and forward integration by merger and acquisition.

This model for nanotechnology commercialisation has some analogy to the forward integration strategy employed in the early stages of the biotechnology industry, with the acronym FIPCO, standing for "fully integrated pharmaceutical company".

However, venture capital firms wishing to invest in nanotechnology ventures may find themselves at a disadvantage compared with investing in biotechnology or other, more mature industries such as aviation and aerospace, computer and IT. The reason being that nanotechnology commercialisation does not have irrefutable indicators for the removal of product development risk or heuristics to measure the increase in value when a risk is abrogated. This makes the process of predicting verifiable, unambiguously defined product development milestones, the time between those milestones and the increase in value of the nanotechnology venture upon achievement of a milestone extremely challenging.

By comparison, the more mature industries have their own indicators and measures of product development. Companies in aviation and aerospace use the Technology Readiness Level (TRL) framework, which is a measure to assess the maturity of the evolving technologies (for example, materials, components, devices) prior to incorporating that technology into a system or subsystem. Generally speaking, when a new technology is first invented or conceptualised, it is not suitable for immediate application. Instead, new technologies are usually subjected to experimentation, refinement, and increasingly realistic testing. Once the technology is sufficiently proven, it can be incorporated into a system or subsystem. There are nine technology readiness levels, although there were previously seven in the original definitions as developed by NASA in the 1980s [30].

In the computer and IT industry, the stages of development are generally accepted to be alpha, beta and advanced beta stage testing before product freeze and release. It is a linear sequence of stages with movement from one stage to the next dependent upon the recognition of value attained by one or more independent third parties such as vendors, customers and end-users.

In the biotechnology industry, the universally accepted signals of an increase in technology value are from the country's impartial health regulator. In the United States, it is the Food and Drug Administration (FDA) which is responsible for assessing the safety and efficacy of candidate medical therapies, devices and diagnostics before they commence the clinical trial process. For example, only if a candidate drug is found to be safe in a Phase 1 clinical trial then the FDA will announce that it can progress to a Phase 2 clinical trial to assess its health benefit for patients. With increasing rigor, the clinical efficacy and health cost-benefit are assessed in the following Phase 3 trial clinical trial then market approval determination by the FDA. Biotechnology and pharmaceutical companies must follow this linear stage-gate process. As an impartial customer-oriented entity, the FDA makes a series of "yes/no" decisions which the financial market has elected to use as product development milestones.

It follows that passing through each of these FDA outcome stagegates is a major determinant of the stepwise increase in the value of a biotechnology company. Accordingly, they are milestones which justify and enable the sequential rounds of financial investment to achieve the next milestone on the path towards a product launch. Schematically illustrated in Fig. 5.3, this relationship between regulatory approval and technology value is the cornerstone of financing in the biotechnology industry. Furthermore, the "universal" nature of these milestones as stages in biotechnology product development with sufficient years of market data has provided deal comparables and heuristics for determining the value of a biotechnology venture dependent, amongst other things, the stage of product development and the nature of the market need it aims to address [14].

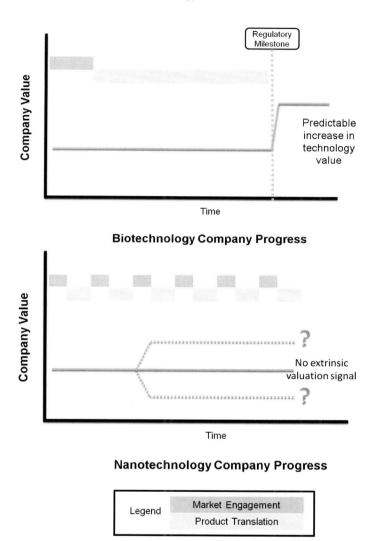

Figure 5.3 Biotechnology and nanotechnology commercialisation progress and valuation "signals".

In stark comparison, nanotechnology does not possess universal milestones for its product development. Nanotechnology is not regulated by the FDA (unless its application is a medical use, which is outside of the terms of reference of this chapter) or any other similar governmental body (other than nanotoxicology and safety regulators). There are no impartial product development milestones along the product development pathway that are universally accepted by the market as generic "signals" of attaining an increase in technology value, let alone an accepted way to measure the absolute value of the nanotechnology commercialisation progress. Hence, a nanotechnology company may use spiral product development and be making significant progress towards a successful nanotech-enabled whole product but have no extrinsic signal to justify a commensurate increase in company value. As shown schematically in Fig. 5.3, it is debatable whether the nanotechnology venture has attained an increase in value (or decreased). We believe this to be a root cause for the lack of funding to traverse the valley of death for this class of technology.

The definition of product development milestones and the acceptance that they have been attained by commercialisation partners (investors, corporate partners and the financial markets) is critical for successful nanotechnology commercialisation. Retention of the commercialisation partners is important because their capital is necessary to fund the next stage of product development to move closer towards product launch. Hence, it was somewhat surprising that our review of the nanotechnology commercialisation literature regarding valuation milestone events found only the paper by Stewart [34].

Stewart [34] suggested that there are two milestone events which have the potential to act as an extrinsic signal of an increase in nanotechnology (or firm) value. The first event relates to nanotechnology intellectual property receiving government R&D grants for product development and commercialisation; the second is the achievement of a significant product development milestone in a corporate sponsored R&D program. Later in this section, we propose our own schema for a nanotechnology investment model, with milestone events, which is integrally linked with end user led product development and market demand.

Before proposing our own schema for nanotechnology valuation milestone events, we first need to describe the dynamic forces which we believe will shape the availability of finance for nanotechnology commercialisation. The authors believe that over time, the venture capital industry will move towards valuation milestones for nanotechnology ventures which are dependent on lead user feedback. It may have hallmarks of the computer and IT industry with generally accepted stages of development such as alpha, beta and advanced beta which are predicated on market feedback and acceptance by the commercialisation partners. The subjective nature of the milestones will mean that the nanotechnology commercialisation partners will need to deeply understand the market, the product development process and the underlying nanotechnology to determine the worthiness of further product development in light of the external market feedback and ability of the nanotechnology venture's management team. It will require considerable time for such investor experience and nano-technology commercialisation successes to emerge before the industry reaches a tipping point and the philosophy of market-correcting milestones in concert with spiral product development becomes the new norm in nanotechnology investing.

A consequence of the nanotechnology commercialisation venture needing to develop its technology and demonstrate its value in the form of the nanotech-enabled whole product means that they will attract interest from multiple large corporate companies already in the marketplace. Such companies are competing in the marketplace and their product could benefit from the inclusion of the nanotechnology into their whole product. This is a desirable scenario of the nanotechnology commercialisation venture as it creates competitive tension amongst the large corporate companies and is likely to maximise the value of the nanotechnology venture in the scenario of a corporate acquisition.

By deduction, we predict that large companies with existing whole products will attempt to make strategic corporate investments in a number of late stage nanotechnology commercialisation ventures. In this manner, the investing company gains a strategic advantage over its rivals as it has privileged knowledge of the nanotechnology, placing it in pole position for a future acquisition of the nanotechnology venture. We suspect, however, that in many cases venture capital investors who invested in the company at an early stage of development will resist such late-stage corporate investment. The venture capital investors may be wary that a corporate investment may preclude other potential acquirers of the company, and in doing so impair company valuation at exit and reduce the venture capital firms' return on capital.

As a consequence, we believe that market forces will lead the dominant nanotechnology commercialisation model towards venture capital investment (and possibly non-controlling corporate investment) through to acquisition, with the milestones for investment anchored in the outcomes of continuous market engagement, beachhead application selection and spiral product development.

5.2.5 Mitigation of Nanotechnology-Specific Technology Risks

For emerging nanotechnologies and new nanomaterials there are numerous challenges and bottlenecks facing both innovators and adopters. These range from scale-up and system optimisation to metrology, quality control, occupational hygiene, and health and safety concerns. Inventors and innovators need to start addressing these challenges early if they are to successfully traverse the valley of death.

5.2.5.1 Manufacturing methods and scale-up

Every day the scientific journal literature publishes work describing a raft of new nanostructured permutations and combinations in the field of nanomaterials. The very high degree of "versatility and tenability" of nanostructured objects, ensembles and systems also typically represents significant challenges when it comes to manufacture and scale-up. Nanomaterials, whether they be particulates, membranes, composites or complex combinations designed to interface with biological systems, can be highly sensitive to variables such as its thermal history, moisture levels, shear rates in processing or the purity of precursor materials. The temptation for researchers is to explore this seemingly endless array of variables to achieve better performance.

However, the reality for scale-up is that "fixating" solely on the most promising recipes or formulations earlier in the innovation process, and placing more of an emphasis on things such as quality, reproducibility, and the development of more "forgiving", cost effective and industrially acceptable manufacturing methods and characterisation tools will ultimately lead to better commercial outcomes. Our experience has been that in many cases, each individual customer or end-user will be need to adjust the technology to suit their own specific performance criteria in any case, so having a robust, industrialised technology which demonstrates utility is more important that having a finely optimised technology that is unreliable.

5.2.5.2 Quality control and specification tolerance of nanotechnology in the whole product

From a manufacturing perspective, a dilemma of a different kind faces manufacturers of nano-enabled products. In an excellent article, Richman and Hutchison [29] explained that in order for nanotechnology discoveries to progress fully to mature applications, more efficient high-throughput characterisation tools with multiple capabilities are required in order to assure customers of quality, performance and safety. If you compare a new technology introduced in a mature field (e.g. the semiconductor industry) contrasted with new nanotechnology implementation, a much slower "rate of decrease in level of metrology" is required for the nanomaterial-enabled product assurance. For example, for nanoparticles, exotic techniques such as dynamic light scattering and transmission electron microscopy are often required for characterisation; however this equipment is far too sophisticated and expensive for all but the most advanced small to medium scale manufacturing enterprises. This relationship goes some way towards explaining the tremendous lag between early stage nanotechnologies and workable regulatory frameworks. A particularly pertinent example is again carbon nanotubes, which are arguably the most inherently variable nanomaterials available.

5.2.5.3 Occupational and environmental health and safety

As nanotechnology discovery and innovation marches rapidly ahead, there is a critical nexus which exists between the science, public opinion and government. Generation of new toxicological scientific data and methodologies, rate-limited by finite R&D funding for "risk-related" nanomaterials research, are balanced and influenced more by political agenda. Regulators and policy decision makers are working hard to arrive at more workable frameworks. In a collaborative paper summarising conclusions from a 2008 NATO workshop on nanomaterial regulation, a multidisciplinary working group reviewed, compared and contrasted an exhaustive amount of information under the categories of (1) science and research aspects, (2) legal and regulatory aspects, (3) social engagements and partnerships and (4) leadership and governance [17]. Clearly, such in-depth international information sharing is going to be required to more efficiently address the existing bottle neck.

Choi and colleagues estimated that if all nanomaterials required long-term *in vivo* testing (comprehensive precautionary approach), in the United States alone, it would cost $1.18 billion and require up to 53 years if all existing nanomaterials were to be thoroughly tested [8]. They strongly supported a more tiered risk assessment strategy, similar to the European Union's 2008 REACH legislation for regulating chemicals. In such a system, initial screening comprising simple and inexpensive tests is used to prioritise substances for further more complex and expensive tests, with increasing degrees of selectivity for adverse effect.

Clearly, for this to happen most effectively, risk assessors need to be engaged very early in the nanomaterial research, development and manufacturing processes, so that lowest risk options or formulations can be selected without compromising ultimate product performance, and so that researchers and workers handling the nanomaterials can be properly protected from the outset.

5.3   Conclusions

In summary, the five critical success factors for nanotechnology commercialisation are as follows:

(i)   presenting the technology in a market-oriented whole product form to readily demonstrate its value to a would-be buyer (product orientation)

(ii)   focusing on the development of the single most commercially viable and attainable product opportunity once known, to the exclusion of all else, until it is successfully commercialised or it fails (beachhead application)

(iii)   engaging with lead users and utilising iterative product development to increase the probability that the product meets the customer's wants and needs (spiral product development)

(iv) attracting and retaining investors and corporate partners through effective sharing of context-rich technology–market knowledge to reduce the risk from an absence of universal product development milestones that represents an increase in value for the nanotechnology venture to justify (commercialisation partners)

(v) identifying and addressing technical risks which may exist in relation to cost, quality control and tolerance of the nanotechnology in the manufacture and use of the whole product for market supply, including occupational and environmental health and safety considerations (mitigation of nanotechnology-specific technology risks).

In conclusion, we propose to bring the critical success factors for nanotechnology commercialisation into a single coherent strategic framework: the nanotechnology commercialisation diamond in Fig. 5.4. It is based on the Strategy Diamond of Hambrick and

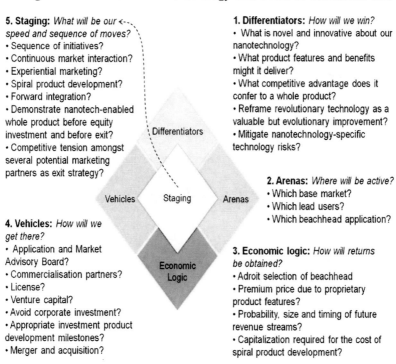

5. Staging: *What will be our speed and sequence of moves?*
• Sequence of initiatives?
• Continuous market interaction?
• Experiential marketing?
• Spiral product development?
• Forward integration?
• Demonstrate nanotech-enabled whole product before equity investment and before exit?
• Competitive tension amongst several potential marketing partners as exit strategy?

1. Differentiators: *How will we win?*
• What is novel and innovative about our nanotechnology?
• What product features and benefits might it deliver?
• What competitive advantage does it confer to a whole product?
• Reframe revolutionary technology as a valuable but evolutionary improvement?
• Mitigate nanotechnology-specific technology risks?

2. Arenas: *Where will be active?*
• Which base market?
• Which lead users?
• Which beachhead application?

4. Vehicles: *How will we get there?*
• Application and Market Advisory Board?
• Commercialisation partners?
• License?
• Venture capital?
• Avoid corporate investment?
• Appropriate investment product development milestones?
• Merger and acquisition?

3. Economic logic: *How will returns be obtained?*
• Adroit selection of beachhead
• Premium price due to proprietary product features?
• Probability, size and timing of future revenue streams?
• Capitalization required for the cost of spiral product development?

Figure 5.4 Nanotechnology commercialisation diamond. Adapted from Hambrick and Fredrickson [13].

Fredrickson [13] but adapts their five key business questions (in relation to differentiators, arenas, economic logic, vehicles and staging) to what we believe are the critical success factors for nanotechnology commercialisation. It can serve as a nanotechnology commercialisation checklist and as a means to ensure the elements of the strategy mutually reinforce each other and fit within the macro-environment.

References

1. Bennett, R. P. (2008). Commercialization-friendly innovation/R&D model for superior exploitation of nanotechnologies. *Nanotech. Law. Bus.*, **5**, pp. 441–446.

2. Chesbrough, H. (2003). *Open Innovation: The New Imperative for Creating and Profiting from Technology.* (Harvard Business School Press, Boston).

3. Christensen, C. M., Musso, C. S., and Anthony, S. D. (2004). Maximizing the returns from research, *Research and Technology Management*, July–August, pp. 12–18.

4. Cohen, W. M., and Levinthal, D. A. (1990). Absorptive capacity: A new perspective on learning and innovation, *Admin. Sci. Quart.*, **35**, pp. 128–152.

5. Cooper, R. G. (1988). Predevelopment activities determine new product success, *Ind. Market. Manag.*, **17**, pp. 237–247.

6. Cooper, R. G. (1990). Stage-gate systems: A new tool for managing new products, *Business Horizons*, May-June, pp. 44–54.

7. Cooper, R. G. (2008). The stage-gate idea-to-launch process–Update, what's new and nextgen systems, *J. Prod. Innovat. Manag.*, **25**, pp. 213–232.

8. Choi, J. Y., Ramachandran, G., and Kandlikar, M. (2009). The impact of toxicity testing costs on nanomaterial regulation, *Environ. Sci. Technol.*, **43**, pp. 3030–3034.

9. Daily, M. R., and Sumpter, C. W. (2005). A structured process for transitioning new technology into fieldable products, *Proceedings 2005 International Engineering Management Conference*, IEEE, pp. 77–81.

10. Frank, C., Sink, C., Mynatt, L., Rogers, R., and Rappazzo A. (1996). Surviving the "valley of death": A comparative analysis, *J. Tech. Transfer.*, **21**, pp. 61–69.

11. Greenwood, R., and Hinings, C. R. (1996). Understanding radical organizational change: Bringing together the old and the new institutionalism, *Acad. Manage. Rev.*, **21**, pp. 1022–1054.

12. Greenwood, R., and Suddaby, R. (2006). Institutional entrepreneurship in mature fields: The big five accounting firms, *Acad. Manage. J.*, **49**, pp. 27–48.

13. Hambrick, D. C., and Fredrickson, J. W. (2001). Are you sure you have a strategy? *Acad. Manage. Exec.*, **19**, pp. 51–62.

14. Kalamas, J., and Pinkus, G. (2003). The optimum time for drug licensing. *Nat. Rev. Drug Discov.*, **2**, pp. 691–692.

15. Kaulio, M. A. (1998). Customer, consumer and user involvement in product development: A framework and a review of selected methods, *Total Qual. Manage.*, **9**, pp. 141–149.

16. Leonard, D. (1995). Wellsprings of knowledge: Building and sustaining the sources of innovation. (HBS Press, Cambridge MA).

17. Linkov, I., Steevens, J., Adlakha-Hutcheon, G., Bennett, E., Chappell, M., Colvin, V., Davis, J. M., Davis, T., Elder, A., Foss Hansen, S., Hakkinen, P. B., Hussain, S. M., Karkan, D., Korenstein, R., Lynch, I., Metcalfe, C., Ramadan, A. B., and Satterstrom, F. K. (2009). Emerging methods and tools for environmental risk assessment, decision-making, and policy for nanomaterials: summary of NATO Advanced Research Workshop, *J. Nanopart. Res.*, **11**, pp. 513–527.

18. Maine, E. M. A., and Ashby, M. F. (2002). An investment methodology for materials. *Mater. Des.*, **23**, pp. 297–306.

19. Maine, E., and Garnsey, E. (2004). Challenges facing new firms commercializing nanomaterials, *Proceedings 9th International Conference on the Commercialization of Micro and Nano Systems*, Edmonton, Alberta, Canada.

20. Maine, E., and Garnsey, E. (2006). Commercializing generic technology: The case of advanced materials ventures, *Res. Policy*, **35**, pp. 375–393.

21. Maine, E., Lubik, S., and Garnsey, E. (2011). Process-based *vs.* product-based innovation: Value creation by nanotech ventures, *Technovation,* doi:10.1016/j.technovation.2011.10.003.

22. Markham, S. (2005). Product champions: Crossing the valley of death, *PDMA Online Seminar Series*.

23. Mazzola, L. (2003). Commercializing nanotechnology, *Nat. Biotechnol.*, **21**, pp. 1137–1143.

24. McGahn, D. P. (2005). Commercializing a new technology in six easy pieces: It all starts with focus, *Nanotech. Law. Bus.*, **2**, pp. 90–94.

25. Moore, G. A. (1991). *Crossing the Chasm*. (Harper Business Press, New York).

26. Morrison, P., Roberts, J., and von Hippel, E. (2000). Determinants of user innovation and innovation sharing in a local market, *Manage. Sci.*, **46**(12), pp. 1513–1527.

27. Murphy, S. A., and Kumar, V. (1997). The front end of new product development: a Canadian survey, *R&D Manage.*, **27**, pp. 5–15.

28. Osman, T. M., Rardon, D. E., Friedman, L. B., and Vega, L. F. (2006). The commercialization of nanomaterials: Today and tomorrow, *J. Manuf.*, **58**, pp. 21–24.

29. Richman, E. K., and Hutchison, J. E. (2009). The nanomaterial characterization bottleneck, *ACS Nano*, **3**, pp. 2441–2446.

30. Sadin, S. R., Povinelli, F. P., and Rosen, R. (1989). The NASA technology push towards future space mission systems. *Acta Astronaut.*, **20**, pp. 73–77.

31. Serrato, R., and Chen, K. (2007). Mergers and acquisitions of nanotechnology companies: A review of the Unidym and CCNI merger, *Nanotech. Law. Bus.*, **4**, pp. 205–210.

32. Smith, P. G., and Reinertsen, D. G. (1998). *Developing Products in Half the Time*, 2nd Ed. (Van Nostrand Reinhold Book, New York).

33. Stevens, G., Burley, J., and Divine, R. (1999). Creativity + Business discipline = Higher profits faster from new product development. *J. Prod. Innov. Manag.*, **16**, pp. 455–468.

34. Stewart, J. (2005). The nanotech university spinout company: Strategies for licensing, developing, commercializing and financing nanotechnology, *Nanotech. Law. Bus.*, **2**, pp. 365–375.

35. Unger, D. W., and Eppinger, S. D. (2002). Planning design iterations, *SMA Working Paper*.

36. Urban, G. L., and von Hippel, E. (1988). Lead user analyses for the development of new industrial products, *Manage. Sci.*, **34**, pp. 569–582.

37. von Hippel, E. (1986). Lead users: A source of novel product concepts, *Manage. Sci.*, **32**, pp. 791–805.

Chapter 6

Overcoming Nanotechnology Commercialisation Challenges: Case Studies of Nanotechnology Ventures

Elicia Maine

Beedie School of Business, Simon Fraser University,
500 Granville Street, Vancouver, BC, V6C 1W6, Canada
emaine@sfu.ca

6.1 Introduction

Nanotechnology ventures, as science-based businesses, are very different from more widely studied technology ventures, such as those built to exploit technology developments in electronics, computers, and software. Yet to date, very little has been written about the commercialisation strategies of nanotech ventures. Science-based businesses, such as biotech, nanotech, and advanced materials ventures, are differentiated from other types of technology ventures by the level of uncertainty they face over long time periods before commercialisation [26]. For instance, new

Nanotechnology Commercialisation
Edited by Takuya Tsuzuki
Copyright © 2013 Pan Stanford Publishing Pte. Ltd.
ISBN 978-981-4303-28-6 (Hardcover), 978-981-4303-29-3 (eBook)
www.panstanford.com

technology ventures have an average gestation period, defined as technology idea to first commercial sales, of 2 years [4], whereas advanced materials, nanotech and biotech ventures have a gestation period in excess of 10 years [20,26]. This prolonged uncertainty, and the high level of capital investment required, is leading to business model innovation, which can be most usefully studied at the level of individual ventures. And, although the distinction between science-based businesses and nearer-to-market technology ventures is critical, nanotech ventures are different from biotech ventures in important ways.

Like biotech ventures, nanotech and advanced materials ventures face high technological and market uncertainty over long time frames [3,18,26]. Nanotechnology, like advanced materials, also comprises a set of technologically diverse enabling technologies based on process innovation [17,25,28]. However, unlike biotech, nanotech ventures have to choose between a vast range of unrelated markets with entirely different alliance partners, regulatory systems, and performance attributes. Nanotech ventures also have different marketing uncertainty challenges, without the step function reduction in uncertainty provided by clinical trials, or the same likelihood that, having resolved technological uncertainty and passed regulatory hurdles, the product will be adopted.

A nanotech venture will benefit from adjusting their commercialisation strategy to address the combined factors of the radical, generic, upstream, slow, process-based and capital-intensive nature of nanotech innovation [22]. Through three case studies, this chapter explores innovative commercialisation strategies which could address these challenges and enable new ventures to exploit nanotechnology more effectively. The ventures are then analysed as to their commercialisation challenges and to how they enact the critical success factors of nanotech commercialisation presented in the previous chapter.

6.2 Case Studies

Case studies of three successful nanotech ventures are explored in this chapter. Each of the nanotech ventures developed and commercialised engineered nanoparticles. Two of the ventures were independent start-ups, commercialising scientific discoveries from university labs after a long process of development within the start-up firm. The third case study is of an internal corporate venture, demonstrating the environment within an incumbent firm most conducive to commercialising radical generic technology such as nanomaterials. The case studies are based on multiple interviews with CEOs and/or operational leaders at both firms, and a wide range of secondary source information sourced from news archives, company websites, financial databases, and the US patenting office.

6.2.1   Hyperion Catalysis[1]

Hyperion Catalysis was formed in 1981 with funding from a Silicon Valley venture capitalist who judged that the advanced materials sector offered outstanding long-term value potential. He brought together a scientific advisory board to help him select an appropriate focus within the advanced materials sector. This board, consisting mainly of scientists from MIT and Harvard, advised on carbon microfilaments, subject to resolving technical uncertainty about synthesis. One employee, a retiring industrial chemist, was hired to start conducting research on this area. With some encouraging results, Hyperion was incorporated in 1982, locating in Cambridge, MA, because of the existing location of their key employee and most of the scientific board. Their goal was to develop a radical innovation in advanced materials technology; if successful, the potential for long-term value creation was enormous, as such an innovation could improve products across most industrial sectors.

From 1982 to 1989, Hyperion focused on developing the first viable multi-walled carbon nanotube (MWCNT) product and process, with patient capital provided by their founder and owner. Their key breakthrough was their 1983 synthesis of MWCNTs, which Hyperion protected by filing for a patent in 1984. This patent, which was issued in 1987, is the first US carbon nanotube patent[2] and became key to Hyperion's patent portfolio (US Pat No. 4,663,230). From 1984 to 1989, Hyperion's scientific team developed their

---

[1]With permission from Elsevier, this case study is reproduced from ref. [20].

[2]Carbon nanotubes have generated considerable interest as they enable radical improvement in the performance attributes of composite materials as well as enabling entirely new products.

technology from a laboratory process to a production process with numerous patents filed on improvements in the reactor design and the development of a continuous manufacturing process. The output of this vapour deposition process is their key intermediate product, MWCNTs, later trademarked FIBRIL™. Hyperion's commercialisation timeline is depicted in Fig. 6.1.

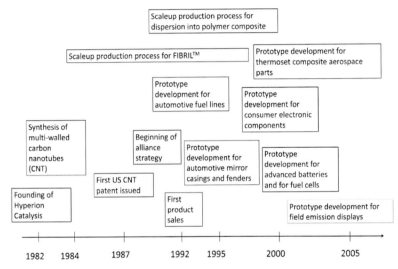

Figure 6.1   Hyperion Catalysis and its commercialisation timeline.

By 1989, Hyperion had achieved their technical objectives, which included learning how to make these MWCNTs in large-scale production volumes and to a high level of purity. When they began focusing on commercialisation, they wanted to follow an in-house manufacturing business model but struggled to choose between the many potential uses for their advanced materials product and process inventions, including potential uses in the automotive, aerospace and power generation industries. Hyperion did not yet have prototypes suitable to demonstrate feasibility to these markets. Hence, they publicised their technical achievements widely, in the hopes of attracting potential customers and/or alliance partners. This strategy proved successful, as it resulted in the approach of their first alliance partner.

This partner, a European-owned resin supplier, thought that Hyperion's technology would solve their own problem with an automotive application. The resin supplier had been attempting to displace steel fuel lines and had established a solid production cost advantage but needed to make their polymeric fuel lines conductive for safety reasons. The resin supplier had already identified the resin, Nylon 12, and was confident that Hyperion's MWCNTs could be compounded with that resin to make conductive composite automotive fuel lines. The resin supplier's compounding and manufacturing equipment, along with their contacts into the automotive industry, were key to Hyperion successfully selling into the automotive market, since automotive original equipment manufacturers (OEMs) and Tier 1 suppliers rarely pay for any prototype development. In successfully developing a prototype, Hyperion developed a process to disperse their interim product of billions of intertwined MWCNTs into individual nanotubes throughout a polymeric resin. In order to have their composite fuel line specified in the development stages of an automotive model, Hyperion also needed to scale up their process to make tonnes of the product. Hyperion filed several patents over 3 years of development, and achieved their first product sales in 1992.

After this first successful product development, Hyperion moved to larger facilities to have room for commercial-scale production equipment and further growth. Hyperion then concentrated on developing prototypes and specifying their product for other automotive applications. In the mid-1990s, Hyperion partnered with GE Plastics to develop further automotive product applications. First they developed conductive polymer composite automotive mirror casings for Ford and other automotive OEMs which could be electrostatically painted (along with the rest of the metallic portions of the car). Next they jointly developed conductive polymer composite fenders, which met or surpassed metallic alternatives, giving advantages of weight-savings and styling options. Most of their materials sales for polymer composite fenders have been for European car models, as weight savings have been more highly valued in the European market.

During this time, Hyperion also continued to scale up their process and developed a high-tonnage nanotube reactor. In 1998, an MIT graduate with technology product development experience was hired as Director of Business Development. He went on to have a major influence on Hyperion's product expansion and commercialisation strategy. Some of his initiatives included expanding their sales presence globally and moving slightly further down the value chain, by compounding resins in-house in order to have control over the dispersion of their MWCNT product. Hyperion's growth was rapid but could have been even more so with additional external financing. And, although their product development efforts were largely successful, they did not meet with universal success. For example, Hyperion's R&D team had been working on developing their product for structural composite aerospace parts. This involved dispersing their nanotube product into the thermoset resins most suitable for aerospace structural parts. Their efforts at demonstrating enhanced value in these applications have been largely unsuccessful to date.

Hyperion's first successful product development outside of the automotive market was in consumer electronics. In this instance, Hyperion was approached by a consumer electronics OEM who valued their material's attributes. Hyperion found consumer electronics OEMs to be far more open to collaboration on product development than automotive OEMs. Hyperion was able to create strategic alliances with consumer electronic OEMs and co-developed several components which took advantage of their static dissipation properties and the integrity and cleanliness of their composite materials. These products, including internal disc drive components, handling trays and devices for manufacturing disk drive components and test sockets for integrated circuits, have become a major product revenue stream for Hyperion.

In the late 1990s and into the early 2000s, Hyperion's R&D team also developed products which used their material in advanced batteries for the power generation industry. Hyperion received competitive Small Business Innovation Research (SBIR) grants from the US Department of Defense (DoD) from 1996 to 1999 to develop MWCNT electrodes for electrochemical capacitors, and issued several patents from this work. Concurrently, they were developing composites with non-polymeric matrix materials. From 2000 to 2004, Hyperion developed their MWCNT product as a catalyst support, which has power generation and emerging alternative automotive applications. They also filed a patent on the use of their product for the emerging application of field emission displays. Hyperion has found IP protection to be critical to their ability to capture value, both in negotiating with large strategic alliance partners and in discouraging new entrants. Hence, they have filed over 100 patents and actively expand and extend their patent portfolio.

Hyperion's product line consists predominantly of composites of their MWCNT product, dispersed into thermoplastic resins. They are continuing to grow their products and revenues into the automotive, electronics, power generation and communication segments, and are looking to expand their sales into other market verticals, as well as "staking out" IP in emerging markets. They are the oldest and, arguably, the most successful dedicated nano-materials venture in the world to date, achieving between $20 and $50 million in annual revenues[3]; yet, to achieve that success, Hyperion needed patient capital, alliance partners and an early focus on substitution rather than emerging markets.

6.2.2 NanoGram/NeoPhotonics Corp.[4]

NanoGram was founded in 1996 by Dr Nobuyuki Kambe and Dr Xiangxin Bi to commercialise their novel nanoparticle manu-facturing process which produced small (5–200 nm), uniform, high-purity nanoparticles for a broad range of materials. Bi had developed the process as a post-doc student at MIT. Kambe had been in the same lab years earlier, having completed his PhD in electrical engineering at MIT, and they met through that con-nection. Kambe was also a successful businessman with a strong network in his native Japan and managerial experience at Nippon Telephone and Telegraph and at the International Center for Materials Research in Japan. Together, Bi and Kambe further developed the process at the University of Kentucky, where Bi had done his PhD, aided by funding from the International Center for Materials Research. NanoGram relocated from Kentucky to Silicon Valley, California, in late 1996, in part because of better access to outside financing.

NanoGram had developed a powerful, flexible nanoparticle manufacturing process (later trademarked as NPM^{TM}) and aimed to produce nanostructured materials for a broad range of emerging markets, including fuel cells, solar cells, solid-state

[3]Revenue estimate obtained from Reference USA.
[4]This case study was based on four primary source interviews conducted on 9 August 2005, 19 August 2005, 26 February 2006 and 13 April 2006, from secondary source information from company websites http://www.NanoGram.com and http://www.neophotonics.com and from other sources cited within the case study.

lighting, portable electronic devices, high-resolution imaging, medical devices, and optical devices for communications networks. Part of the reason that NanoGram's process was so flexible was because it did not require high vacuum, making it far easier to change compositions and thus to produce a wide range of materials. They began developing their process for this broad range of potential applications, helped by $1 million in seed financing from Institutional Venture Partners, who installed experienced technology entrepreneur Mike Hodges as interim CEO. In 1998, Tim Jenks, a Stanford MBA and MIT engineering post-graduate who had been a vice president at Raychem Corporation, joined NanoGram as CEO and began to carefully match NanoGram's enabling product attributes with specific market needs, and to develop a strategy for creating strategic alliances.

The initial rational for an in-house manufacturing revenue model was that customer adoption was the most important issue for NanoGram, and the adoption of its technology was initially limited by demand, as it had not yet been able to demonstrate its value in specific applications to potential customers. NanoGram felt the need to manufacture product to demonstrate its potential to customers. However, the decline of the technology sector in 2000 and 2001, and the subsequent uncertainty after 11 September 2001, led to a massive decline in the availability of investment capital. Most ventures, including NanoGram, re-examined their revenue models and looked for ways to lower operating costs and capital intensity. Jenks recalls, "We knew it would be impossible to fund all of our applications—a narrow focus was essential, yet we knew we had a very powerful platform." Thus, NanoGram was forced to make tough decisions between maintaining sufficient resources and exploiting the breadth of its technology.

NanoGram first responded by narrowing its focus to its technology competencies with films and the application development of novel circuits for optical networks in the telecommunications industry. Through precise control of nanoparticle size and subsequent optimisation of refractive index, NanoGram's technology could enable lower cost, high-efficiency circuits. Jenks liked this application of NanoGram's technology because it was not the focus of large incumbent players, as the portable electronic device component markets they had targeted earlier had been. To reflect this altered focus, NanoGram changed its name to NeoPhotonics Corporation.

Shortly after this name change, more favourable market conditions and a "desire to diversify their risk" led them to spin out another applications company, NanoGram Devices Corporation, in November 2002, for the development and commercialisation of medical battery devices. For the medical device market, NanoGram's NPM^{TM} materials production process and its laser reactive deposition LRD^{TM} process enabled batteries with much faster recharging time than any other battery available. This process also enabled a wide range of batteries with desirable characteristics including long runtimes and high power output per unit weight. At this stage, there were two application companies, NeoPhotonics and NanoGram Devices Corporation (NDC), both demonstrating value through improved performance in existing applications in established markets.

As these new applications companies had the potential for conflict over common IP, an IP company was also formed with a licensing revenue model which would focus on technology incubation and fuller exploitation of the broad platform of technologies which had applications across multiple industries. Somewhat confusingly, this new company was named NanoGram Corporation, the original name of NeoPhotonics. Jenks describes the revenue model change as follows:

> Then, two market-focused companies were sharing common intellectual property—IP that defined a powerful multi-facet platform. The licensing model derived from these two decisions—we formed NanoGram as a platform company with initial application licenses to the two market-focused companies, thereby enabling the two market focused companies to succeed. NanoGram has been highly successful since, continuing forward as a platform IP and discovery company. (Tim Jenks, 26 February 2006)

Thus, although NanoGram was founded in 1996, the current form of the family of related ventures (see Fig. 6.2) was established in 2002. NanoGram raised three rounds of financing into the IP company, $7 million in 2004, $18.7 million in 2006, and $32 million at the beginning of 2008, the latter earmarked for their wholly owned solar-focused business.

The two application companies, NeoPhotonics and NDC, were successful in raising funding independently and in developing alliances specific to their industry, although there were further bumps along the way. NeoPhotonics (the primary venture), with Tim Jenks as president and CEO, raised $25 million in 2002,

NanoGram Corporation (2003 IP venture)

Commercialising Generic Nanomaterials Technology
through R&D, Alliance Generation and Licensing

NeoPhotonics Corp. (1996 original venture)	**NanoGram Devices Corp. (2003)**	**Kainos Energy Corporation (2003)/ NanoGram Solar (2008)**
This is the original NanoGram Corp, renamed NeoPhotonics in 2002, reflecting its altered focus on enabling lower cost and higher efficiency circuits for optical telecom networks	Applied NanoGram's technology to enable smaller and longer life Medical Devices through enhanced batteries	Applications venture which remained wholly owned subsidiary of NanoGram Corp
		Kainos applied NanoGram's technology to alternative energy products (solid oxide fuel cell cost reduction and later solar cell cost reduction)
Raised $115 million in VC financing	Raised $10 million in VC financing as separate entity	Raised $32 million of financing through NanoGram
Growth and acquisition of complementary assets through mergers & acquisitions	Acquired for $45 million in 2004 by partner/customer Wilson GreatBatch Technologies	NanoGram unsuccessfully attempted to sell equity in Kainos to raise the further $75 million to develop a plant.
IPO in 2011		Acquired in 2010 by alliance partner Teijin Ltd for its potential to enable solar cell and consumer electronics applications

Figure 6.2 NanoGram/NeoPhotonics IP and application ventures.

acquired a photonics integrated circuit manufacturer in 2003, reorganised under Chapter 11 bankruptcy, emerged and raised a further $40 million in 2004. After an alliance with a Chinese photo diode manufacturer, Photon Technology Co., to integrate Neo-Photonic's novel optical glass devices with Photon Technology's products, the two firms merged in 2005. This merger resulted in a sharp upturn in customers, revenue and employees.[5] The combined entity, still called NeoPhotonics and still led by CEO Tim Jenks, continued to grow rapidly, their growth partially fuelled by several strategic acquisitions of optical device manufacturers, and was responsible the majority of revenue generation from the family of ventures depicted in Fig. 6.2. NeoPhotonics collected growth distinctions during this time, including being listed in Deloitte's "Silicon Valley Fast 50", the "Inc. 500" Fastest Growing Private Companies, the Red Herring "North America 100" of Most Promising Tech Companies, and Deloitte's 500 Asia Pacific 2009.

---

[5]At the time of the merger, NeoPhotonics was estimated to have 100 employees and approximately $10 million in revenues, whereas Photon Technology Co. had 1000 employees and $40 million in revenues [12].

NeoPhotonics raised $75 million in a third round of VC investment in 2006, and a further $30 million in VC investment in 2008. Their nanomaterials technology enabled higher-speed, higher bandwidth data transmission, resulting in product sales to all of the largest optical network hardware vendors globally [24].

NanoGram Devices Corporation (NDC), the other applications company to exploit the nanomaterials technology developed by NanoGram, spun out in January 2003 and quickly raised $10 million in venture financing. NanoGram's process was especially flexible for metal matrix oxides, and ideal for enabling novel cathode materials. Thus, looking for high margin substitution applications, new CEO Barry Cheskin approached the dominant incumbent in the medical power market, Greatbatch Technologies, in 2003, proposing that NDC supply novel cathode material to WGT. Successful development and commercialisation of NDC's high performance nanoparticle silver vanadium oxide battery technology, in the form of implantable batteries for medical devices, led to NDC's acquisition by Greatbatch for $45 million in March of 2004. Greatbatch used NDC's process to develop new higher energy density batteries enabling smaller and longer life implantable medical devices utilising NDC's technology [11].

In August 2003, NanoGram created a wholly owned subsidiary company, Kainos Energy Corp., primarily for the development of low-cost membranes for solid oxide fuel cells. NanoGram's nanomaterials processes also enabled a wide range of batteries with exceptionally fast recharging rates. Kainos exploited this broad technology platform in developing fuel cell, solar cell, and hybrid automotive battery prototypes. Kainos obtained a Phase I SBIR government grant for further R&D into fuel cells, but, after deciding that the economics for fuel cells were not attractive yet, shifted its focus exclusively to the solar cells and was merged back into NanoGram. The IP part of the company, NanoGram, was focused on licensing the technology more broadly, scaling up its processes for higher volume production, and developing technology for emerging markets. After NeoPhotonics' merger with Photon Technology Co. was completed in 2005, Jenks realised that he could no longer run both NeoPhotonics and NanoGram. Jenks searched for a replacement CEO who could improve NanoGram's alliance creation and licensing capabilities by understanding customers' mindsets. He recruited Kiernan Drain, who brought relevant experience from his role as vice president of Avery Dennison Corporation's worldwide performance polymers division.

Drain put more resources into NanoGram's alliance creation strategy, AccessNano™ and networked actively himself. He also created a strategic marketing team and made NanoGram much more proactive at identifying good opportunities for its technology, desirable alliance partners and promising value networks. As part of reaching out to potential alliance partners and customers in Asia, NanoGram opened a Japanese national subsidiary (Nano-Gram KK) and a Korean national subsidiary (NanoGram Korea) in 2006 with experienced and well-connected presidents hired to develop regional business opportunities. Drain also focused on maximising profits through NanoGram's licensing model by helping its customers scale up volume.

NanoGram was an upstream R&D and licensing company broadly exploiting their engineered nanoparticle platform technology, but simultaneously attempted to develop and manufacture products in alternative energy and consumer electronics markets through their wholly owned subsidiary Kainos Energy Corp (later renamed NanoGram Solar). Focusing on a novel solar cell fabrication approach as NanoGram Solar, the company raised $32 million in venture capital financing in January 2008 and used that money to build a pilot solar cell manufacturing facility, having developed yet another process innovation, SilFoil™, to reduce the cost of solar cell production. NanoGram Solar actively pursued applications and alliance partnership in the cleantech energy sector. A key alliance, with materials supplier Nagase & Co. of Japan, was initiated in late 2006. By the end of 2007, NanoGram had installed its NPM™ system at Nagase ChemteX Center in Tatsuno, Japan, and they jointly announced higher volume nanoparticles and nanomaterials production ability to serve global customers for optical, electrical and energy applications. Nagase became an equity partner in NanoGram through investment in the third round of equity financing in 2008. In late 2008 and most of 2009, NanoGram attempted to raise a further $75 million to scale up to a high-volume manufacturing plant for solar cells [13]. NanoGram signed a technology development deal with Tokyo Electron Ltd., a corporate investor in the 2008 financing round, but were unable to raise further financing or proceed with an IPO in the difficult market conditions.

R&D on consumer electronics applications had also been ongoing at NanoGram and its alliance partners. The alliance formed in 2006 with Nagase in solar cells also extended into the consumer electronics field, as Nagase also manufactured LCD screens for consumer electronics products. NanoGram's process technology offered promise of lower cost transistors and display backplanes. In early 2009, NanoGram formed an R&D alliance with Teijin Ltd., a Japanese materials multinational, to adapt Nano-Gram's processes to economically produce and disperse crystalline silicon nanoparticles (referred to as "silicon ink") in high volumes to be "printed" into flexible electronics applications. Nanogram and Teijin were able to demonstrate that NanoGram's silicon ink materials could be used in the conventional manufacturing facilities existing printing equipment of electronic device manufacturers.

NanoGram was acquired by Teijin in 2010, with the intention of further developing NanoGram's NPM™ process to make low-cost, lightweight, flexible electronics such as thin-film transistors for liquid crystalline displays. Teijin also aims to supply the silicon ink material and sintered silicon nanoparticle films to customers manufacturing flexible displays, semiconductors, and solar cells.

Thus, of the four entities formed to commercialise Nano-Gram's engineered nanoparticle manufacturing process, only NeoPhotonics (the original venture and the most successful in terms of value creation) remains as an independent company, and they succeeded largely by partnering and then acquiring another photonics component manufacturer to enable enhanced telecom components. NeoPhotonics completed a successful IPO in February, 2011, raising $82.5 million [7]. NeoPhotonics continues to grow revenues and profits in the growing data transmission markets, with key customers such as Alcatel-Lucent, Cisco Systems and Nokia Siemens Networks.

6.2.3 Degussa Advanced Nanomaterials (AdNano)[6]

Degussa AG, now Evonik Degussa, conducted in-house nano-materials R&D and had a long tradition of technology and produc-tion competencies in nanoparticles such as catalysts, pigments,

---

[6]With permission from Wiley, this case study is reproduced from ref. [21].

and fumed oxides. However, R&D performed within their regular business units was generally near term, and R&D performed within their corporate R&D unit had been traditionally constrained by their existing core competencies. They pursued two avenues to increase their exposure to the growth potential of more radical nanomaterials advances. First, they formed an internal nano-materials venture within their corporate R&D and innovation unit. Second, they participated in external monitoring of nanomaterials developments and importation of nanomaterials ideas through their Business Ventures group.

The internally focused project houses were created from platform technology opportunities categorised as "medium risk" and complementary to Degussa's capabilities. These project houses were staffed by employees from multiple Degussa business units and from within Creavis, the strategic R&D and new product development (NPD) subsidiary of Degussa. Each project house was given a window of 3 years in which to "start a new business start-up and/or enhance the existing capabilities of its business units, whether it is through new products or through access to new markets" [8]. AdNano was Degussa's and Creavis' first project house, and its success led to the creation of five additional project houses, each staffed by 20–30 scientists [9].

Project house projects which were still too risky or too long term for a business unit after the expiration of their 3-year project house time period could become internal ventures. Internal ventures "incubate the business from inception as a start-up to sales and profitability of a mid-sized business, after which the business is typically transferred to a Degussa business unit" [8]. After breaking a new path as Degussa's first project house, AdNano also became Degussa's first internal venture and serves as a case study exemplar of an incumbent's internal nano-materials venture. AdNano won an external award for leadership in nanomaterials commercialisation and successfully developed several new products and processes which were incorporated into Degussa's Aerosils & Silanes business unit in 2007.

Degussa's internal nanomaterials venture, AdNano, was conceived in 1998 as an idea from a group of five persons working on gas-phase processing of ultra fine particles within Degussa's Central Process Technology Group. The Central Process Technology Group was a corporate services unit which served all of Degussa. In this instance, the team, led by Dr Andreas Gutsch, was collaborating with two of Degussa's business units: Advanced Fillers & Pigments and Aerosil & Silanes.

Gutsch's research group began looking for funding to work on their ideas, which were too high risk for the business units to be interested in funding directly. However, Germany has a government funding program called Deutsche Forschungsgemeinschaft, similar to the United States' National Science Foundation, which was funding innovative nanomaterials R&D. Gutsch, along with Geoffrey Varga (at that time an internal consultant from Degussa's Electronic Materials subsidiary in New Jersey) proposed that Degussa create a "project house" [1] for risky but high-potential nanomaterials R&D within Degussa, which would be subsidised by the German government. They applied for and were successful in obtaining government funding, conditional on matching funding from a corporate sponsor. Degussa corporate agreed to match most of the funding, with a smaller portion contributed through additional funding from both relevant business units. The funding coincided with the inception of Creavis, and the team implemented the formation of Degussa's first project house.

In 2000, Degussa Project House Nanomaterials (PH Nanomaterials), wholly owned by Degussa AG, began with €13 million in total funding. This consisted of €6 million from Deutsche Forschungsgemeinschaft, €4 million from Degussa corporate, €1 million each from the two relevant Degussa business units, and €1 million from the German Ministry for Education and Research. Degussa and Creavis management decided from the start to limit the lifetime of Project Houses to 3 years. Given their overall funding, Gutsch, as the first leader of PH Nanomaterials, compiled a team of 17 technician and professional staff from various parts of Degussa, which included the original Central Process Technology Group team, Varga, who relocated to Germany full time in 1999 to help initiate PH Nanomaterials, and staff from Degussa's Advanced Fillers & Pigments and Aerosil & Silanes business units. Gutsch, who had previous experience in managing a strategic development group at Degussa, designed PH Nanomaterials to consist of an interdisciplinary team of researchers with marketing input right from the beginning. He intentionally brought together employees with experience in materials science, chemistry, physics, industrial design, production and marketing and had them work in close proximity within one building. Beyond the physical proximity of diverse R&D and functional specialists, Gutsch also established routines and incentives to encourage free communication.

PH Nanomaterials was considered an experiment of how to innovate in a large firm. So, although it was only a 17-person project within a firm of over 50,000 employees, it reported to a steering committee which closely followed its progress. For the first 12–18 months, they were in experimental mode and did not have a lot of market focus. Between 12–18 months they became more focused on market applications. PH Nanomaterials enjoyed some early technical successes in areas which were also seen to be highly relevant to Degussa's customers. During its 3-year life-span, PH Nanomaterials developed promising new nano-structured materials offering distinctive functionalities.

Around the 18-month mark, Gutsch and Varga realised that PH Nanomaterials had internal support and some momentum but would soon be facing the notorious "Valley of Death." Varga had experience with technology start-up firms in the United States and knew some of the challenges they were likely to face. He also knew that their product development team would face potentially insurmountable management and incentive problems if they moved directly into an established Degussa business unit before they incubated the technology independently. Varga was further convinced that an alternative organisational structure was required through reading *Corporate Venturing* [2], which warned of the dangers of subjecting a new venture team to the management routines and incentive systems of a large firm too soon. Gutsch and Varga pitched their ideas to the steering committee of Creavis, and, after proposing that their alternative was to spin out the venture externally, won management agreement to start the internal corporate venture, AdNano. In early 2002, after securing agreement to begin Degussa's first internal corporate venture in a year's time, Gutsch was promoted to the head of Creavis, and Varga was promoted to the head of PH Nanomaterials and the director of AdNano.

Gutsch and Varga had secured an agreement that AdNano would have 4 years of "political protection" as an internal venture during which time it would incubate the new technology and product processes and start the integration process into one or more of Degussa's existing business units. Degussa Corporate, Creavis and the Aerosil & Silanes business unit agreed to contribute €25 million to the internal venture. After 4 years' time, if successful, they would become completely integrated within Degussa's Aerosil & Silanes business unit (Fig. 6.3). Although Varga had proposed the idea of spinning the venture out externally in order to convince Degussa management to agree to the internal venture, he was always convinced that the resources and capabilities of Degussa's Aerosil & Silanes business unit would be an immense help to AdNano in their development. This proved to be the case, with AdNano drawing extensively on the marketing and production expertise of the Aerosil & Silanes business unit.

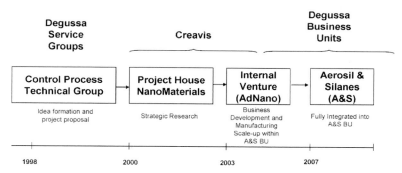

Figure 6.3   Evolution of internal venture Degussa Advanced Nano-materials (AdNano). Reprinted with permission from ref. [21]. Copyright 2008 Wiley.

Degussa's first internal corporate venture, AdNano, was formed in 2003. As with PH Nanomaterials, it remained wholly owned by Degussa AG. Most of the personnel from PH Nanomaterials continued with AdNano and 50% more staff were brought into the venture. All employees continued to be on Degussa's payroll but were given the option to exchange their standard Degussa bonus incentives for AdNano sales and earning incentives. Thus, AdNano employees had financial and career incentives to help make AdNano a success, without the personal risks associated with an independent venture.

AdNano's stated goal was "to produce innovative nanomaterials and tap into new business segments in attractive markets" [10].

Their steering committee put in place both financial and non-financial milestones to determine success or failure. The financial milestones were in the areas of target sales, profit, and capital expenditures. The non-financial milestones were in the areas of resource building, internal execution, customer acceptance of technical performance, and external perception. The steering committee made it clear that AdNano would be abandoned if the significant milestones were not achieved.

AdNano met this goal by developing nanostructured materials and dispersion systems, including indium tin oxide, zinc oxide, ceria and various composites. AdNano's biggest successes were in the area of fumed oxides, making them a much more natural fit with the Aerosil & Silanes business unit than the Advanced Fillers & Pigments business unit. AdNano's steering committee determined that AdNano had met enough milestones by the time that their 4-year window of incubation expired in 2007 to be fully integrated into Degussa's Aerosil & Silanes business unit (Fig. 6.3). AdNano and now the Aerosil & Silanes business unit have been customising these products for key customers in the automotive, consumer electronics, chemical, energy, and cosmetics industries. AdNano focused on product and process development work which can be considered variations of standard Degussa technology. AdNano developed a diversified portfolio of product applications, about 80% of which had some existing market pull. The other 20% of their efforts were focused on creating new markets.

AdNano encountered political resistance when they began developing products with the potential to cannibalise portions of existing product lines within the Aerosil & Silanes business unit. However, with the support of their high-level steering committee, they were able to overcome this natural organisational resistance. AdNano's Director, Geoffrey Varga, explained:

> Two of our leading 4 materials in fact have the potential to partially substitute existing offerings in the market from the Aerosil & Silanes business unit. These two materials both have multiple applications, and several of those have some overlap with established products. Of course business unit product line management initially had some concerns about this, but with support from upper management we were able to move forward. Eventually it became clear to everyone involved that developing

replacements or alternatives for your existing offerings in the market for both present and potential future applications is preferable to waiting for your competitors to do it first. These are the kinds of issues where the support of very high-level management can help to keep things on track and guide activities in the proper direction, providing much needed political protection at critical points in time. (Geoffrey Varga, February 26, 2006)

Thus, AdNano was able to avoid one of the common constraints encountered by internal corporate ventures.

AdNano heavily leveraged Degussa's expertise and established production and marketing and sales competencies. AdNano had their own pilot production and used it to "sample potential customers with materials, conduct Beta-testing, as well as providing multi-ton quantities for proof-of-concept in the market while the production facilities are being designed & built".[7] Given the economies of scale required to profitably manufacture and sell these products, AdNano did not generate substantial product revenues before integrating into the Aerosil & Silanes business unit of Degussa. Varga was wary of attempting to rapidly integrate into the Aerosil & Silanes business unit at the end of AdNano's incubation time and decided instead to smooth this process over the year prior to full integration, gradually transferring people and ideas into the Aerosil & Silanes business unit. AdNano also worked closely with Aerosil & Silanes design and production engineering group to develop their own higher volume manufacturing facilities.

6.3   Analysis of Case Study Commercialisation Challenges

All three case studies demonstrate the challenges of nanotech innovation: radical generic technology, along with upstream, capital-intensive process innovation. These factors lead to sustained high levels of technology and market uncertainty and often make it difficult for nanotech ventures to raise the required financing. The challenges faced by Hyperion Catalysis, NanoGram/ NeoPhotonics and AdNano are discussed here and summarised in Table 6.1.

---

[7]Primary source interview with Geoffrey Varga, February 24, 2006.

Table 6.1 Commercialisation challenges and strategies of Hyperion, NanoGram/NeoPhotonics, and AdNano

Metric	Hyperion catalysis	NanoGram	AdNano
Founding Year	1982	1996	2000 (for Project House)
Technology	Fullerenes/Fullerenes Dispersed in Resin	Nanoparticle manufacturing process NPM™	Nanoparticle processing
Ownership	Privately owned venture without VC financing	Privately owned venture with VC financing	Fully owned internal venture of Degussa AG
Business Models	In-house Manufacturing	Licensing, In-house Manufacturing	In-house Manufacturing
Target Markets	Automotive, Consumer Electronics, Power Generation, Aerospace, Transportation	Medical Devices, Optical Devices for Communications Networks, Fuel Cells, Solar cells, Solid State Lighting, Portable Electronic Devices, High Resolution Imaging	Automotive, Consumer Electronics, Chemical, Energy, Cosmetics
Technological Uncertainty at Founding	High (radical technology; established substitute products; need for process innovations; multiple markets)	High (radical technology; established substitute products; need for process innovations; multiple markets)	High (radical technology; established substitute products; need for process innovations; multiple markets)
Market Uncertainty	High (upstream position in value chain; need for complementary	High (upstream position in value chain; need for complementary	High (upstream position in value chain; need for

at Founding	innovations; lack of continuity, observability, trialability; multiple markets)	innovations; lack of continuity, observability, trialability; multiple markets)	complementary innovations; lack of continuity, observability, trialability; multiple markets)
Number of Patents	>100	>100	>10
Access to Complementary Assets	Through alliance partners	Through alliance partners, market vertical subsidiaries, and merged companies	Through parent company
Availability of Finance	Patient angel investor, SBIR, retained earnings, VC financing purposefully avoided	SBIR, angel investment, $230 million raised over 11 years from VCs	€ 38 million over 7 years from German Government, Degussa Corporate, business units
Time to First Product Commercialisation	10 year (1992 fuel lines)	10 years (telecommunications components)	7 years (substitution into coatings)
National/Regional Support for Experimentation	SBIR grants, Knowledge workers from Boston region. Key employee(s) from MIT. Scientific advisory board from MIT and Harvard	SBIR grants, Knowledge workers and VCs from Silicon Valley	Grants from government, incubated within Degussa

Hyperion's technology was radical in that it was protected by over 100 patents, enabled substantial cost reduction and weight savings in automotive components, new combinations of performance attributes for consumer electronics components, and could potentially enable entirely new consumer electronics and alternative energy products. Hyperion's technology was generic in that they actively developed it across five target markets. Their process technology was upstream in all of their targeted industry value chains, both in the manufacturing of carbon nanotubes and in the dispersion of these nanotubes into polymer resins. They enabled enhanced components but did not manufacture them. Their nanomaterials technology was capital intensive, although the exact amounts are unknown: Hyperion funded development through patient angel capital, SBIR grants, partnership with suppliers and customers and retained earnings.

NanoGram's technology was radical in that it was protected by over 100 patents, dramatically increased the performance of cathode materials, and substantially lowered costs and increased performance of optical network components. It was generic in that it was prototyped across 7 target markets, and actively developed for 4 target markets. Their key IP was process-based, rather than product-based, and this key IP was upstream when creating nanoparticles and dispersing them into a nanomaterial and midstream when the nanoparticles were printed directly onto a component. The development and scale-up of these novel processes were very capital intensive: an example is the proposed scale-up of NanoGram Solar's process into a solar cell manufacturing facility which was estimated to cost $75 million [21].

AdNano's technology was radical and generic in that it was protected by over 10 patents, substantially improved performance attributes in existing Degussa product applications across the automotive, consumer electronics, chemical, energy, and cosmetics markets, and had the potential to enable new applications. It was upstream in that AdNano manufactured the nanoparticles and the nanostructured composite materials, but not any components or whole products. It was capital intensive in that, up until integration into Degussa's Aerosil and Silanes business unit (i.e. before production scale-up), AdNano and its predecessor project house had already spent 38 million Euros over 7 years of research and process and product development.

6.4 Approaches to Nanotech Commercialisation Critical Success Factors

To overcome these commercialisation challenges, these largely successful case studies demonstrate many of the success factors identified in the previous chapter as being critical for nanotech commercialisation. How they do so is described in this section.

6.4.1 Product Orientation (and Not Technology Admiration)

Commercialising radical, generic technology is not a linear or rapid process; however, all three case study ventures developed a product orientation within 5 years of founding and prioritised substitution applications over emerging applications: there was enough technology and market uncertainty in developing nanomaterials for substitution applications without adding the additional uncertainties associated with emerging applications. And all three ventures were able to clearly articulate the performance enhancements and attribute combinations enabled by their technologies. IP was critical to both independent ventures and important for AdNano. It was their unique IP applied to enable enhanced or novel performance attributes in specific existing product applications which drew strategic alliance partners to both independent ventures. A European-owned, multinational resin supplier with existing automotive development partners approached Hyperion with a product in mind–composite automotive fuel lines–which Hyperion's carbon nanotube technology could enable. For Nano-Gram, their medical devices venture, NDC, was partnered with (and eventually acquired by) the leading multinational medical device battery OEM once NDC demonstrated superior product attributes in prototype nanotech enabled batteries. As argued in the previous chapter, the focus needed to be on a "nanotech enabled whole product", in this case, rapid recharge rates for small medical devices, before the technology was deemed valuable.

Yet this product orientation had to wait until the platform nanomaterials technology was sufficiently developed from the lab scale to a consistent, well characterised material with the potential to be manufactured at commercial scales. Hyperion had the longest period of platform technology development after the venture was founded—5 years—and communicated its unique technology attributes through white papers and conferences, before being approached by the resin supplier with the composite automotive fuel line product development proposal. AdNano's predecessors also took 5 years of platform technology development time, but AdNano was product focused from founding as an internal corporate venture. NanoGram became product focused with the arrival of Tim Jenks, 2 years after founding, but it was 6 years after the founding of NanoGram before the renamed company, NeoPhotonics, focused exclusively on enabling telecom components. As NeoPhotonics raised VC financing, and it looked as though finance could be raised through other spinoffs, the family of NanoGram ventures again began to exploit the platform technology. Subsequently, each application venture focused exclusively on product development in a single target market: NeoPhotonics in telecom, NDC in medical devices, and Kainos/NanoGram Solar in alternative energy generation.

6.4.2 Continuous Market Interaction and Selection of a Beachhead Application

All three ventures faced challenges with prioritising target markets for their generic nanomaterials technology. Significant R&D needs to be undertaken before the viability of an application is clear, multiple applications may look equally promising, and the presence of a strong alliance partner often shifts the balance in market and application selection [19,20]. Because of nanotech ventures' need for complementary assets through alliance partnerships, the beachhead market is generally a substitution application, rather than an emerging market. Each of these nanomaterials ventures first commercialised their technology by displacing another material in an existing component for an established application, even though all had the potential to create entirely new applications for emerging markets.

The applicability of Hyperion's technology to multiple applications across several industries added to the potential for value creation but made demonstration of value in any one specific application more challenging. Hyperion divided their R&D and business development focus between applications in the automotive industry, the aerospace industry, the consumer electronics industry, and the power generation industry. Exploration of each of these industries required the development of relationships with different customers and made it necessary to engage in unique process R&D and additional complementary innovations. The need for industry-specific regulatory changes and education of designers also contributed to the market uncertainty involved with a focus on multiple markets. Hyperion's beachhead application emerged through an established resin manufacturer, which needed the performance attributes of Hyperion's MWCNTs in order to meet safety regulations by making composite automotive fuel lines conductive. The application development involved innovation at several levels along the value chain. Hyperion's alliance partner needed to match a suitable resin to Hyperion's nanotubes to enable good composite properties, adequate dispersion, and good secondary processability. Next, Hyperion's tier one automotive customer needed to develop design and process changes to take advantage of composite material strengths. In the automotive fuel line, this involved altering the powertrain design, with new fasteners and assembly methods, and had the benefit of elimi-nating the multiple forming steps required to make steel fuel line. These additional steps led to the relatively long prototype development time indicated in Hyperion's commercialisation timeline (Fig. 6.1).

NanoGram also took advantage of the generic and radical nature of its nanomaterials process to establish 4 target markets. These emerged over time through an iterative market matching process. NanoGram's CEO used the analogy that NanoGram had:

> discovered a cure and [were] looking for a disease. The first "disease" we identified and targeted was batteries for cell phones and laptops. However, the disease and the cure were not an ideal match. In this case, the need associated with the "disease" was energy time (overall battery time), whereas NanoGram's "cure" was rate capability (how quickly a battery can discharge or re-charge), so there was not a perfect fit.... Additionally, Nano-Gram was flexible enough to realize when the right "disease" came along that we had the "cure". [In their second application], NanoGram saw that within the medical devices space, that discharge rate was very important for implantable devices. From these markets, NanoGram was now able to "springboard" to other

battery applications. For example, the batteries powering hybrid electric vehicles are rate dependent, and thus a good target market for NanoGram. (Kieren Drain, 13 April 2006)

NanoGram's leadership used this matching of its unique technology attributes with market needs for an initial selection of markets and applications, and then considered incumbent competitors, potential customers, and potential alliance partners within each value network. Unsuccessful experiments, such as that of fuel cells, were abandoned when it became evident that other target markets were more promising.

AdNano's market selection was more constrained than that ofthe two independent ventures, in that its clear goal was to integrate into an existing Degussa business unit. That still left AdNano targeting applications in the automotive, consumer electronics, chemical, energy and cosmetics markets but allowed them to leverage all of the existing application-specific knowledge and established customer relationships of their affiliated business unit. Once generic technical goals were achieved, AdNano developed and beta-tested their nanoparticle products with existing customers of Degussa's Aerosil & Silanes business unit.

6.4.3 Application of Spiral Product Development Methodology

An early and frequent interaction between marketing and R&D has long been advocated for effective product development [6,14]. However, new product development must be tailored to the level of technology and market uncertainty [16]. An appropriate new product development strategy is also contingent on the breadth and depth of a technology, as well as the position of the commercialising venture in industry value chains [20,22].

Spiral product development involves repeatedly testing ideas, prototypes, and beta products with potential customers in between each stage in a StageGate® new product development process. It also allows for shifting the objectives and constraints of a stage as uncertainties are resolved. Spiral product development is particularly useful where product development cycles are short, and early prototypes can be relatively easily trialled on potential customers, such as in the software and information technology sectors. Spiral product development is far more difficult for a radical generic technology commercialised from an upstream position in several industry value chains. To achieve its spiral product development, while keeping the focus on nanotech enabled whole products, computer modelling and very rough prototypes must be employed until a customer or alliance partner has committed to buying or funding the product. These types of testing of the market are what Chesbrough refers to as low fidelity experiments [5].

For technology ventures in environments of high uncertainty, Chesbrough (ref. [5], p. 362) recommends "high fidelity, low cost, quick performing, and usefully informative experiments". All three nanomaterials case study ventures, however, faced a trade-off between "high fidelity... and usefully informative experiments" and "low cost, quick performing" experiments. Many technology ventures can have both sets of attributes in their experiments, whereas nanotechnology ventures have to either give up fidelity in order to have a low cost in a reasonably short timeframe or spend a lot of money over a long time period to achieve high fidelity in their experimentation. The pursuit of high-fidelity experimentation in new product development can lead to costly failures, as in the case of NanoGram Solar's large investment in a scaled-up solar cell manufacturing facility.

6.4.4   Attraction and Maintenance of Commercialisation Partners

For nanomaterials ventures, commercialisation partners are essential to value creation [18,20,23]. Commercialisation partners were essential to Hyperion Catalysis' new product development. Their policy was "to partner with a larger/established player somewhere along the value chain in each industry vertical they pursue[d]". Hyperion's partnership with a resin supplier for their automotive fuel line application provided Hyperion with access to technological assets for compounding, co-extrusion and injection moulding, full scale production facilities and fuel line prototype development, and with marketing relationships with an automotive Tier 1 supplier and OEM. In the consumer electronics industry, Hyperion partnered with several consumer electronic OEMs, which helped them access more varied markets.

Likewise, alliance creation was integral to NanoGram's commercialisation strategy. NanoGram searched broadly and proactively selected promising target markets and key alliance partners and fostered deep partnerships through its AccessNano™ partnership program. Within the value network of each target market, the application spinoffs/subsidiary were able to focus specifically on product development, complementary innovation, and financing within one target market. In all four markets of telecom, medical devices, alternative energy and consumer electronics, NanoGram and its spinoffs partnered in such a way as to access required complementary assets, such as design, regulatory and application-specific production capabilities. Such alliances also give a technology venture some influence over the development of required complementary innovations.

AdNano had less need of commercialisation partners, given the complementary assets already available through Degussa's business units, and their early decision to focus 80% of their efforts on existing markets already served by Degussa. AdNano enjoyed direct access within Degussa's Aerosil & Silanes business unit to design and production capabilities, marketing and established customer relationships. Of course, AdNano still needed to beta test products on customers but could leverage Degussa's existing partners and strong customer base. Maintaining some exploration of emerging markets allowed AdNano to build the IP and market awareness for future growth opportunities. Exploitation of these emerging markets would require commercialisation partners.

6.4.5 Mitigation of Nanotechnology-Specific Technology Risks

All three ventures faced the challenges associated with commercialising a radical process innovation from an upstream position in multiple industry value chains [17,20,22]. Manufacturing scale-up of a nanomaterials process technology is inherently challenging, as the materials characteristics are co-dependent on the process conditions. Furthermore, although each nanomaterials venture had developed a platform technology, their process needed to be customised to each application and regulation differed in each industry. These three ventures demonstrate strategies to over come such nanotech-specific technology risks.

Hyperion was a pioneer in nanotechnology. It faced all of the nanotech-specific commercialisation challenges but was unusual in that it enjoyed patient angel capital and did not have to be constrained by the timelines of venture capitalists. That was fortunate, because, as shown in Fig. 6.1, Hyperion had long periods of refining and scaling up production processes—first for their FIBRIL^{TM} carbon nanotube production, then for the controlled dispersion of their nanotubes throughout a polymeric resin—along with subsequent application-specific product development and regulatory hurdles. As a pioneer, and having widely publicised their technical achievements in patents, journal papers and white papers and at conferences, Hyperion's leadership was also fortunate to be approached by alliance partners who were excited by the performance attributes enabled by Hyperion's technology, and who guided the application-specific product development of the composite automotive fuel line. The successful commercialisation of their first product, 10 years after founding, helped fund Hyperion's new prototype development R&D for additional automotive applications, as well as applications in consumer electronics, aerospace, and power generation.

NanoGram utilised organisational innovations to overcome challenges to nanomaterials commercialisation. It had the chance to be successful because, through organisational innovation and managerial strategies, it creatively reduced risk, accessed financing and complementary assets, and demonstrated value in specific applications. NanoGram's main organisational innovation was commercialising a radical generic process technology through a broad IP development company and market oriented application spin-offs, which allowed for reduction of market risk and additional opportunities for financing and alliance creation. NanoGram's value chain positioning represented a purposeful shift from the exclusively upstream position earlier nanomaterials ventures had adopted. Its CEO, Jenks, decided to forward integrate to a midstream position in each of NanoGram's target markets, but to do so in a manner which allowed it to maintain a broad but focused strategy for technology and product development, alliance creation and financing. To accomplish this, Jenks decided that NanoGram would compete both in the upstream market for technology **and** in the upstream and midstream product markets. NanoGram's innovative business model encompassed all three of the technology commercialisation business models identified by Pries and Guild [27]: creating a manufacturing or service company, selling the technology, and broad licensing.

AdNano's strategy to reduce nanotech commercialisation risks included beta testing of their products with existing Degussa customers, drawing on the production and marketing expertise of Degussa, and focusing their product development on existing markets while still exploring emerging markets. AdNano's steering committee also employed active risk management techniques, including creating financial milestones and timelines in the areas of target sales, profit, and capital expenditures, and non-financial milestones in the areas of resource building, internal execution, customer acceptance of technical performance, and external perception. Lastly, their persistence in and high-level support for that developing new products with the potential to cannibalise portions of Degussa's existing product lines mitigated a major drawback in developing radical technology within internal corporate ventures.

6.4.6 Licensing vs. Manufacturing Decision

Although a choice for science-based ventures is to pursue a less capital-intensive pure licensing revenue model [15], for nano-technology ventures this choice is problematic in that they may have to create a market for their technology. If NanoGram had pursued solely a licensing model without manufacturing subsidiaries/spinoffs, it may not have been able to sufficiently demonstrate the potential of its process technology. Additionally, the overall value created by NanoGram Corporation and its subsidiary and spinoff ventures through forward integration into component manufacturing was far higher than if NanoGram had solely produced or licensed nanoparticles. However, the large amount of capital investment required to forward integrate into manufacturing components in several industry value chains usually necessitates venture capital funding. Relying heavily on VC investment also necessitates an exit or an IPO within a 5–7-year timeframe. The other alternatives are to stay solely with a licensing model, which captures less of the value created by a nano-technology innovation, or to be fortunate enough to enjoy either patient angel capital or some other form of long-term, non-dilutive financing. Hyperion Catalysis demonstrates the feasibility of a manufacturing revenue model with an organic growth strategy: however, patient angel capital was required for Hyperion's survival and eventual success.

6.5 Conclusion

As the most developed sub-sector of nanotechnology, nano-materials commercialisation activity is providing initial evidence of the emerging commercialisation strategies that could prove successful in the wider exploitation of nanotechnology. The detailed evidence and analysis of successful nanomaterials ventures in this chapter reveal potential commercialisation strategies for other nanotech ventures. In particular, nanotech ventures can create value through radical IP, a product focus on predominantly sub-stitution applications, multiple target markets each with a narrow operational focus, forward integration, and substantive capabilities in market selection and prioritisation, strategic alliance creation, and raising financing. The critical success factors for nanotechno-logy commercialisation (proposed in Chapter 6) are also demon-strated through the strategies of these ventures.

Commercialisation strategies in the nanotech sector are limited by the radical, generic, upstream, slow, process-based and capital-intensive nature of nanotech innovation. In nanotech, and possibly for all science-based business, it is not possible to perform the "high fidelity, low cost, quick performing and usefully infor-mative experiments" recommended by Chesbrough [5] for techno-logy ventures. Nanotech ventures have to either give up fidelity in order to have a low cost in a reasonably short timeframe, or spend large amounts of money over a long time period to achieve high fidelity in their experimentation.

Acknowledgements

The author gratefully acknowledges the financial support of the Social Sciences and Humanities Research Council of Canada through an Initiatives in the New Economy grant.

References

1. Azom.com, (2005). Degussa starts new project house, online article 16 March 2005, http://www.azom.com/news.asp?newsID=2700/, last accessed on 12 February 2006.

2. Block, Z., and MacMillan, I. C. (1993). *Corporate Venturing: Creating New Business within the Firm* (Harvard Business School Press, Boston, MA, USA).

3. Braunschweig, C., (2003). Nano nonsense, *Venture Capital Journal*, 1 January 2003, http://www.vcjnews.com/.

4. Carter, N. M., Gartner, W. B., Shaver, K. G., and Gatewood, E. J. (2003). The career reasons of nascent entrepreneurs, *J. Bus. Venturing*, **18**(1), pp. 13–19.

5. Chesbrough, H. (2010). Business model innovation: Opportunities and barriers, *Long Range Plann.*, **43**(2/3), pp. 354–363.

6. Cooper, R. (1990). Stage-gate systems: A new tool for managing new products, *Business Horizons*, May–June, pp. 44–54.

7. Cowan, L. (2011). NeoPhotonics, Epocrates jump after IPOs, *Wall Street Journal*, February 2, 2011, accessed at http://online.wsj.com/article/SB10001424052748703960804576120002547766070.html/.

8. Creavis Technologies & Innovation, http://www.creavis.com/site_creavis/en/default.cfm?content=download&cat=14/, last accessed on 10 February 2006.

9. Creavis Technologies & Innovation, http://www.creavis.com/site_creavis/en/default.cfm?content=bto/mission/, last accessed on 12 February 2006.

10. Degussa AG, online article, http://www.degussa.com/degussa/en/press/news/details?NewsID=1033/, last accessed on 12 February 2006.

11. European Medical Device Technology, (2007). Improved batteries developed for implantable applications, *Euro, Med, Device Technol.*, online article March 1, 2007, http://www.emdt.co.uk/article/improved-batteries-developed-implantable-applications/, last accessed on 30 October 2010.

12. Fitzgerald, M. (2005). Anatomy of a merger. *Inc. Magazine*, Nov. 2005.

13. Flanagan, J. (2007), Nanotechnology near the point when it's time to go public, *The New York Times*, 24 December 2007.

14. Freeman, C. (1982). *The Economics of Industrial Innovation.* (Pinter, London, UK).

15. Gans, J., and Stern, S. (1993). The product market and the market for "ideas": Commercialization strategies for technology entrepreneurs. *Research Policy,* **32**(2), pp. 333–350.

16. Leonard, D. (1995). *Wellsprings of Knowledge: Building and Sustaining the Source of Innovation.* (Harvard Business School Press, Boston, MA, USA).

17. Linton, J., and Walsh, S. (2008). A theory of innovation for process-based innovations such as nanotechnology, *Technol. Forecast. Soc. Change,* **75**(5), pp. 583–594.

18. Maine, E., and Garnsey, E. (2004). Challenges facing firms commercializing nanomaterials. *Proceedings of the 9th International Conference on the Commercialization of Micro and Nano Systems.* Edmonton, Alberta, Canada.

19. Maine, E., Probert, D., and Ashby, M. (2005). Investing in new materials: A tool for technology managers, *Technovation,* **25**(1), pp. 15–23.

20. Maine, E., and Garnsey, E. (2006). Commercializing generic technology: The case of advanced materials ventures, *Research Policy,* **35**(3), pp. 375–393.

21. Maine, E. (2008). Radical innovation through internal corporate venturing: Degussa's commercialization of nanomaterials, *R&D Management,* **38**(4), pp. 359–371.

22. Maine, E., Lubik, S., and Garnsey, E. (2012). Process-based *vs.* product-based innovation: Value creation by nanotech ventures, *Technovation,* **32**(3/4), 179–192.

23. McGahn, D. (2005). Commercializing a new technology in six easy pieces: It all starts with focus. *Nanotech. Law. Bus.,* **2**(1), pp. 90–94.

24. NeoPhotonics IPO prospectus, p. 80, http://www.sec.gov/Archives/edgar/data/1227025/000119312511012096/ds1a.htm#toc59690_10/, last accessed on 13 April 2011.

25. Pandza, K., and Holt, R. (2007). Absorptive and transformative capacities in nanotechnology innovation systems. *J. Eng. Technol. Manage.,* **24**(4), pp. 347–365.

26. Pisano, G. (2010). The evolution of science-based business. *Ind. Corp. Change,* **19**(2), pp. 465–482.

27. Pries, F., and Guild, P. (2011). Commercializing inventions resulting from university research: Analyzing the impact of technology characteristics on subsequent business models. *Technovation*, **31**(4), pp. 151–160.

28. Shea, C. (2005). Future management research directions in nanotechnology: A case study. *J. Eng. Technol. Manage.*, **22**(3), pp. 185–200.

Chapter 7

Intellectual Property and Nanomaterials: Trend and Strategy

Daisuke Kanama

Department of Business and Information Systems,
Hokkaido Information University, Nishinopporo 59–2,
Ebetsu, Hokkaido 069-8585, Japan

dkanama@do-johodai.ac.jp

7.1 Introduction

Nanotechnology and nanoscience have received increased attention in Japan and other countries, following the launch of the National Nanotechnology Initiative in the United States as the first national initiative in the world. Many governments are introducing various measures to give a national priority to this area. Matters exhibit unique characteristics on a nanometre scale, different to those in the bulk state or at a molecular level. Their unique properties are expected to have scientific novelty as well as potential for many applications (see Chapters 1 and 2). In other words, nanotechnology and nanoscience are regarded as an area of science and technology full of promise that can have a significant impact on our society and economy in the future.

Nanotechnology Commercialisation
Edited by Takuya Tsuzuki
Copyright © 2013 Pan Stanford Publishing Pte. Ltd.
ISBN 978-981-4303-28-6 (Hardcover), 978-981-4303-29-3 (eBook)
www.panstanford.com

A wide range of analytical activities have been undertaken to understand the nature of nanotechnology [1–3,6–8,10]. A common recognition of nanotechnology is that it is an interdisciplinary research field. Huang *et al.* conducted a large-scale text-based analysis on the titles and abstracts of nanotechnology-related patents in the United States Patent and Trademark Office (USPTO) and identified major research topics in the field of nanotechnology and nanoscience [3]. Meyer identified four large clusters of nanotechnology invention areas through the analysis of co-occurrences of the International Patent Classification (IPC) of nanotechnology patents in the USPTO [7]. It was pointed out that nanotechnology and nanoscience are a set of inter-related and overlapping but not necessarily merging technologies. Several studies on scientific publications have also been undertaken. Schummer investigated multi- and inter-disciplinary nanotechnology and nanoscience by applying a co-author analysis to the eight existing "nano journals". The analysis showed that current nanometre scale research reveals no particular patterns or degrees of inter-disciplinarity and that the multi-disciplinarity derives from loose interaction between traditional scholarly fields [9].

A recent study of scientific publications via co-citation analysis identified approximately 30 research areas related to nanoscience and materials, and illustrated their multi-disciplinary character [4]. A mapping of science, drawn using co-citation analysis, clearly showed the emergence of a precursor to nano-bioscience and provided a picture of its evolving nature. The analysis of relative specialisation indices in various countries indicated the substantial and increasing presence of China in nanoscience and materials research [5].

7.2 Background: Patent Application Trends within Each Strategic Priority Area in Japan

In the Third Science and Technology Basic Plan, the eight strategic priority areas designated in the Second Science and Technology Basic Plan were reinforced through further selection and concentration processes. In this Third Basic Plan, the priority areas have been re-designated into "four prioritised promotion areas" and "four promotion areas" (some changes have been made to the names given to those areas covered). Information on the number of patent applications in these eight areas is available from the Japan Patent Office (JPO) and these figures show where nanotechnology is positioned among the eight priority areas. Figure 7.1 shows patent applications by technological area, based on the number of patent applications to the Japanese, US and European patent offices in 2007. It is clear that Japan is far behind the United States and Europe in terms of the relative volume of applications in the area of life science. The areas in which Japan is ahead of the United States and Europe are environmental science and social infrastructure, although the patent applications in these areas account for only a fraction of the total number of applications lodged with the JPO. Among the three patent offices, the USPTO received the largest number of patent applications in the information and communication technology area. In Europe, the relative volume of patent applications in the life science area is higher than in the United States and Japan. The area of nanotechnology and materials accounts for approximately 20% of the total patent applications at each of the three patent offices, with Japan having the highest figure.

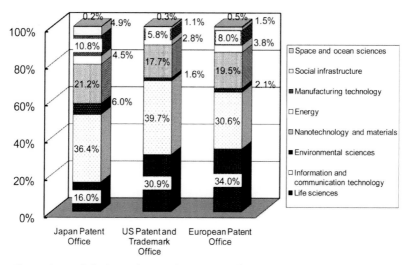

Figure 7.1 Relative volume of patent applications in the eight strategic priority areas that were lodged to patent offices in Japan, the United States and Europe in 2007.

7.3 Trend in Patent Applications in the Area of Nanotechnology

7.3.1 Classification

The Nanotechnology Researchers Network Center of Japan identifies the world's four largest patent organisations as the JPO, the USPTO, the European Patent Office (EPO) and the World Intellectual Property Organization (WIPO) [Note 1]. The centre extracts nanotechnology-related patents from monthly patent publications released by these patent organisations, using preset keywords. Extracted patents are then categorised according to nine defined technology areas. The technology areas of interest are materials, medicine and life sciences, electronic devices, information and communications, optoelectronics, measurement and testing, environment and energy, processing, printing, and photography. These nine areas cover almost all the potential fields of nanotechnology. Table 7.1 shows the technologies designated for each of the nine areas. The centre compiles a database that includes the name of the inventor, invention, applicants and other information.

In this chapter, the screening of nanotechnology-related patents (hereinafter referred to as nanotechnology patents) was carried out based on the following principles:

(a) Nanotechnology patents were divided into two main categories: the technology to alter and develop *materials* at the atomic or molecular level to add new characteristics to existing materials, and the technology to *process* materials and fabricate nanostructures. For the purposes of this article, both of these areas have been screened.

(b) Nanotechnology patents that included nano-*scale* manipulation or processing, in terms of "time", "wavelength", "mass" and "volume" was also screened. An example of this is the nanotechnology patent that proposes a method of using a picogram amount of protein to screen crystallisation conditions.

(c) Patents that selectively utilise nanotechnology-related techniques and materials as a part of the new invention were also included. For example, patents concerning conductive polyamide compounds that included electrically

Table 7.1 Classification of technology areas related to nanotechnology

Classification number	Technology area	International patent classification	Technology content
1	Materials	B01J	Catalysts/colloid science (scientific physical method)/hydrophobic magnetic particles
		B81B	Microstructure devices and systems /carbon nanotubes
		B82B	Microstructure techniques and nanotechnology/carbon nanotubes/ functional nanostructures
		C01B	Carbon structure/manufacturing of fullerenes/manufacturing of carbon nanotubes /synthetic porous crystalline substances
		C01G	Metal-bearing compounds/metal particles
		C03B	Manufacturing, moulding or supplementary processes
		C03C	Glass or glassy enamels
		C04	Artificial stone/ceramics
		C07	Organic chemistry
		C08	Organic polymer compounds/biopolymer nanoparticles/conductive polyamide compounds/toughened polymers through introduction of carbon nanotubes/photopolymers
		C09	Inks/dyes/resins/adhesive
		C22	Metals/Iron or non-ferrous alloys, and their processing

(Continued)

Table 7.1 (*Continued*)

Classification number	Technology area	International patent classification	Technology content
		C23C	Coatings/dispersion across surfaces/surface finishing through chemical transduction or substitution/diamond coating/nanoparticles coating
		C30	Crystal growth/synthesis of organic nanotubes/synthesis of ultra-thin nanowires
2	Medicine and Life Science	A61	Medical science/cosmetics containing electrochemically and biologically active particles/biodegradable nanocapsules/stents coated with nanoparticles/the use of the optical contrast factor consisting of quantum dots/optically active nanoparticles for treatment and diagnosis/cancer drugs/personalised medicines
		C12	Microbiology/enzymology/genetic engineering/determination of nucleic acid molecule sequence/measuring equipment
3	Electronic Devices	H01L	Basic electric elements/semiconductor equipment/patterning of silicon nanoparticles/membrane sensors consisting of semiconductor film containing nanocrystals/quantum dot phosphor/monoelectron transistors
		H01J	Field-emission-type electron source
4	Information and Communication	G06N	Signalling polymers/quantum computers

Classification number	Technology area	International patent classification	Technology content
5	Optoelectronics	G11	Information storage/memory with nanomagnets/memory media with nanometer-order memory layer
		G02	Microstructure optical fibres/accumulation-type photonic circuits/microlens EUV lithography/silicon nanoparticle luminescent devices/optical waveguide that creates a core and dad with nano-porous materials
		H01S	Optic amplifiers and lasers formed on the surface of semiconductor nanocrystals
6	Measurement and Testing	G01	Method of analysis that uses nanocrystal index/nanopumps/gene sequencers/manufacturing of DNA chips/ultra micro liquid dispensers/nanothermometers
7	Environment and Energy	C02F	Treatment of water, wastewater, sewage or sludge/treatment of exhaust gas
		H01M	Betteries/positive electrode of a rechargeable lithium battery
8	Processing	B01	Separation/mixing/manufacturing of self-cleaning surfaces
		B21	Processing/forming/diamond polishing of coated layers
		B23	Machine tools/use of femtosecond lasers/forming of silicon nano-scale dots
		B32B	Laminated bodies
9	Printing and Photography	B41J	Printing/ink jet heads/forming of nano-thickness images of goods
		G03	Photogrphs/electronic photographs

conductive particulate materials, were screened when one or more of the electrically conductive particulate materials were nano-materials such as carbon nanofibres.

(d) In the areas of micro-electro-mechanical systems (MEMS), superlattice structures, photonic crystals and quantum wells, a large number of inventions such as machines with micro and nanostructured components (electrical elements and lights) have been proposed. For these areas, the screening decision was made based on whether the patents used the nanotechnology described in (b) as a material or as a part of the fabrication process.

7.3.2 Trends in Nanotechnology Patents in the Four Largest Patent Organisations

This section outlines the number of nanotechnology applications lodged with the world's four largest patent organisations, screened and classified according to the above screening criteria. As shown in Fig. 7.2, all four patent organisations saw a significant increase in the number of nanotechnology patent applications between 2005 and 2007. In recent years, the total number of applications lodged at the JPO has remained at a level slightly exceeding 400,000. Nanotechnology patents thus accounted for approximately 1% of the total number of patent applications submitted to the JPO in 2007. Similarly, the number for the USPTO was approximately 1.5%, that for the EPO was approximately 1% and that for the WIPO was approximately 2.5%.

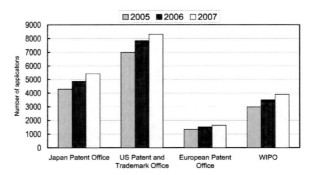

Figure 7.2 Number of the nanotechnology patent applications submitted to the four largest patent organisations.

7.3.3 Trends in Nanotechnology Patent Application According to the Applicant's Nationality

This section examines the nationality of the applicants who filed nanotechnology patent applications at the world's four largest patent organisations (Note 2). Figure 7.3 compares the nationality of applicants who lodged nano-related patents to the four largest patent organisations in 2007. It can be seen that US applicants outnumber the others with approximately 6,600 patent applications, which is about 1.6 times the number of applications filed by Japanese applicants (ranked second) and approximately 5 times that by German applicants (ranked third).

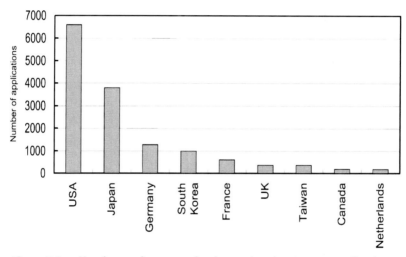

Figure 7.3 Number of nanotechnology-related patent applications submitted to the four largest patent organisations, sorted according to the nationality of the applicants (2007).

Figure 7.4 shows the classification results for the nationality of applicants who filed patent applications to the world's four largest patent organisations in 2007. In Japan, approximately 72% of applications submitted to the JPO were filed by Japanese applicants. In comparison, only about 62% of the applications to the USPTO were filed by US applicants. According to the report released by the JPO, similar trends were observed in other fields of patents. In order to manufacture or sell goods in a foreign country, it is necessary to obtain a patent in that particular country. In this context, filing applications to a patent organisation in a foreign country may reflect the strong intention of the applicants to develop, manufacture and sell goods in the particular country. Similar analysis of the patent applications from Asian countries revealed thatSouth Korea and Taiwan submitted a large number of applications to the USPTO (Fig. 7.5 a,b).

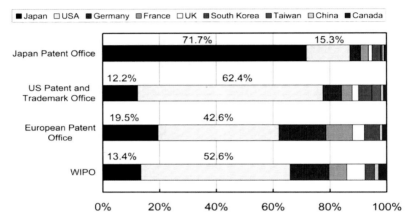

Figure 7.4   Breakdown of the nanotechnology patent applications submitted to the four largest patent organisations by nationality (2007).

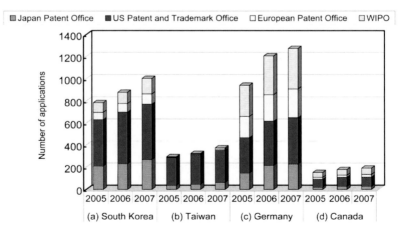

Figure 7.5   Nanotechnology patent applications by nationality (South Korea, Taiwan, Germany and Canada) (2007).

In contrast, the number of applications to the EPO and the WIPO from these countries was small. These countries are striving to strengthen their competitiveness in the field of nano-technology, particularly in the fields of information and com-munications technology (ICT) and electronics. The United States is a leading force for the industries in these fields, which is probably one of the factors that determine their application behavi-our. Therefore, when analysing the trends in patent application in terms of nationality, it is necessary to consider more specific sub-areas of the technology concerned.

For reference, the sections (c) and (d) in Fig. 7.5 show the number of patent applications that were lodged at the world's four largest patent organisations by German and Canadian applicants. A high proportion of patent applications from these countries were made to the USPTO and the WIPO. More interestingly, patent applications from these countries continued to increase until 2005, when they suddenly either levelled off or began to decrease. This trend was also observed for the number of patent applications made by applicants from other European countries.

7.3.4 Number of Nanotechnology Patent Applications by Corporations, Universities and Public Research Organisations

This section shows the results of an analysis of the relative volume of nanotechnology patent applications by three sectors: corpora-tions, universities and public research organisations. Figure 7.6 shows the breakdown of nanotechnology patent applications according to sector, for the top 10 countries in terms of the number of patent applications in 2007. Corporations filed the largest number of nanotechnology patent applications, regardless of country. Overall, corporations filed more than 80% of the total number of nanotechnology patent applications. This trend is expected to continue, with figures showing a small but steady increase between 2003 (81%) and the first half of 2005 (83.3%). Interestingly, while universities filed the second largest number of applications in countries such as the United States, the United Kingdom, Canada and the Netherlands, it was public research organisations that filed the second largest number of applications in Japan, Germany, France and South Korea. Hence, two major trends can be observed: the US-type trend and the Japanese-type trend. The figure also shows that the relative volume of applications from corporations was approximately 63% in Taiwan, which was the lowest among these countries.

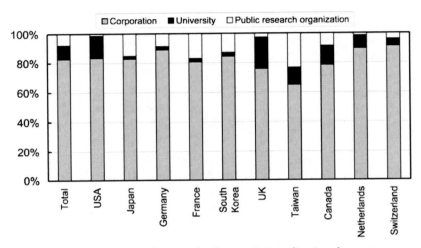

Figure 7.6   Percentage of nanotechnology patent applications by sector.

Figures 7.7 a,b,c show the relative volume of nanotechnology patent applications by country, for each applicant sector. In 2007, the United States accounted for the largest proportion of nanotechnology patent applications by corporations, followed by Japan. These results correspond to the overall ranking.

The figures also reveal that the United States accounted for an overwhelming percentage of the nanotechnology patent applications by universities, which is totally different from the trend observed for corporations. In this sector, the United States headed up the table, followed by the United Kingdom and then Japan. It is worth noting that China and Israel were ranked among the top 10 countries in this sector, although they were not among the top 10 countries in the overall ranking. In particular, Israel is filing an increasing number of nanotechnology patent applications in the field of medicine and life sciences.

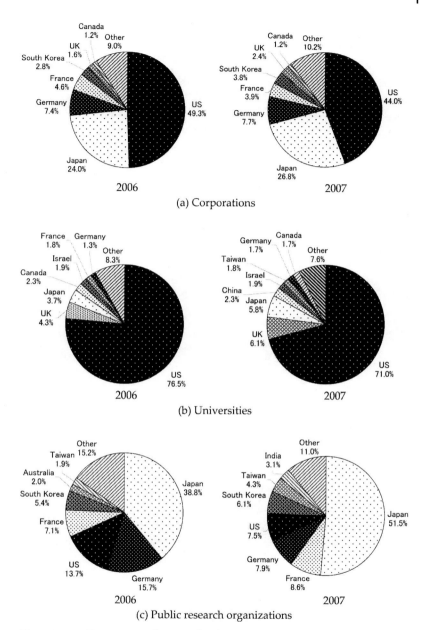

(a) Corporations

(b) Universities

(c) Public research organizations

Figure 7.7 Percentage of nanotechnology patent applications by sector.

The number of nanotechnology patent applications by public research organisations shows that, unlike the figures for the corporate sector, Japan accounted for the largest percentage of applications, followed by France, Germany and the United States. It is worth noting that many applications from Japan were filed by public research organisations such as the Japan Science and Technology Agency and the National Institute of Advanced Industrial Science and Technology. Looking at changes over the period from 2006 to 2007, there were no noticeable shifts in rankings for the corporate and university sectors. However, there was a slight decrease in the relative volume of applications from the United States. In the sector of public research organisations, Japan upped its percentage sharply, while Germany and the United States saw their percentage cut in half. Such organisations as the Max-Planck Institute and the Fraunhofer Gesellschaft filed many applications in Germany, as did the Centre National de la Recherche Scientifique (CNRS) in France.

7.3.5 International Comparison of Nanotechnology-Related Patents in Nine Designated Technology Areas

Finally, nanotechnology patent applications were categorised into nine specific areas of technology in order to compare trends by country (see Table 7.1 for the details of classification). Figure 7.8 shows the results of classification for all nanotechnology patent applications submitted to the world's four largest patent organisations. The largest number of patent applications was found in the field of materials, followed by electronic devices, then medicine and life sciences. The trends of patent applications within these nine areas vary significantly from one patent organisation to another. In the case of the JPO, the percentage of applications in the medicine and life sciences area was small. By contrast, the USPTO received a small percentage of applications from the materials field, which was offset by a large percentage of applications from the electronics device area. Patent applications lodged with the EPO and the WIPO showed the same tendency.

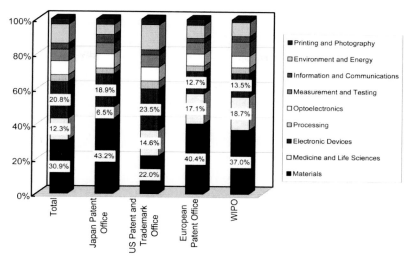

Figure 7.8 Percentage of nanotechnology patent applications by nine designated areas of technology (2007).

The three specific areas of technology, namely materials, electronic devices, medicine and life sciences, constituted a large proportion of the total number of applications (Fig. 7.9 a,b,c). Hence, the number of nanotechnology patent applications filed in those three specific areas in 2007 was compared among countries. In the area of materials, US patents accounted for the largest percentage, with Japanese patents falling a little short of the American figure. Together, US and Japanese patents accounted for approximately 70% of the total number of patent applications in the materials area. US and Japanese patents also led others in the electronic device area, in which South Korea and Taiwan were ranked third and fifth, respectively. The United States dominated in the area of medicine and life sciences, with Japan accounting for only a fraction of the applications submitted in this area. It is important to remember that Ireland and Israel were ranked among the top 10 in this field. Ireland achieved a remarkable breakthrough when it was ranked fifth in the first half of 2007, compared with its ranking of 13th in 2005. This leap in the rankings reflects how much importance the country places on the areas of medicine and life sciences.

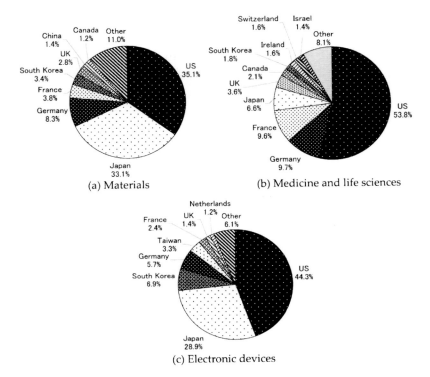

Figure 7.9 Percentage of nanotechnology patent applications by country in the three major areas of technology (2007).

7.4 Examples of Applied Nanotechnology: Carbon Nanotube Technology

Technology that is related to carbon nanotubes (CNT) can be divided into two categories: production technology and applied technology. This section explains the state of development of these technologies based on the number of patent applications.

7.4.1 Growth in the Patent Applications Related to CNT

Figure 7.10 shows the number of Japanese patent applications related to the manufacture, fabrication and processing of CNT as well as the application of CNT, published in the period from 1993 to 2006. Both numbers started to increase around 1997, with the number of patents relating to the application of CNT showing a sharper rise.

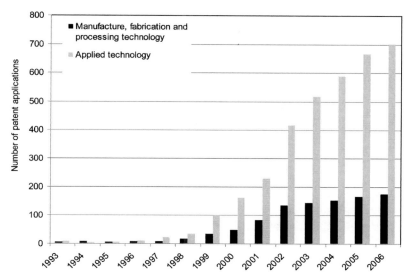

Figure 7.10   Trends in the number of patent applications related to CNT manufacture, fabrication and processing technology and the number of those related to the applied technology of CNT (2006).

7.4.2   Level of the Maturity of CNT Technology from the Perspective of Patent Trends

In general, as development in a field of technology advances, more corporations apply for patents and the number of patent applications increases accordingly. Then, as the technology in this field matures, the number of patent applications begins to decrease gradually. With the accumulated number of patent applications placed on the horizontal axis and the year of filing placed on the vertical axis, such a relation forms an S-shaped curve, known as a growth curve, as shown in Fig. 7.11. The left side of the S indicates the state where patent applications start to emerge, i.e., the embryonic period of the technology.

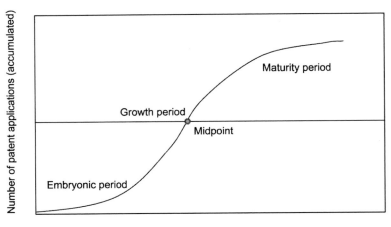

Figure 7.11   Growth curve model.

By plotting the growth curve of a technology, we can see at which stage the technology stands. Figure 7.12 shows the growth curve of the number of patent applications related to the manufacture, fabrication and processing of CNT. As for the data for 2007 and thereafter, the graph uses figures calculated with on the number of patent applications filed until 2006, instead of the real figures in 2007, as Japanese patent applications are not published until about 18 months elapse following the date of filing, and it takes some time to integrate the relevant data into the database. The growth curve suggests that the manufacture, fabrication and processing technology of CNT passed the midpoint around 2005 and currently stands at the later stage of the growth period. Hence, CNT manufacturing technology seems to be established, especially as a number of corporations have already started to sell CNT products based on the manufacturing technology patents they filed.

Figure 7.13 shows the growth curve of the number of patent applications related to the applied technology of CNT. The technology passed the midpoint of the growth curve after 2004, a little earlier than the manufacture, fabrication and processing technology of CNT. The technology is currently at a later stage of the growth period. This finding is consistent with our actual recognition of the current state where CNT technology has started to be used in some commercial products, e.g. sporting equipment.

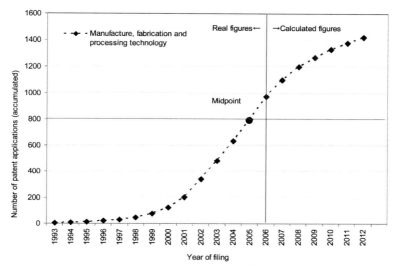

Figure 7.12 Growth curve of the number of patent applications related to CNT manufacture, fabrication and processing technology.

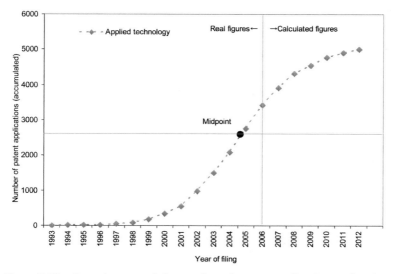

Figure 7.13 Growth curve of the number of patent applications related to the applied technology of CNT.

7.4.3 Trends in Patent Application Related to CNT, Based on the Type of Technological Fields

While the applied technology of CNT has already been used for sporting equipment, it has not reached that same mature stage in other application fields, especially electronic materials, where the method to control the chirality of single-walled CNT has not yet been established. The growth curve of the number of patent applications also suggests this situation. CNT study is currently in the growth period for the manufacture, fabrication and processing technology and for applied technology, and more products involving the use of CNT are expected to emerge in the future. Table 7.2 shows the number of patent applications categorised by the type of applied technology and by manufacture/fabrication/processing technology.

7.5 Conclusion: Intellectual Property Strategy in the Field of Nanotechnology

7.5.1 IP Strategy at the Stage of Basic Research

As with other fields of science and engineering, basic research in the field of nanotechnology and nanomaterials is time-consuming. Nonetheless, steady basic research can bring quantum leaps in technological development beyond the extension of conventional technology. These discontinuous technological changes lead to the development of innovative materials and the discovery of unexpected, new applications of technology. Hence, it is extremely important to carry out basic research activities for the benefit of society.

In the context of intellectual property, however, it is very difficult to evaluate the industrial applicability of the outcomes of basic research in realistic terms at the very early stage of technology development, even if the outcomes offer great potential for a wide range of applications in the future. It is also difficult to identify which technology among those achievements can be put into practice, and when and how this can be realised.

Table 7.2 Number of patent applications by type of applied technology and by manufacture/fabrication/processing

	Year of filing	1993	1994	1995	1996	1997	1998	1999	2000	2001	2002	2003	2004	2005	2006	Total	Share
Applied technology	Display	1		1	2	12	18	67	70	81	95	98	108	121	140	814	19.1%
	Semiconductor	1			1	1	1	2	11	26	36	58	63	68	77	345	8.1%
	Electronic materials					1	1	2	10	9	33	29	44	60	66	255	6.0%
	Optical device								2	2	6	6	21	30	31	98	2.3%
	Storage medium								1	3	11	12	2	7	7	43	1.0%
	Capacitor					2	1			3	4	12	9	9	17	57	1.3%
	Fuel cell				1			7	16	17	49	40	38	45	59	272	6.4%
	Secondary battery	2		1	1		3	2	3	5	2	12	6	22	20	79	1.9%
	Photo-electric cell									3		5	5	6	4	23	0.5%
	Sensor, etc.					1	3	7	7	12	29	29	24	29	22	163	3.8%
	Lighting, etc.								1	1	3	3	3	5	5	21	0.5%
	X-ray, electron beam					1			3			2	1	1	3	11	0.3%
	Conductive materials		1				1	1	4	10	32	19	28	33	25	154	3.6%
	Photography, photocopy							1	11	8	14	17	7	4	10	72	1.7%

(Continued)

Table 7.2 (Continued)

Year of filing		1993	1994	1995	1996	1997	1998	1999	2000	2001	2002	2003	2004	2005	2006	Total	Share
Inkjet				5		1			2	2	2	7	1	3	3	26	0.6%
Composite				2	4	1	3	4	8	19	47	75	67	80	85	395	9.3%
Automobile, etc.									5	2	9	17	17	25	41	116	2.7%
Sporting equipment									1		4	2	2	2	2	11	0.3%
Machine tool									1	1	2	2	1	2	3	10	0.2%
Catalyst support		1							2	4	7	9	5	7	10	45	1.1%
Medicine, biotechnology								1		1	5	14	13	10	10	54	1.3%
Food, etc.										1	1		2	2	2	8	0.2%
Nanofiber		2	1					2	1	5	11	14	14	15	9	73	1.7%
Filter							1		1	1	5	2	3	2	3	18	0.4%
Micro-machine								2	2	12	13	34	23	22	24	132	3.1%
Manufacture, fabrication & processing	Manufacturing method	3	6	4	6	8	16	30	46	77	127	131	138	151	144	887	20.8%
	Fabrication, processing	1	1			1	1	3	3	3	7	12	14	15	18	75	1.8%
Total		11	9	13	15	28	49	131	208	307	551	660	659	776	840	4257	100

However, it is important to secure basic patents in anticipation of the practical application of the target technologies in the future at this stage. Hence, there is no choice but to apply for patents even when the prospects for commercialisation cannot be clearly seen. This is also true for the filing of patent applications overseas. The applicants should also be aware that, as it takes a long time in the field of nanotechnology to commercialise the outcomes of basic research and development, even if basic patents based on those outcomes are successfully obtained, it is highly likely that the patents could expire or almost expire by the time the patented technologies are put into practical applications. For instance, CNT was discovered in 1991 yet their applied technology has only recently bloomed on a large scale, and the expiration of the basic patents for CNT is quickly approaching.

7.5.2 IP Strategy at the Stage of Application Development

Success in the field of nanotechnology depends on whether the outcomes of basic research can be linked with specific uses or markets and be commercially feasible in the form of new products. Corporations normally carry out product development while taking account of specific market needs. Efforts made to develop practical applications, as a phase between basic research and product development, have an important role to play in the commercialisation of nanotechnology.

At this stage of commercial development, corporations taking on the task of commercialisation usually identify market trends and needs from a long-term perspective, and carry out technological development through collaborative research with universities and other research institutes that engage in the relevant basic research. In this process, a wide variety of intellectual property, including know-how, is created through the fusion of essential technologies brought together for commercialisation at a level unanticipated in laboratories.

As mentioned earlier, considering that it is highly likely that basic patents will expire by the time the target technologies reach the mature stage of practical application, it is insufficient to secure patents relating to basic inventions and discoveries. To overcome this problem, it is important to obtain patents at home and abroad, and to manage knowledge with respect to each of the basic technologies and applied technologies, as well as the technologies for manufacturing products created through collaborative research.

In addition to securing patent rights and other intellectual property rights, it is also important to make preparations for international standardisation, in anticipation of the practical applications of the patented technologies in the future. International standardisation activities in the field of nanotechnology have gradually increased since 2002. The International Organization for Standardization (ISO) set up the TC 229 (nanotechnology) in 2005. Since the first meeting of the TC 229 in December of 2005 in London, the TC 229 has been held twice a year with more than 200 participants from many countries. The IP strategy and the standardisation strategy are closely related. By anticipating the strong economic growth in the nanotechnology industry, each organisation has a strong incentive to incorporate their IP at the basic and applied stages into international standardisations. This is vital as the technologies associated with materials, measurement, testing and energy will have a long-term impact on economy.

7.5.3 Connecting Basic Research and Application Development

If corporations cannot make use of the achievements established in basic research in their product development activities, they will fall behind in the development of new business opportunities. To avoid such a situation, it is extremely important to connect with the universities and other research institutes engaging in the relevant basic research and to promote the development of practical applications through this collaboration.

The intermediary role for connecting the two parties can be performed by technology licensing organisations (TLOs), the intellectual property departments of universities, and venture businesses. Productive collaboration among three parties, that is, the industrial sector, the academic sector, and the intermediary between the two, is an essential factor for securing competitiveness in the fields of nanotechnology and nanomaterials.

7.6 Notes

7.6.1 Note 1

When applying for an international patent, one may follow what is called the Paris Convention route by directly applying to the patent agencies of various countries, or the Patent Cooperation Treaty (PCT) route using a unified international procedure. Filing the patent application through the PCT route is equivalent to filing the patent application in each member country, but one cannot obtain an actual patent right from the PCT. In order to obtain the patent right, the assessment of the patent application must be done by each member country where the patent is filed. Patent applications filed through the PCT route are published by the WIPO. The EPO serves the same function for its member countries as the PCT does. Unlike the PCT, however, the EPO has the authority to grant patent rights.

7.6.2 Note 2

The "nationality of applicants" is defined herein as the nationality of the chief inventor. The cases, where inventors file the patent applications to foreign countries via the local patent offices in their own countries, are also counted by the nationality of the chief inventor.

References

1. Glänzel, W., Meyer, M., Du Plessis, M., Thijs, B., Magerman, T., Schlemmer, B., Debackere, K., and Veugelers, R. (2003). Nanotechnology, analysis of an emerging domain of scientific and technological endeavour, *Steunpunt O&O.*

2. Heinze, T. (2004). Nanoscience and nanotechnology in Europe: analysis of publications and patent applications including comparisons with the United States. *Nanotechnol. Law Bus.*, **1**(4), pp. 427–445.

3. Huang, Z., Chen, H., Chen, Z. K., and Roco, M. C. (2004). International nanotechnology development in 2003: country, institution, and technology field analysis based on USPTO Patent Database, *J. Nanopart. Res.*, **6**(4), pp. 325–254.

4. Igami, M., and Okazaki, T. (2007a). Capturing nanotechnology's current state of development via analysis of patents, *OECD STI Working Paper*, 2007/4.

5. Igami, M., and Saka, A. (2007b). Capturing the evolving nature of science, development of new scientific indicators and mapping of science. *OECD STI Working Paper*, 2007/1.

6. Meyer, M. (2006). Are patenting scientists the better scholars? An exploratory comparison of inventor-authors with their non-inventing peers in nano-science and technology, *Res. Policy*, **35**, pp. 1646–1662.

7. Meyer, M. (2006). What do we know about innovation in nanotechnology? Some propositions about an emerging field between hype and path-dependency. Paper presented at *SPRU 40th Anniversary Conference. The Future of Science, Technology and Innovation Policy*, SPRU, Brighton, East Sussex, United Kingdom.

8. Scheu, M., Veefkind, V., Verbandt, Y., Molina Galan, E., Absalom, R., and Förster, W. (2006). Mapping nanotechnology patents: The EPO approach. *World Patent Inform.*, **28**, pp. 204–211.

9. Schummer, J. (2004). Multidisciplinarity, interdisciplinarity and patterns of research collaboration in nanoscience and nano-technology. *Scientometrics*, **59**, pp. 425–465.

10. Zucker, L. G., Darby, M. R., Furner, J., Liu, R. C., and Ma, H. (2006). Minerva unbound: knowledge stocks, knowledge flows and new knowledge production, *NBER Working Paper*, No. 12669.

Chapter 8

Government Regulation of Nanotechnologies

Diana M. Bowman[a,b] and Joel D'Silva[c]

[a]*Risk Science Center and Department of Health Management & Policy,*
School of Public Health, The University of Michigan,
1420 Washington Heights, Ann Arbor MI 48109-2029, USA
[b]*Department of International and European Law,*
Faculty of Law, KU Leuven, B-3000 Leuven, Belgium
[c]*Independent Legal Researcher, B-3000 Leuven, Belgium*

dibowman@umich.edu, joel.dsilva@telenet.be

Regulation shapes decision making and human behaviour. However, its impact—as it has often been argued—can encourage or stifle innovation, especially in relation to emerging technologies such as nanotechnologies. It is therefore not surprising that the regulatory debate in relation to nanotechnologies is highly contested. As an increasing number of products incorporating nanomaterials and nanoparticles make their way onto the market in an environmental of imperfect information, such debates have only intensified. The aim of this chapter is to introduce readers to the notion of regulation and illustrate how regulation in its many guises is currently being employed to regulate nanotechnologies.

Nanotechnology Commercialisation
Edited by Takuya Tsuzuki
Copyright © 2013 Pan Stanford Publishing Pte. Ltd.
ISBN 978-981-4303-28-6 (Hardcover), 978-981-4303-29-3 (eBook)
www.panstanford.com

8.1 Introduction

Global efforts to develop enhanced and new products containing engineered nanomaterials, or produced using nanotechnology-based processes, within the public and private sectors have escalated over the last decade. Products containing nanomaterials are now a ubiquitous feature in the fields of cosmetics, sporting goods and textiles, with emerging applications now evident within the medical sector, agri-food and automobile industries. The degree to which the technology, and specific applications there of, will impact our daily lives seems likely to only amplify over the coming years.

The ever-increasing production of engineered nanomaterials and their use in various consumer products pose a number of challenges to governments and health and safety regulators who are vested with the role of safeguarding human and/or environmental health and safety. Although the discussion on the challenges posed by regulating engineered nanomaterials has moved from the abstract into more specific and concrete areas of concern, there is still little consensus as to the potential risks—if any—of the technology [39]. This in itself is not new, nor specific to nanotechnologies, as the emergence of any new technology brings with it the promise of both benefits and risks, and the need to minimise the latter [46]. However, unlike the introduction of many earlier technologies, the state of scientific art in relation to potential risks is littered with significant gaps in knowledge [2,47,59].

Despite the anticipated benefits, recognition of the paucity of toxicological and ecotoxicological data in relation to certain nanomaterials has resulted in increasing debate over the effectiveness of current regulatory regimes for managing potential risks posed by the technology. The maturation of nanotechnologies has resulted in this debate shifting to consider both the need for, and the possible form(s) of nano-specific regulation, as well as wider ethical and societal concerns [30]. With the European Parliament having passed the first nano-specific piece of national or supra-national legislation through the *Cosmetics Regulation (Regulation (EC) No 1223/2009)* questions relating to risk management and the governance of nanotechnologies are only likely to intensify in the short to medium term.

With this in mind, the aim of this chapter is to outline how the regulatory and governance challenges posed by nanotechnologies are currently being confronted and met by governments and other relevant stakeholders. In doing so, Section 8.2 begins by defining what is meant by regulation, its objectives, and the strengths and weaknesses adopted by different regulatory approaches. With the regulation of nanotechnologies having largely occurred under a range of existing regulatory regimes, Section 8.3 provides an overview of a number of the regulatory reviews undertaken to date within jurisdictions such as Australia, the European Union and the United States of America, and the key findings and implications. Multi-lateral and multi-party activities aimed at ensuring the benefits of the technology and minimising potential risks are considered in Section 8.4. Key challenges, common issues and conclusions are brought together in Section 8.5.

8.2   The World of Regulation: Unpacking Different Regulatory Models

The regulation of various new technologies is constantly in the public eye as it is often felt that new technologies are outpacing regulation. It is often expressed by many that the need is for improved legislation and implementation so that it is more efficient and can work in a rapidly changing environment. However, this is not the case as in order to "design and implement a legal system able to cope with rapid changes in technology, a broader perspective is required" [52]. The regulatory environment for new technologies in general can be said to comprise "a fluid, evolving collection of inter-related and overlapping measures, with jurisdictions at local, national and international levels" [32]. According to Moses [52], there are four potential problems that may result from the failure of law and regulation to keep pace with new technologies. These may be summarised as follows:

(1) failure in developing oversight approaches and appropriate legal restrictions and precautions to control risks associated with new technologies

(2) uncertainty in the application of existing frameworks to new technologies

(3) existing rules might under or over-regulate the technology and finally, and

(4) the potential for the technology to make existing rules obsolete (see also, 46).

While the notion of regulation has traditionally referred to legislative instruments enacted through formal processes of Parliament or Congress, as well as developments in the common law [10], regulation is today considered to be much broader than simply "command and control" or "black letter law". Julia Black [5] has defined this contemporary view of regulation as,

> the sustained and focused attempt to alter the behaviour of others according to defined standards or purposes with the intention of producing a broadly identified outcome or outcomes.

According to Black's view, the primary aim of regulation is to influence behaviour of individuals or groups more generally [5]. As noted by Brownsword [10] "'regulation' refers [therefore] to any instrument (legal or non-legal in its character, governmental or non-governmental in its source, direct or indirect in its operation... that is designed to channel behaviour". Regulation may, therefore, be undertaken by a range of different actors, and not just government, in order to elicit or shape a range of different outcomes, albeit positive or negative. While regulation may often be perceived as a way to minimise or manage risk of type of harm, it may also be used as a way to encourage behaviour by encouraging a particular type of activity in order to promote economic interests or innovation [44].

8.2.1 State-Based Regulation

Despite claims that nanotechnologies are "essentially unregulated" (see, for example, [24]), there can be little doubt that nano-based products and processes fall under existing regulatory frameworks. These frameworks are established by, for example, Acts, Regulations and Directives. These traditional state-based "command and control" forms of regulation have considerable legitimacy with the public despite their perceived imperfections. Their compulsory nature, the appearance of strong accountability and higher certainty are all characteristics that appeal to voters. Moreover Ludlow *et al.* [44] have suggested that,

> clear and consistent state-based regulation provides many advantages compared with no regulation. In many instances industry prefers this

form of regulation because it provides a level playing field, as well as providing protection against short-cutting competitors. It also provides certainty, assisting in securing capital finance and insurance.

However, in order to achieve these objectives, Ludlow *et al.* [44] have argued in relation to nanotechnologies that the instruments employed must also be appropriate for the subject matter being regulated. "Appropriate" is, however, an ambiguous term, informed by past experiences, different views on risk, benefits, innovation and broader societal considerations, and the degree to which the instruments favourably or unfavourably impact upon your behaviour or that of others [44].

So while state-based regulation may provide clarity, this traditional form of regulation suffers from a number of perceived shortcomings [1,37,77]. It is often viewed as being slow, cumbersome, rigid and involving high transaction costs [51,70]. These limitations may be amplified in the context of rapidly evolving fields or areas where the regulatory requirements are more dynamic and information is imperfect. Given this it is unsurprising that some commentators have suggested that in the short term, nano-specific legislation may not be the most appropriate or effective way to regulate specific areas of the technology [61].

8.2.2 Civil-Based Regulation

In light of the broadly accepted limitations associated with command and control there has been increasing interest among stakeholders to develop alternative regulatory tools [70]. "Soft law"—or civil regulation—mechanisms are part of the regulatory continuum and include various forms of self-regulation, regulation and third-party regulation, as well hybrid arrangements [37,77]. Importantly, "civil regulation extends regulatory authority 'sideways' beyond the state to civil society and to non-state actors", thereby removing the potential for legally binding standards [77]. Although under these mechanisms government no longer functions as the regulatory authority, the very broad nature of this regulatory category may enable the state to have some level of involvement in the broader governance framework [37].

The use of different types of civil regulation, such as codes of conduct, risk management frameworks and industry codes, has steadily increased. High-profile activities have included the chemical industry's Responsible Care program [42,62]. It has been argued that these voluntary forms of regulation are less resource intensive to develop and administer and have the ability to evolve and respond to the changing environment quicker than state-based regulation [8,70]. It has also been suggested that these approaches provide breathing space for innovation and creativity to occur [78]. Arguably it is these perceived strengths that have inspired a number of different organisations, across different sectors and jurisdictions to develop and implement voluntary codes of conduct and risk management frameworks. These have included the "Responsible NanoCode", "NanoSafe", the "NanoRisk Framework" and the "Principles for the Oversight of Nanotechnology" [14,22, 41,66].

These programs operate within the shadow of formal regulatory obligations and do so by, for example, incorporating a precautionary approach to the manufacturing of nanoscale materials, by implementing innovative processes and state of the art scientific knowledge into current practices, or through the data gathering processes which can then be used to develop risk appropriate management practices. These approaches therefore work in combination with other forms of regulation in order to govern the safe and responsible development of nanotechnologies. This contention sits comfortably with Black's [6] broad notion of regulation in that it is something more than merely government enacted legislation.

Civil regulation, however, also has the potential to suffer from a range of perceived limitations [6]. Self-regulation, in particular has been accused of serving the interests of industry above that of society by circumventing market forces, having variable standards of enforcement and lacking the accountability and legitimacy of government regulation [9,78]. Gunningham and Rees [37] also note that in relation to self-regulation "the effectiveness (or ineffectiveness) of self-regulation [can] var[y] enormously among industries...".

8.2.3 Co-Regulation

Co-regulation may be generally defined, as the "use of a panoply of tools and actors, formal and informal, governmental and nongovernmental, national and international" [53]. Existing at the interface of state-based and civil regulation, this approach has the ability to draw upon the strengths of these other models while avoiding a number of their potential weaknesses. Given the flexibility of this regulatory approach, it is therefore not surprising that government have viewed co-regulation as a fertile and productive area for regulatory innovation [3,76].

Evidence of co-regulation has already emerged in relation to managing the challenges presented by nanotechnologies with a number of governments having implemented—in consultation with industry and other stakeholders—voluntary reporting schemes or data stewardship programs [see, for example, 11,18,23,55,56]. The development of the European Commission's voluntary Code of Conduct for Responsible Nanosciences and Nanotechnologies Research [26] provides a further example of how government may work with industry and civil society to shape behaviour.

According to Gunningham and Sinclair [38], various approaches are complementary and "'single instrument' or 'single strategy' approaches are misguided, since all instruments have strengths and weaknesses, and none are sufficiently flexible and resilient to be able to address all environmental problems in all context[s]". As noted by Ludlow *et al.* [44], "the ongoing challenge for stakeholders will [therefore] be finding the acceptable balance between the different approaches". This is likely to be dependent in part, for example, on the evolving state of scientific knowledge, the degree of legitimacy attached to any one approach, the jurisdiction in which it is employed, and the way in which benefits and risks of particular areas are perceived within those jurisdictions [44].

In summary, there will be no "magic bullet" or single approach that will solve the regulatory challenges posed by nanotechnologies or indeed, any new technology—and even when suitable solutions are found it is inevitable that the proposed response will not be acceptable to all those affected by its implementation. Finding a workable and effective balance, which has the necessary legitimacy with the public, and other key stakeholders will be an ongoing and evolving task.

8.3 Current Regulatory Frameworks and Their Effectiveness for Nanotechnologies

Faced with uncertainties and increasing scrutiny, governments in a number of jurisdictions have initiated reviews of their regulatory frameworks in order to examine their appropriateness for nanotechnologies. Reviews have, for instance, been undertaken in Australia, the European Union, the United Kingdom, the United States, and New Zealand [13,23,27,28,35,36,43,65]. These reviews have varied in their scope and focus. What have been the key lessons and recommendation to come out of the various reviews undertaken to date?

There are several, many of which are horizontal in nature and therefore common to multiple jurisdictions. For instance, each review highlighted the fact that just like their conventional counterparts, nanotechnology-based products and process fall within the ambit of statutory instruments and are regulated accordingly. They "inherit" such regimes. As such the instruments were sufficiently broad in their application so as to capture the emerging technology. Several commentators have, however, highlighted the fact that the implementation of existing instruments and the principles underpinning them and their application to nanotechnologies are potentially problematic [13,43]. These reviews have highlighted how current scientific uncertainties and knowledge gaps have the capacity to impact upon the operation and effectiveness of the frameworks.

The situation in terms of the scientific and technical issues posed by these uncertainties in relation to the operation and implementation of statutory frameworks was arguably best summarised by the European Commission when it stated that,

> [w]hilst the Community legislative framework generally covers nano-materials, implementation of legislation needs further elaboration. The scientific basis to fully understand all properties and risks of nanomaterials is not sufficiently available at this point in time. [28]

The various government-initiated reports have also been complemented by a large number of regulatory and policy assessments by members of the research community, non-governmental organisations and other stakeholders [16,17,39,71,72].

Against this backdrop, a number of Royal Commissions and Parliamentary Inquiries have also been initiated [40,54,64]. A number of scientific opinions concerning the potential risks associated with nanotechnologies have also been published [29,67–69]. While the findings and recommendations vary they nonetheless highlight the need for "risk-focused research" in order to address the knowledge gaps currently challenging regulators and risk managers

[47]. Moreover, the increasing number of such reports suggests that concerns over the adequacy of currently regulatory arrangements for governing different facets of the technology, including worker safety and food safety, are unlikely to be resolved in the short to medium term [50]. A theme running through most of the reports is also the need for coordination not only between jurisdictions, but also between other key stakeholders.

Two other issues that are also being deliberated upon have been the issue of finding suitable and accepted definitions and the need for labelling of nanomaterials. Attempts at securing a suitable and generally accepted definition of a "nanomaterial" have been going on in Europe and worldwide for some time [7,19]. Definitions play an important role in the regulatory sphere as "they assist in establishing the subject matter and scope of what is to be regulated" [7]. However, considering the complexities associated with nanoscience and also differing opinions on various definitions, this has not been an easy task. Reaching an agreement on fundamental terms and definitions such as "nanotechnology", "nanoparticle", "nanoscale" and a "nanomaterial" is both complex and contested. To date, there is no internationally recognised and accepted definition of a "nanomaterial". This has not, however, stopped the European Commission itself from adopting such a definition for its own purposes. The central challenge in developing a definition for nanomaterials that can be used for regulatory purposes is for it "to be broad enough that it encompasses those materials that plausibly may exhibit the risk-creating properties that have led to the calls for regulation of nanomaterials in the first place without sweeping in many materials that plausibly cannot be expected to display such properties" [15]. According to Bergeson and Auer [4], "The definitional void is frustrating stakeholders as the lack of definitional clarity invites commercial, legal, and compliance uncertainties. So also, the lack of a regulatory definition has created uncertainty and possibly complexities for regulators" [63]. There is also no denying that definitional issues are as much a political endeavour as it is a scientific one [60]. Since it is likely to "take years before adequate nano-standards and nanotoxicological models have been developed, the political question is how to regulate nanomaterials in the meantime" [60]. Adopting a preliminary or broad over-arching definition is crucial to this process. At the same time, a definition by itself should not–and cannot—be seen as the end point. The continuous re-examination and revision of terminology, testing and measurement standards and other requirements will be necessary as the state of the scientific art advances [19].

As illustrated by Table 8.1, the European Union—in 2009—introduced legislation that would mandate the labelling of nano-scale ingredients incorporated into cosmetics as part of the new regulatory framework for cosmetics. Similar requirements were proposed as part of the recast of the Novel Foods Regulation [7], the negotiations for which have since collapsed. Such action by the European Parliament has fostered debate on the need for product labelling for nanomaterials in consumer products, as well as the nature and form that any such labelling should take [21,31]. As shown by these discussions, labelling of products in general is not only a highly sensitive area, but also a highly contested one. Consumer product labels have always been a point of contention between regulatory agencies, business and the public [21,31].

Table 8.1 Overview of regulatory initiatives for nanotechnologies

Year	Initiative
2000	Foresight Nanotech Institute releases the Foresight Guidelines on Molecular Nanotechnology; voluntary guidelines for the responsible development of nanotechnology.
2003	Luna Innovations initiated the development of NanoSAFE, a nanotechnology risk governance framework.
2004	Initiation of the NanoMark certification system
	Code of Conduct for Nanotechnology published by BASF
2005	International Organizations for Standardization (ISO) technical committee, ISO/TC 229 *Nanotechnologies* hosts its first meeting.
	OECD Chemical Committee hosted the first OECD Workshop on the Safety of Manufactured Nanomaterials.
2006	Australia's industrial chemical regulator launch its voluntary call for data on nanomaterials
	DEFRA launch the UK Voluntary Reporting Scheme for engineered nanoscale materials.
	City of Berkeley, California, amends the Hazardous Materials and Waste Management sections of the Berkeley Municipal Code to regulate "manufactured nanoparticles".
2007	The OECD's Committee for Science and Technology Policy establishes a Working Party on Nanotechnology

	Coalition of Non-Governmental Organizations publish their "Principles for the Oversight of Nanotechnologies and Nanomaterials"
	Environmental Defence and DuPont publish their "Nano Risk Framework"
2008	EPA formally launches its two year "Nanoscale Materials Stewardship Program"
	European Commission launches its voluntary "Code of Conduct for Responsible Nanoscience and Nanotechnologies Research"
	Australia's National Industrial Chemical Notification and Assessment Scheme launch its second voluntary call for data on nanomaterials
	Federal Register notice by the EPA to inform manufacturers that carbon nanotubes are to be registered
	Swiss Retailer's Association publish a "Code of Conduct for Nanotechnologies"
	Launch of the Responsible Nano Code and the Benchmarking Framework
2009	Californian Department of Toxic Substances Control announce their mandatory data "call-in" for carbon nanotubes
	European Parliament vote in favour of a report relating to the recast of the Novel Foods Directive; the proposed Novel Foods Regulation put forward by the Parliament specifically includes nanomaterials within its scope and provides a definition of nanomaterials for the purpose of the Regulation (negotiations collapse in 2011)
	Adoption of the final text of the Cosmetic Regulation by the European Parliament and Council, which incorporates mandatory provisions – including labelling requirements — relating to the regulation of cosmetic products containing nanomaterials.
2010	EC publish the a consultation document on the definition of 'nanomaterial'
	Australia's industrial chemical regulator amends its regulatory framework in order to explicitly take into account industrial nanomaterials
2011	US EPA publish a "New Policy for Nanotechnology in Pesticides"
	US Government publishes a Policy document for the regulation and oversight of nanotechnologies.
	European Commission publish their definition of a "nanomaterial"
	Australia's National Industrial Chemical Notification and Assessment Scheme adopts a working definition of an "industrial nanomaterial"
	Executive Office of the President publishes a Memorandum for the Heads of Executive Departments and Agencies: Policy Principles for the U.S. Decision-Making Concerning Regulation and Oversight of Applications of Nanotechnology and Nanomaterials
	Adoption of the final text of the Regulation (EU) No 1169/2011 on the provision of food information to consumers (the Food Labeling Regulation), which contains nano-labelling requirements
2012	FDA releases draft guidance material for industry in relation to the use of nanotechnologies in food ingredients and food contact substances

While labelling policies can provide workers and consumers with information in order to assist them in assessing relative risk and exercising an informed choice, disclosure of such information cannot be employed instead of instruments which assess, for example, the safety and efficacy of drugs or the safety of foods [12,21,33]. In this respect, labelling policies can be used to communicate and inform consumers of relative risks, but they remain a component of a larger regulatory regime and are, therefore, dependent on the operation of other regulatory instruments and policies. The debate over the need for labelling regimes for products containing nanomaterials and nanoparticles is likely to intensify as the number of products available on the market increases [33]. Fundamental to these debates will be the questions of the purpose of such a regime and the nature it should take in order to meet its objectives. Each of these questions is highly contentious dependent on culture, and is likely to represent significant battlegrounds, not only across stakeholder groups but also within stakeholder groups like various governments, given the diversity of interests at play and their economic implications [21].

8.4 Multi-Lateral and Multi-Party Initiatives

There can be little doubt that dialogue on the regulation of nanotechnologies at the international level has also gathered pace over the last five to ten years with inter-governmental bodies, such as the Organization for Economic Cooperation and Development (OECD), the United Nations (UN) and the World Health Organization (WHO) in partnership with the Food and Agricultural Organization (FAO) of the UN, having turned their attention to nanotechnologies. The activities and initiatives of these organisations, some of which are highlighted in Table 8.1, have not been designed to develop regulatory instruments; rather they have instead largely focused on the safety aspects of nanotechnologies and in identifying the challenges faced by, for example, safety regulators. The International Organization for Standardization (ISO), for example is supporting governments and industry through the development of definitions, common nomenclature and standards for classification and testing of nanotechnology and nanomaterials [49]. Also of importance, industry and civil society have been key players within many of these discussions and have assisted in shaping their outcomes.

The OECD has established two Working Parties to consider different aspects of nanotechnologies: the Working Party on Manufactured Nanomaterials (WPMN) and the Working Party on Nanotechnology (WPN). The WPMN has implemented eight projects including the creation of a database on human health and environmental safety research (publically launched April 2009), safety testing of a representative set of manufactured nano-materials, for example, fullerenes, single-walled carbon nano-tubes, and titanium dioxide, to which a number of countries and private sector participants have contributed to, and cooperation on risk assessment schemes in order to determine their suitability for nanomaterials [58].

The WPN has focused on six specific project areas, which include public outreach activities, documenting the ways in which nanotechnologies may be employed to assist in addressing global challenges, and acting as a facilitator for international cooperation and collaboration on research activities [57,58]. While the objec-tives of the WPN are clearly ambitious, the proposed outputs should provide governments with well-informed insights into the potential impacts—both positive and negative—and provide them with fundamental knowledge and tools to move forward with policy developments.

Committees and agencies of the UN have similarly been engaged in discussions about nanotechnologies on the interna-tional stage, as outlined by the work agenda of UNESCO [74,75], and more recently the FAO—in partnership with the WHO [34]—and the United Nations Committee of Experts on the Transport of Dangerous Goods and on the Globally Harmonized System of Classification and Labelling of Chemicals [73]. UNESCO, has focused their efforts on "ethical reflection… to address the potential benefits and harms of nanotechnologies but even more important is assessing and publicly discussing the goals for which these technologies will be used" [75]. It has recognised the scientific challenges and current knowledge deficit associated with nano-technologies. In order to address the numerous scientific and social challenges posed by the technology, UNESCO has advocated for voluntary ethical guidelines for nanotechnologies. While aspirational in nature, the principles and approach advocated by UNESCO highlight how longer-term regulatory instruments within this area may not only focus on the scientific risks, but also address deeper social and political considerations.

The recognition of the increasing commercialisation of food and feed products processed with nanotechnologies, or incorporating engineered nanomaterials, and potential safety implications there of prompted the FAO and WHO to convene an "Expert Meeting" in June 2009. Issue identification and capacity building of national food safety regulators to meet the potential challenges posed by the use of nanotechnologies in the agri-food sector appear to be paramount considerations for the FAO and WHO.

Against this backdrop, a number of multi-party initiatives focused on responsible development and governance of nano-technologies have occurred. These have included, the International Dialogue on Responsible Research and Development of Nano-technology, the establishment of the International Council on Nanotechnology, and the International Risk Governance Council's (IRGC) project on risk governance of nanotechnology.

Moving forward, it is clear that some sectors or actors will be better positioned to influence the development of policies for nanotechnologies than others, especially those with represen-tation in multiple forums. Given the diverse interests and economic value involved, it is argued that this sphere is likely to become increasingly populated by parties with vested interests pushing their own policy and regulatory agendas. The uncertainty here lies in the ultimate impact that these activities will have at the national level and the policy and regulatory frameworks that will be fashioned as a result.

8.5 Conclusion: Acknowledging the Elephant in the Room

Seven years ago Marchant and Sylvester [45] made the following statement:

> [m]any other questions continue to nip at nanotechnology's heels, not the least of which are debates about what is and is not technically feasible. Despite these uncertainties, we can have complete confidence in one aspect of nanotechnology's future—it will be subject to a host of regulations.

There can be little doubt now about the accuracy of their words. And while uncertainties still persist, there appears to be some consensus that change is needed under the broader regulatory matrix. While the 2009 recast of the European Union's Cosmetics Directive may have been the first national or supranational legislative instrument to be passed with nano-specific provisions, as highlighted by the passage of the Food Information Regulation, it will certainly not be the last. The European Parliament is still to vote on the inclusion of nano-specific amendments to the *Registration, Evaluation, Authorisation and Restriction of Chemicals (REACH) Regulation (Regulation 1907/2006)* but such amendments now appear inevitable. While such measures are jurisdiction specific, their impact will be felt more widely: non-European parties wishing to place products onto the European market will be required to comply with such instruments. As such, regulatory action within any one jurisdiction will have an impact beyond their territorial boundaries.

At present, there is growing speculation that existing voluntary reporting or data collection activities may be hardened in a number of jurisdictions [50]. Such actions, if they transpire suggest that some governments are willing to enact regulations despite the many uncertainties and the challenges thereof. Such action will also have wide ranging implications for regulators, not least that they will be required to stay on top of the changes within a dynamic regulatory environment.

A number of key challenges and issues must be addressed to ensure that the predicted benefits of the technology are fully realised. Arguably the most pressing of these is the need to tackle the known scientific uncertainties in a systematic and strategic manner [48]. In doing so, the continuation and expansion of the current multi-party and multi-jurisdiction collaborations will be most beneficial in identifying and managing potential risks. However, uncertainty and the need for more scientifically sound data should not in itself give rise to paralysis in relation to the broader governance frame-work. Whilst voluntary codes of conduct, risk management frameworks and certification schemes may not be perfect, they nevertheless provide a framework for managing potential risks. In this sense, these initiatives illustrate that innovative, consent-based governance regimes can be developed and implemented by institutions despite the fast moving pace of the technologies. Frameworks of this type can be easily refined over time in order to reflect the evolving scientific state. Importantly though, these types of regulatory approaches will only be part of the broader regulatory environment. Public (consumer) participation is also being increasing seen as an important tool in restoring public faith in innovation and in regulatory oversight of science and technology [20]. The number of projects that encourage the public to engage with nanotechnology governance issues is growing. The European Commission in its communication on a European strategy for nanotechnology emphasises an open and proactive approach to the governance of nanotechnology to ensure public awareness and confidence, as well as dialogue with citizens to promote informed judgements and consumer decisions [25]. Hence, even though public and consumer fears cannot be always quelled by mere participation in nanotechnology related decisions; public engagement has become a serious component of nanotechnology policy.

The regulation of nanotechnologies represents a difficult and ongoing challenge, and as with the science itself, there are likely to be many events, contributions and players within the international and national spheres that will shape the regulatory environment. The question hence to be considered is when seeking the best method of achieving a policy or regulatory aim, is traditional governmental regulation the best method, or can alternative models provide desired results as well or better? In an area in which the pace of evolution is so rapid, thinking beyond conventional boundaries and beyond pure government regulation would appear to be relevant to considering how to govern such a dynamic and multifaceted technology. This will not be an easy task, but it is pivotal to ensuring the safety and success of nanotechnologies.

References

1. Aalders, M., and Wilthagen, T. (1997). Moving beyond command and control: reflexivity in the regulation of occupational safety and health and the environment, *Law Policy*, **19**(4), pp. 415–443.

2. Aitken, R. J., Creely, K. S., and Tran, L. (2004) *Nanoparticles: An Occupational Hygiene Review* (Institute of Occupational Medicine, Edinburgh).

3. Ayres, I., and Braithwaite, J. (1992) *Responsive Regulation: Transcending the Deregulation Debate* (Oxford University Press, New York).

4. Bergeson, L., and Auer, C. (2011). Nano disclosures: Too small to matter or too big to ignore?, *Nat. Resources Environ.*, **25**(3), pp. 26–30.

5. Black, J. (2002). Critical reflections on regulation, *Austl. J. Legal Philos.*, **27**, pp. 1–36.

6. Black, J. (1996). Constitutionalising self-regulation, *Mod. Law Rev.*, **59**(1), pp. 24–55.

7. Bowman, D. M., D'Silva, J., and Van Calster, G. (2010). Defining nanomaterials for the purpose of regulation within the European Union, *Eur. J. Risk Regul.*, **1**(2), pp. 115–122.

8. Braithwaite, J. (1982). Enforced self-regulation: a new strategy for corporate crime Control, *Mich. Law Rev.*, **80**(7), pp. 1466–1507.

9. Braithwaite, J. (1993), *Business Regulation and Australia's Future,* eds, Grabosky, P. N. and Braithwaite, J., Chapter 6 "Responsive Regulation for Australia", (Australian Institute of Criminology, Canberra) pp. 81–96.

10. Brownsword, R. (2010). *International Handbook on Regulating Nano-technologies,* eds. Hodge, G. A., Bowman, D. M., and Maynard, A. D., Chapter 4 "The Age of Regulatory Governance and Nanotechnologies", (Edward Elgar, Cheltenham), pp. 60–80.

11. Californian Department of Toxic Substances Control (2009), *Chemical Information Call-In: Carbon Nanotubes,* (DTSC, Sacremento).

12. Caswell, J. (1998). How labeling of safety and process attributes affects markets for food, *Agr. Resource Econ. Rev.*, **27**(2), pp. 151–158.

13. Chaudhry, Q., Blackburn, J., Floyd, P., George, C., Nwaogu, T., Boxall, A., and Aitken, R. (2006), *Final Report: A Scoping Study to Identify Gaps in Environmental Regulation for the Products and Applications of Nanotechnologies,* (Defra, London).

14. Coalition of Non-Governmental Organizations (2007). *Principles for the Oversight of Nanotechnologies and Nanomaterials.* Available at: http://www.foeeurope.org/activities/nanotechnology/Documents/Principles_Oversight_Nano.pdf/.

15. Dana, D. (2010), *The Nanotechnology Challenge,* Northwestern Public Law Research Paper No. 10–83, ed. Dana, D., Chapter 1 "Can the Law track Scientific Risk and Technological Innovation? The Problem of Regulatory definitions and Nanotechnology", (Cambridge University Press, 2010), p. 1, http://ssrn.com/ abstract = 1710928/.

16. Davies, J. C. (2006). *Managing the Effects of Nanotechnology* (Project on Emerging Nanotechnologies, Washington, DC).

17. Davies, J. C. (2009). *Nanotechnology Oversight: An Agenda for the New Administration* (Project on Emerging Nanotechnologies, Washington, DC).

18. Department of Environment Food and Rural Affairs (2006). *UK Voluntary Reporting Scheme for Engineered Nanoscale Materials* (Defra, London).

19. D'Silva, J. (2011). What's in a name?—Defining a "nanomaterial" for regulatory purposes in Europe, *Eur. J. Risk Regul.*, **1**, pp. 85–91.

20. D'Silva, J., and Van Calster, G. (2010). For me to know and you to find out? Participatory mechanisms, the Aarhus Convention and new technologies, *Stud. Ethics, Law Technol.*, **4**(2), pp. 1–34.

21. D'Silva, J., and Bowman, D. (2010). To label or not to label?—It's more than a nano-sized question?, *European J. Risk Regul.*, **4**, pp. 420–427.

22. Environmental Defense and DuPont (2007). *Nano Risk Framework* (EDF, New York).

23. Environmental Protection Agency (2008). Notice: nanoscale materials stewardship program, *Federal Register*, **73**(18), pp. 4861–4866.

24. ETC Group (2004). *News Release—Nanotech: Unpredictable and Un-Regulated: New Report from the ETC Group* (ETC Group, Ottawa).

25. European Commission (2004). *Towards a European Strategy for Nanotechnology* (EC, Brussels).

26. European Commission (2007). *Towards a Code of Conduct for Responsible Nanosciences and Nanotechnologies Research—Consultation Paper* (EC, Brussels).

27. European Commission (2008a). *Regulatory Aspects of Nanomaterials: Summary of Legislation in Relation to Health, Safety and Environment Aspects of Nanomaterials, Regulatory Research Needs and Related Measures* (EC, Brussels).

28. European Commission (2008b). *Regulatory Aspects of Nanomaterials* (EC, Brussels).

29. European Food and Safety Authority (2009). *Scientific Opinion: The Potential Risks Arising from Nanoscience and Nanotechnologies on Food and Feed Safety—Scientific Opinion of the Scientific Committee* (EFSA, Brussels).

30. European Group on Ethics in Science and New Technologies (2007). *Ethical Aspects of Nanomedicine* (EC, Brussels).

31. European Parliament (2009). *Press Release: MEPs Approve New Rules on Safer Cosmetics*, 24 March (EP, Brussels).

32. Faulkner A. (2004). *Regulation and Governance of Tissue Engineered Technologies in the United Kingdom and Europe: Social Science Research, Summary of Results.* Available at: http://www.uwe.ac.uk/hlss/research/pdfTERG-FinalReportSummary04.pdf/.

33. Falkner, R., Breggin, L., Jasper, N., Pendergrass, J., and Porter, R. (2009). *Consumer Labelling of Nanomaterials in the EU and US: Convergence or Divergence?*, EERG Briefing Paper (Chatham House, London).

34. Food and Agriculture Organization and World Health Organization (2008). *Joint FAO/WHO Expert Meeting on the Application of Nanotechnologies in the Food and Agriculture Sectors: Potential Food Safety Implications* (FAO and WHO, Rome).

35. Food and Drug Administration (2007). *Nanotechnology—A Report of the U.S. Food and Drug Administration Nanotechnology Task Force* (FDA, Washington, DC).

36. Food Safety Authority of Ireland (2008). *The Relevance for Food Safety of Applications of Nanotechnology in the Food and Feed Industries* (FSAI, Dublin).

37. Gunningham, N., and Rees, R. (1997). Industry self-regulation: an institutional perspective, *Law Policy,* **19**(4), pp. 363–414.

38. Gunningham, N., and Sinclair, D. (1999). Regulatory pluralism: designing policy mixes for environmental protection, *Law Policy*, **21**(1), pp. 49–76.

39. Hodge, G. A., Bowman, D. M., and Maynard, A. D. (2010). *International Handbook on Regulating Nanotechnologies* (Edward Elgar, Cheltenham).

40. House of Lords Science and Technology Committee (2010). *Nanotechnologies and Food* (UK Parliament, London).

41. Hull, M. (2010). *Nanotechnology Environmental Health and Safety: Risks, Regulation and Management,* eds. Hull, M., and Bowman, D. M., Chapter 7 "Nanotechnology Risk Management and Small Business: A Case Study on the NanoSafe Framework", (Springer, London), pp. 247–294.

42. King, A. A., and Lenox, M. J. (2000). Industry self-regulation without sanctions: the chemical industry's responsible care program, *Acad. Manage. J.*, **43**(4), pp. 698–716.

43. Ludlow, K., Bowman, D. M., and Hodge, G. A. (2007). *A Review of Possible Impacts of Nanotechnology on Australia's Regulatory Framework* (Monash Centre for Regulatory Studies, Melbourne).

44. Ludlow, K., Bowman, D. M., and Kirk, D. (2009). Hitting the mark or falling short with nanotechnology regulation?, *Trends Biotechnol.*, **27**(11), pp. 615–620.

45. Marchant, G. E., and Sylvester, D. J. (2006). Transnational models for regulation of nanotechnology, *J. Law Med. Ethics*, **34**(4), pp. 714–725.

46. Marchant, G. E., Abbot, K. W., and Sylvester, D. J. (2009). What does the history of technology regulation teach us about nano oversight?, *J. Law Med. Ethics*, **37**, pp. 724–731.

47. Maynard, A. D., Aitken, R., Butz, T., Colvin, V. L., Donaldson, K., Oberdörster, G. Philbert, M. A., Ryan, J., Seaton, A., Stone, V., Tinkle, S. S., Tran, L., Walker, N. J., and Warheit, D. (2006). Safe handling of nanotechnology, *Nature*, **444**, pp. 267–269.

48. Maynard, A. D., Bowman, D. M., and Hodge, G. A. (2010). *International Handbook on Regulating Nanotechnologies*, eds. Hodge, G. A., Bowman, D. M. and Maynard, A. D., Chapter 26 *"Conclusions: Triggers, Gaps, Risk and Trust"*, (Edward Elgar, Cheltenham) pp. 573–586.

49. Miles, J. (2010). *International Handbook on Regulating Nano-technologies*, eds. Hodge, G. A., Bowman, D. M., and Maynard, A. D., Chapter 5 "Nanotechnology Captured", (Edward Elgar, Cheltenham) pp. 83–106.

50. Monica, J. C., and Van Calster, G. (2009). *Nanotechnology Environmental Health and Safety: Risks, Regulation and Management,* eds. Hull, M. and Bowman, D. M., Chaper 4 "A Nanotechnology Legal Framework", (Springer, London) pp. 97–142.

51. Moran, A. (1995). *Markets, the State and the Environment: Towards Integration,* ed. Eckersley, R., Chapter 3 "Tools of Environmental Policy: Market Instruments Versus Command-and-Control", (Macmillan Education, South Melbourne) pp. 73–85.

52. Moses, L. B. (2007). Recurring dilemmas: the law's race to keep up with technological change, *University of Illinois Law, Technology & Policy*, **2**, pp. 239–285.

53. National Research Council (2001). *Global Networks and Local Vales— A Comparative Look at Germany and the United States* (National Academy Press, Washington, DC).

54. New South Wales Legislative Council Standing Committee on State Development (2008). *Nanotechnology in NSW* (NSW Legislative Council, Sydney).

55. National Industrial Chemical Notification and Assessment Scheme (2007). *Summary of Call for Information and the Use of Nanomaterials* (Australian Government, Canberra).

56. National Industrial Chemical Notification and Assessment Scheme (2008). Industrial nanomaterials: voluntary call for information 2008, *Australian Government Gazette*, **C10** (7 October), pp. 25–38.

57. Organization for Economic Cooperation and Development (2007). *OECD Working Party on Nanotechnology (WPN): Vision Statement* (OECD, Paris).

58. Organization for Economic Cooperation and Development (2008). *Nanotechnologies at the OECD* (OECD, Paris).

59. Poland, C. A., Duffin, R., Kinlock, I., Maynard, A. D., Wallace, W., Seaton, A., Stone, V. Brown, S., MacNee, W., and Donaldson, K. (2008). Carbon nanotubes introduced into the abdominal cavity of mice show asbestos like pathogenicity in a pilot study, *Nat. Nanotechnol.*, **3**, pp. 423–428.

60. Rathenau Instituut (2010) *Reply to the Consultation on the Proposal for a Commission definition of the term 'nanomaterial'.* Available at: http://www.rathenau.nl/fileadmin/user_upload/rathenau/NanoDialoog/Nieuws/reply_form_nanomaterials_-_Rathenau_Institute.pdf/.

61. Renn, O., and Roco, M. C. (2006). Nanotechnology and the need for risk governance, *J. Nanopart. Res.*, **8**, pp. 153–191.

62. Rees, J. (1997). Development of communitarian regulation in the chemical industry, *Law Policy*, **19**(4), pp. 477–528.

63. Rizzuto, P., and Pritchard, B. (2010). Industry developing nanoengineered goods frustrated by regulators' lack of definitions, *Daily Environ. Rep.* **93**, pp. B-1.

64. Royal Commission on Environmental Pollution (2008). *Novel Materials in the Environment: the Case of Nanotechnology* (UK Parliament, London).

65. Royal Society and Royal Academy of Engineering (2004). *Nanoscience and Nanotechnologies: Opportunities and Uncertainties* (RS-RAE, London).

66. Royal Society, Insight Investment, Nanotechnology Industries Association, Nanotechnology Knowledge Transfer Network (2007). *Responsible Nanotechnologies Code: Consultation Draft—17 September 2007* (Version 5) (Responsible NanoCode Working Group, London).

67. Scientific Committee on Emerging and Newly Identified Health Risks (2006). *Modified Opinion (after Public Consultation) on the Appropriateness of Existing Methodologies to Assess the Potential Risks Associated with Engineered and Adventitious Products of Nanotechnologies* (Directorate General for Health and Consumers, EC, Brussels).

68. Scientific Committee on Emerging and Newly Identified Health Risks (2009). *Risk Assessment of Products of Nanotechnologies* (Directorate General for Health and Consumers, EC, Brussels).

69. Scientific Committee on Consumer Products (2007). *Opinion on Safety of Nanomaterials in Cosmetic Products* (Health and Consumer Protection Directorate-General, EC, Brussels).

70. Sinclair, D. (1997). Self-regulation versus command and control? Beyond false dichotomies, *Law Policy*, **19**(4), pp. 529–559.

71. Taylor, M. R. (2006). *Regulating the Products of Nanotechnology: Does FDA Have the Tools It Needs?*, (Project on Emerging Nanotechnologies, Washington, DC).

72. Taylor, M. R. (2008). *Assuring the Safety of Nanomaterials in Food Packaging: The Regulatory Process and Key Issues* (Project on Emerging Nanotechnologies, Washington, DC).

73. United Nations Committee of Experts on the Transport of Dangerous Goods and on the Globally Harmonized System of Classification and Labelling of Chemicals (2009). *Ongoing Work on the Safety of Nanomaterials* (UN, Geneva).

74. United Nations Educational, Scientific and Cultural Organization (2006). *The Ethics and Politics of Nanotechnology* (UNESCO, Paris).

75. United Nations Educational, Scientific and Cultural Organization (2007). *Nanotechnology and Ethics: Policies and Actions—COMEST Policy Recommendations* (World Commission on the Ethics of Scientific Knowledge and Technology, Paris).

76. Utting, P. (2005). *Rethinking Business Regulation—From Self-Regulation to Social Control* (United Nations Research Institute for Social Development, Geneva).

77. Vogel, D. (2006). *The Private Regulation of Global Corporate Conduct* (Centre for Responsible Business Working Paper Series, University of California, Berkeley).

78. Webb, K. and Morrison, A. (1996). The legal aspects of voluntary codes, paper presented at the *Exploring Voluntary Codes in the Marketplace Symposium*, 12–13 September, Ottawa.

Chapter 9

Metrology, Standards and Measurements Concerning Engineered Nanoparticles

Åsa Jämting and John Miles

National Measurement Institute, Department of Industry, Innovation, Science, Research and Tertiary Education, Bradfield Rd, West Lindfield, NSW 2070, Australia

Asa.Jamting@measurement.gov.au

To be able to characterise the various properties of engineered nanoparticles (ENPs) and to communicate the results from such measurements is crucial in the fast developing field of nanotechnology. It is only through a full understanding of each measured quantity that it is possible to move from a research basis towards fully-fledged commercialisation and control of manufacturing processes, ensuring product quality. International standardisation of measurements, sampling procedures and conditions and the development of reference materials are instrumental in facilitating this understanding. In addition, for an open communication between the stakeholders (manufacturers, researchers, regulators and consumers) to take place, there is a need for an agreed upon nomenclature to be developed and disseminated.

Nanotechnology Commercialisation
Edited by Takuya Tsuzuki
Copyright © 2013 Pan Stanford Publishing Pte. Ltd.
ISBN 978-981-4303-28-6 (Hardcover), 978-981-4303-29-3 (eBook)
www.panstanford.com

This chapter introduces some key concepts in the area of measurement science and highlights some of the international efforts that are carried out to provide a solid foundation for documentary standards and measurements at the nanoscale. Specific issues related to the characterisation of ENPs will be further explored in this chapter and there is also a brief summary of some of the most common techniques currently used for dimensional measurements of ENPs and their principles, advantages and limitations.

9.1 Metrology: The Science of Measurement

The term "metrology" is defined as the science of measurement and its application [26]. It follows that nanometrology is the science of measurement in the nanoscale, where the nanoscale is *the size range from approximately 1 nm to 100 nm* [22].

In a lecture to the Institution of Civil Engineers in London, William Thomson (Lord Kelvin) made the famous remark [38]:

> When you can measure what you are speaking about, you know something about it. But when you cannot measure it, your knowledge is of a meagre and unsatisfactory kind. It may be the beginning of knowledge, but you have scarcely advanced to the stage of science.

A saying commonly added to this remark is that you cannot make what you cannot measure and historically metrology has always been a prerequisite for successful manufacturing and commercialisation.

Some of the key terms and concepts used in the field of metrology, as defined in ref. [26], include the following:

- *Measurand*—"the quantity intended to be measured". This needs to be clearly defined and understood. For example, the measurand for the size of a complex shaped nanoparticle (NP) may involve lengths in three dimensions, the aspect ratio, the temperature and the measuring technique used.
- *Reference*—"a measurement unit, a measurement procedure, a reference material, or a combination of such". For example, the length of a given rod may be 5.34 m, a product of a number and a measurement unit, namely the metre.
- *Calibration*—"an operation that, under specified conditions, in a first step, establishes a relation between the quantity

values with measurement uncertainties provided by measurement standards and corresponding indications with associated measurement uncertainties and, in a second step, use this information to establish a relation for obtaining a measurement result from an indication". This complex definition may be summarised as the operation that relates a measurement standard to the reading of an instrument.

- *Measurement Uncertainty*—"non-negative parameter characterising the dispersion of the quantity values being attributed to a measurand, based on the information used". The measurement uncertainty is an estimate of the range of values within which the true value lies. It is a fundamental parameter, as important as the measurement result itself. For example, the diameter of a nanoparticle may be written as 10 nm ± 2 nm, where ± 2 nm is the measurement uncertainty.

- *Metrological Traceability*—"property of a measurement result whereby the result can be related to a reference through a documented unbroken chain of calibrations, each contributing to the measurement uncertainty". The traceability of measurements typically relates measurements to the International System of Units (Système international d'unités–SI). Establishing metrological traceability is crucial if measurements are to be compared and accepted internationally.

9.2 Standards

9.2.1 Physical Standards

A physical standard is the actual physical realisation of a unit of measurement. For example, the SI unit for length, the metre, is the length of the path travelled by light in vacuum during a time interval of 1/299,792,458 of a second. Many national metrology institutes (NMIs) physically realise the metre using the vacuum wavelength of the 633 nm light from an iodine-stabilised helium-neon laser. This laser is then the primary physical standard for he metre in that country. More recently, optical frequency combs are being used instead of iodine-stabilised helium-neon lasers.

National primary standards are disseminated to the community via a chain of instruments and comparisons, establishing metrological traceability. The quality and international acceptance of measurements made in the factory or store depends on their traceability.

The NMIs of many countries are now developing nano-metrology infrastructures, initially for dimensional (length) measurements. A fundamental step is to transfer the realisation of the primary standard for the metre down to measurements at the nanometre (nm) level. Traceability is typically achieved by using primary length standards to calibrate high magnification microscopes, such as an electron microscope (EM) and an atomic force microscope (AFM), fitted with optical interferometers on the translation axes. These microscopes are then used to calibrate grids, gratings and line scales that may be used to calibrate secondary AFMs or EMs.

9.2.2 Documentary Standards

Physical standards and documentary standards are often confused. Documentary standards are published documents that set out specifications and procedures designed to ensure that a material, product, method or service is fit for its purpose and consistently performs in the way intended. Documentary standards establish a common language that defines quality and establishes criteria. The aim of international standardisation is to facilitate the exchange of goods and services through the elimination of technical barriers to trade. International documentary standards provide a reference framework, or a common technological language, between suppliers and their customers, facilitating trade and the transfer of technology.

Three bodies are foremost in the planning, development and adoption of International Documentary Standards: The International Organization for Standardization (ISO) is responsible for all sectors excluding electrotechnical, which is the responsibility of the International Electrotechnical Committee (IEC), and telecommunications, which is the responsibility of the International Telecommunication Union (ITU).

ISO is the world's largest developer of standards. It is a non-governmental network of the national standards bodies of 157 countries, supported by the Central Secretariat based in Geneva, Switzerland. The principal deliverables of ISO are international standards, embodying the essential principles of global openness and transparency, consensus and technical coherence. ISO standards are developed by experts nominated by the national member bodies contributing to the work of the particular committee responsible for the subject matter under consideration.

The need for international documentary standards for nano-technology has been recognised for some time [34,35]. These standards will play a critical role in ensuring that nanotechnology is safely integrated into society and help create a smooth transition from the laboratory to the marketplace. Standards will be required at all points of the nanotechnology value chain—from nanoscale materials that form the building blocks for components and devices to the integration of these devices into functional systems. They will support growth in productivity by encouraging innovation, value generation, compliance and regulation. The production of well-characterised and controlled nanotechnology-enabled products depends on the availability of documentary standards.

This need for nanotechnology standards resulted in ISO forming the technical committee (TC) TC 229—Nanotechnologies in 2005, with the aim to not only develop new standards but also liaise with other TCs to monitor already existing standards which are relevant to nanotechnology. Today, TC 229 has 34 participating countries, each with mirror groups within their own standards organisations (for example: USA ASTM International Committee E56 on Nanotechnology; Australia SA NT-001 Nanotechnologies; UK BSI Committee for Nanotechnologies (NTI/1); Germany DIN) NA 062-08-17 AA Nanotechnologien; China Working Group for Nanomaterial Standardization plus 11 observing countries. TC 229 also liaises with other organisations such as International Bureau of Weights and Measures (Bureau International des Poids et Measures (BIPM), European Environmental Citizens Organisation for Standardisation (ECOS), European Trade Union Institute (ETUI), European Commission Joint Research Centre (JRC), International Union of Pure and Applied Chemistry (IUPAC), Organisation for Economic Co-operation and Development (OECD) and Versailles Project on Advanced Materials and Standards (VAMAS).

Four categories of standards are being developed by TC 229:

- *Terminology and nomenclature standards* provide a common language for scientific, technical, commercial and regulatory processes.
- *Measurement and characterisation standards* provide an internationally accepted basis for quantitative scientific, commercial and regulatory activities.
- *Health, safety and environmental standards* will improve occupational safety, and consumer and environmental protection, and promote good practice in the production, use and disposal of nanomaterials, nanotechnology products and nanotechnology-enabled systems and products.
- *Materials specification standards* will specify the relevant characteristics of manufactured nanoscale materials for use in specific applications.

Many of the documents produced by TC 229 will be anticipatory (developed ahead of the technology that act as "change agents" and guide the market) and horizontal (provide underlying support to a technology or range of technologies but are not themselves application specific) as nanotechnology is still in the early stages of development and evolution. Table 9.1 lists the standards that has been developed and published by TC 229. The ISO website has the full list of standards developed as well as standards under development by TC 229 [24].

There is a wide variety of different ISO TCs and subcommittees (SCs) which liaise with TC 229. Some of these include TC 24/SC 4—Particle characterization, ISO/TC 35–Paints and varnishes, ISO/TC 150—Implants for surgery, ISO/TC 184/SC 4—Industrial data, ISO/TC 201—Surface chemical analysis, TC 202—Microbeam analysis, ISO/TC 206—Fine ceramics, ISO/TC 207—Environmental management, ISO/TC 209—Cleanrooms and associated controlled environments and the ISO/REMCO—Committee on Reference Materials.

For particle size characterisation, TC 24/SC 4 is involved in developing and maintaining standards which particularly concern the techniques commonly used for particle size measurements. The ISO website has the full list of standards developed by TC 24 [25].

Table 9.1 List of standards published by TC 229 (current as of April 2013) [24]*

ISO/TS 10797:2012—Nanotechnologies—Characterization of single-wall carbon nanotubes using transmission electron microscopy

ISO/TS 10798:2011—Nanotechnologies—Charaterization of single-wall carbon nanotubes using scanning electron microscopy and energy dispersive X-ray spectrometry analysis

ISO 10801:2010—Nanotechnologies—Generation of metal nanoparticles for inhalation toxicity testing using the evaporation/condensation method

ISO 10808:2010—Nanotechnologies—Characterization of nanoparticles in inhalation exposure chambers for inhalation toxicity testing

ISO/TS 10867:2010—Nanotechnologies—Characterization of single-wall carbon nanotubes using near infrared photoluminescence spectroscopy

ISO/TS 10868:2011—Nanotechnologies—Characterization of single-wall carbon nanotubes using ultraviolet-visible-near infrared (UV-Vis-NIR) absorption spectroscopy

ISO/TR 10929:2012—Nanotechnologies—Characterization of multiwall carbon nanotube (MWCNT) samples

ISO/TS 11251:2010—Nanotechnologies—Characterization of volatile components in single-wall carbon nanotube samples using evolved gas analysis/gas chromatograph-mass spectrometry

ISO/TS 11308:2011—Nanotechnologies—Characterization of single-wall carbon nanotubes using thermogravimetric analysis

ISO/TR 11360:2010—Nanotechnologies—Methodology for the classification and categorization of nanomaterials

ISO/TR 11811:2012—Nanotechnologies—Guidance on methods for nano- and microtribology measurements

ISO/TS 11888:2011—Nanotechnologies—Characterization of multiwall carbon nanotubes—Mesoscopic shape factors

ISO/TS 11931:2012—Nanotechnologies—Nanoscale calcium carbonate in powder form—Characteristics and measurement

ISO/TS 11937:2012—Nanotechnologies—Nanoscale titanium dioxide in powder form—Characteristics and measurement

ISO/TS 12025:2012—Nanomaterials—Quantification of nano-object release from powders by generation of aerosols

ISO/TR 12802:2010—Nanotechnologies—Model taxonomic framework for use in developing vocabularies—Core concepts

(Continued)

Table 9.1 *(Continued)*

ISO/TS 12805:2011—Nanotechnologies—Materials specifications—Guidance on specifying nano-objects
ISO/TR 12885:2008—Nanotechnologies—Health and safety practices in occupational settings relevant to nanotechnologies
ISO/TS 12901-1:2012—Nanotechnologies—Occupational risk management applied to engineered nanomaterials—Part 1: Principles and approaches
ISO/TR 13014:2012—Nanotechnologies—Guidance on physico-chemical characterization of engineered nanoscale materials for toxicologic assessment
ISO/TR 13014:2012/Cor 1:2012—
ISO/TR 13121:2011—Nanotechnologies—Nanomaterial risk evaluation
ISO/TS 13278:2011—Nanotechnologies—Determination of elemental impurities in samples of carbon nanotubes using inductively coupled plasma mass spectrometry
ISO/TR 13329:2012—Nanomaterials—Preparation of material safety data sheet (MSDS)
ISO/TS 14101:2012—Surface characterization of gold nanoparticles for nanomaterial specific toxicity screening: FT-IR method
ISO/TS 27687:2008—Nanotechnologies—Terminology and definitions for nano-objects—Nanoparticle—nanofibre and nanoplate
ISO 29701:2010—Nanotechnologies—Endotoxin test on nanomaterial samples for in vitro systems—Limulus amebocyte lysate (LAL) test
IEC/TS 62622:2012—Artificial gratings used in nanotechnology—Description and measurement of dimensional quality parameters
ISO/TS 80004-1:2010—Nanotechnologies—Vocabulary—Part 1: Core terms
ISO/TS 80004-3:2010—Nanotechnologies—Vocabulary—Part 3: Carbon nano-objects
ISO/TS 80004-4:2011—Nanotechnologies—Vocabulary—Part 4: Nanostructured materials
ISO/TS 80004-5:2011—Nanotechnologies—Vocabulary—Part 5: Nano/bio interface
ISO/TS 80004-7:2011—Nanotechnologies—Vocabulary—Part 7: Diagnostics and therapeutics for healthcare

*For explanation of the acronyms used in the table, please see the glossary.

9.2.3 Reference Materials

Reference materials (RMs) may also be used to calibrate or verify nanoscale instruments. Reference materials are "sufficiently homogeneous and stable with reference to specified properties, which has been established to be fit for its intended use in measurement or in examination of nominal properties" [26]. A Certified Reference Material (CRM) is a "reference material accompanied by documentation issued by an authoritative body and providing one or more specified property values with associated uncertainties and traceabilities, using valid procedures" [26]. Metrological traceability of a technique or measurement may be established using CRMs.

A number of NP RMs and CRMs of different composition (gold (Au), silicon dioxide (SiO_2), polystyrene latex (PSL)) is commercially available. Some of these materials, their properties and the suppliers are listed in Table 9.2. The German Federal Institute for Materials Research and Testing (BAM) has developed a database in collaboration with TC 229, which lists RMs with properties at the nanoscale [7].

Most of the documentary standards that are available for particle characterisation recommend that regular verification procedures be put in place. For this, it may be sufficient to use a test material that fulfils the assumptions of the measurement technique that it is being used to verify. A number of well-characterised NP materials are commercially available, some of which are listed in Table 9.3.

Table 9.2 Some available NP reference materials (current as of April 2013)

RM name	Material	Certified value (nm)	Techniques used[*]	Supplier	RM type
GBW 12011	Polystyrene	61	TEM	AQSIQ	RM
ERM-FD 100	Silica; reference material	20	TEM, SEM, DLS, DCS, AFM	IRMM	CRM
ERM-FD 304	Silica; quality control	40	DLS, DCS	IRMM	CRM
STADEX SC-0030-A	Polystyrene	29	DMA	JSR	RM

(Continued)

Table 9.2   (*Continued*)

RM name	Material	Certified value (nm)	Techniques used[*]	Supplier	RM type
STADEX SC-0050-D	Polystyrene	48	DMA	JSR	RM
STADEX SC-0055-D	Polystyrene	55	DMA	JSR	RM
STADEX SC-0060-D	Polystyrene	61	DMA	JSR	RM
STADEX SC-0070-D	Polystyrene	70	DMA	JSR	RM
STADEX SC-0075-D	Polystyrene	76	DMA	JSR	RM
STADEX SC-0080-D	Polystyrene	80	DMA	JSR	RM
STADEX SC-0100-D	Polystyrene	100	DMA	JSR	RM
RM8011	Gold	10	AFM, SEM, TEM, DMA, DLS, SAXS	NIST	RM
RM8012	Gold	30	AFM, SEM, TEM, DMA, DLS, SAXS	NIST	RM
RM8013	Gold	60	AFM, SEM, TEM, DMA, DLS, SAXS	NIST	RM
SRM 1964	Polystyrene	60	DMA	NIST	CRM
SRM1963a	Polystyrene	100	DMA	NIST	CRM

[*]For explanation of the acronyms used in the table, please refer to the glossary.

Table 9.3   Some NP test materials for general laboratory use (current as of April 2013)

Supplier[*]	Materials	Size range (nm)
British Biocell International	Silver, gold	2–200
Sigma-Aldrich	Wide range: metals, metal oxides and polystyrene	Wide range
NanoComposix	Silver, gold, core–shell	1–100
Thermo Scientific (previously Duke Scientific)	Polystyrene latex	20–900
Polysciences Inc.	Polystyrene latex	40–800

[*]Certain company names and products are mentioned in the text in order to identify adequately the product referred to. In no case does such identification imply recommendation or endorsement by the National Measurement Institute, nor does it imply that the products are necessarily the best available for the purpose.

9.2.4 Nanoparticle Metrology

The ability to provide accurate, high-precision, traceable measurements of dimensional and other physicochemical properties of NPs is indispensable and underpins a fuller understanding of each measured quantity. It is only through this understanding that it is possible to move from a research basis towards fully-fledged commercialisation and control of manufacturing processes, ensuring product quality. When focusing on NP measurements, the list of properties is extensive.

9.2.4.1 Nanoparticle properties

The properties of interest that can be measured for a NP are described in a recent report from a workshop on NP metrology as follows [37]:

- Morphology
  - ○ Characteristic length and areas in 2D-projection, e.g. x_{Feret}, x_{perim}, x_{Lmax} ...
  - ○ Parameters describing aggregates/agglomerates, e.g. x_{Gyr}, N_{agg}, ...
  - ○ Shape parameters from morphology data, e.g. sphericity, aspect ratio, fractal dimension for x_{Gyr}
- Size related properties based on hydrodynamics and/or interaction with external fields
  - ○ Diffusion coefficient, hydrodynamic diameter (of translation)
  - ○ Settling velocity, Stokes diameter
  - ○ Aerodynamic diameter
  - ○ Acoustophoretic mobility
  - ○ (Partial) scattering or extinction cross section, x_{Gyr}, $x_{projection}$
- Surface area of the dispersed phase
  - ○ Via adsorption of gases, e.g. BET (after Brunauer, Emmett and Teller [40])
  - ○ Via small angle X-ray scattering (SAXS)
  - ○ Via titration experiments with surfactants, polyelectrolytes, etc.
  - ○ Since these methods are not yet mature techniques, and partly only possible with calibration there is need for pre-normative research and standardisation

- Chemical composition and phase
  - ○ Crystallinity (amorphous fraction versus crystalline fraction)
  - ○ Phase fractions of different crystallographic phases
- Concentration of particles
  - ○ Mass, surface, number concentration
  - ○ Total or fractional concentration
- Interfacial properties (which depends on whether the engineered NP is in polar or non-polar suspendants, or in gas)
  - ○ Surface charge
  - ○ Zeta-potential
  - ○ Surface conductivity
  - ○ Pristine point of zero charge and iso-electric point (for different charge determining ions)
- Interaction with continuum/suspendants
  - ○ Solubility and dissolution kinetics
  - ○ ROS (radical oxidising species) potential
  - ○ Wettability.

There is an extensive range of measurement techniques providing information on these parameters. Techniques such as scanning and transmission electron microscopy (SEM and TEM) and AFM provide information about particle size and shape; dynamic light scattering (DLS) and particle tracking analysis (PTA) can be used for size measurements and aggregation studies whereas separation techniques such as differential centrifugal sedimentation (DCS), field flow fractionation (FFF) and size exclusion chromatography can give information about the size distribution of both simple and complex particle suspensions.

For single particle analysis, TEM can be used in combination with energy-dispersive X-ray spectroscopy (EDS) and selected area electron diffraction (SAED), which can also be used to study crystallographic orientation of small volumes. X-ray diffraction (XRD) can be applied to assess bulk properties.

Particle concentration is also a property of interest, where mass concentration can be measured using FFF, combined with spectroscopy. Number concentration can be measured using techniques such as particle tracking analysis, Coulter counter devices and laser-induced break-down spectroscopy (LIBD).

NP interfacial properties, such as surface charge, zeta potential and surface functionalisation can be measured using potentiometric titrations, electrophoretic mobility and surface plasmon resonance (SPR), respectively. X-ray photoelectron spectroscopy has been used to study the redox state of NPs. For these techniques to be quantitative rather than qualitative, rigorous protocols must be applied to each technique. These protocols should include a clear definition of the measurand and calibration and/or verification procedures of both the instrument and the method used.

9.2.4.2 Nanoparticle size measurements

The choice of size as an important measurand for NPs is based on common agreements. For example the Scientific Committee on Emerging and Newly Identified Health Risks (SCENHIR) [6], concluded that "size is universally applicable to nanomaterials and the most suitable measurand".

Many of the properties that make NPs important are strongly related to their dimensional properties. To adequately describe the size of three-dimensional objects such as NPs poses a challenge. The particle morphology is often a determining factor for how the particle dimensions can be described. If the particles are spherical, their size could be described by one single number but for non-spherical particles, other descriptors have to be used, such as length, width and the combination, aspect ratio. Standard terminology and methodology have been developed specifically for this purpose (ISO 9276–6 [11]). To express the variation of properties within a population of NPs statistically, the distribution of the measurand is often described by its mean and variance (ISO 9276–1 [9] and ISO 9276–2 [10]).

When it comes to applying practical measurement techniques, the most common solution is to equate the particle with a sphere, where the measurand is the diameter. Even then, the range of different measurands is extensive as indicated by the following list:

- Volume diameter, x_v, the diameter of a sphere having the same volume as the particle (measurand in for example laser diffraction measurements)

- Surface volume diameter, x_{sv}, the diameter of a sphere having the same surface to volume ratio as the particle (measurand in for example gas adsorption measurements)

- Stokes diameter, x_{Stk}, the diameter of a sphere having the same settling velocity in the laminar flow region ($Re < 0.2$) as the particle (measurand in for example disc centrifugation measurements)

- Projected area diameter, x_a, the diameter of a circle having the same area as the projected area of the particle resting in a stable position (measurand in for example microscopy based image analysis measurements)

- Projected area diameter, x_p, the diameter of a circle having the same area as the projected area of the particle resting in random orientation (measurand in for example dynamic image analysis measurements)

- Feret's diameter, x_F, the mean value of the distance of parallel tangents of the projected outline of the particle position (measurand in for example microscopy based image analysis measurements)

- Martin's diameter, x_M, the mean chord length of the projectedoutline of the particle (measurand in for example microscopy based image analysis measurements)

- Hydrodynamic diameter, x_h, the diameter of a sphere which has the same diffusion coefficient as the particle (measurand in for example DLS and PTA measurements)

- Radius-of-gyration, x_{Gyr}, is used to describe the radius of the radial mass distribution of an object (often non-spherical) around its centre of gravity (used in multi-angle light scattering (MALS) and SAXS measurements). Also referred to as root mean square (RMS) radius.

9.2.5   Sampling and Dispersion

Prior to measuring any parameter related to a NP it is critical to first achieve a representative sample. Following are some of the issues with sampling:

- ensuring that any sub-sample is representative of the whole sample
- minimising sampling errors
- ensuring that the sample is well-mixed before sampling

- choosing appropriate sampling techniques
- documenting the sampling steps.

If the particles are in the form of a dry powder some of the concerns related to dispersions in liquid are associated with

- choice of a suitable suspendant
- choice of surfactant (if any)
- method of dispersing the powder into the suspendant
- wettability of the dry particles by the suspendant
- suspension stability.

If the sample is a dilute liquid suspension, the sampling process is less complicated but care needs to be taken to ensure that the sample is well dispersed before sampling.

Sample dispersion can also be a challenge, particularly for dry powders. ENPs have a strong tendency to aggregate when dried and it is often required to use some form of energy to break up the aggregates in the suspension, using techniques such as ultrasonication and vortexing. Care has to be taken when using ultrasonication, as high power levels can damage the particles themselves.

There are two ISO standards that deal with the issues related to sampling and dispersion [17,18]. These standards, along with other publications [1,27,30] provide guidelines for sampling and dispersion of particles.

9.3 Measurement Techniques for Nanoparticle Characterisation

There is a range of nanoparticle characterisation instruments available and they typically use a variety of physical principles. Unfortunately, this means that different parameters are measured with each technique. Furthermore, the type of sample may be different as some techniques only allow measurements of particles dispersed in liquid suspensions, whereas others can only be applied to particles dispersed in air and others require particles dried on a substrate.

As a consequence, size measurements give a range of results and will depend on the instrument used. In addition, the physical models used for data analysis involve assumptions and simplifications that can cause large differences to occur even when the physical principle is the same. Hence, when comparing the results of different particle characterisation methods, it is important to take into account the different measurands of each technique and how dimensional properties are derived from them.

Particle characterisation techniques are often separated into three different classes. Ensemble techniques measure a large number of particles and provide an average value for the system as a whole. These techniques provide good statistical representation of the particle system but are often unable to resolve contributions from a small number of particles in a broad particle size distribution (PSD), which may be dominated by particles with larger sizes.

Single particle analysis techniques measure the properties of individual particles. These techniques can resolve PSDs in great detail, but the extent to which the measurements are representative of the PSD can be limited due to the relatively small sample sizes. Although it is possible to increase the number of particles that are measured, the time and expense involved often makes such efforts prohibitive.

The third class of techniques is separation techniques, which are based on a sample separation step before applying detection techniques. The fractionation allows the sample to be separated into smaller volume fractions. These can be detected with either an ensemble technique, which now is capable of detecting contributions to the measurement from each fraction, and/or further analyzed using singe particle analysis techniques.

9.4 Selected Nanoparticle Size Measurement Techniques: Benefits and Limitations

In this section, some common techniques used for size measurements of NPs in liquid suspension will be presented, and their advantages and limitations explored. These techniques complement each other so the use of several techniques assists in building up a comprehensive understanding of NPs.

9.4.1 Dynamic Light Scattering

9.4.1.1 Principles

DLS, also known as photon correlation spectroscopy or quasi-elastic light scattering, is an ensemble technique based on light scattering that measures the diffusion coefficient of suspended particles undergoing Brownian motion [2,5]. To perform a measurement, a particle suspension is placed in a cuvette and illuminated by a laser beam (see Fig. 9.1). The particles scatter the light from the beam, which changes in intensity as the particles move randomly in the sample (small particles move more rapidly than larger particles). A detector is used to monitor how rapidly the fluctuations in the scattered light intensity occur which is presented as a correlation curve. Information about the diffusion behaviour of the particles can be determined from this curve by applying well known models to fit the data [32,36]. For particles in a liquid with a low Reynolds number, the diffusion coefficient can then be related to the hydrodynamic diameter, d_h, via the Einstein–Stokes equation [8].

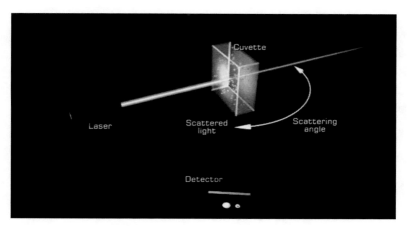

Figure 9.1   Typical DLS configuration: the laser illumination source, the sample under measurement in the cuvette and the detector.

9.4.1.2   Advantages

DLS can provide accurate PSDs for samples that approximate a monodispersed system. The technique is fast and easy to use, and can measure particles with diameters in the range from approximately 1 nm to 5 μm, depending on particle density. The sample concentration is typically limited to 10^7–10^{11} particles/mL. DLS requires only small sample volumes (typically less than 1 mL of suspension at ∼0.1% particle mass fraction). A range of suspendants can be used; keeping in mind that the viscosity and the refractive index of the suspension medium, as well as the temperature of the system, have to be known. Because of its ease of use and quick turnaround, DLS is commonly used in research and for quality control purposes. This technique is non-invasive and the samples can be fully recovered after analysis.

9.4.1.3   Limitations

DLS measurements are sensitive to the quality of the suspension, as the intensity of the scattered light is proportional to the sixth power of the particle size. Thus the scattering signal due to the presence of dust or agglomerated or aggregated particles will obscure the scattering signal from smaller particles. This also makes DLS less suitable for accurate measurements of broad PSDs. For low particle concentrations, there is a risk of number fluctuations due to a low number of particles in the measurement volume. If the particle concentration is too high, errors can occur due to multiple scattering events. For analysis, it is assumed that the particles are spherical and that the sample composition is homogenous. Information on particle shape can only be obtained in instruments equipped with multi-angle or goniometric detector configuration.

9.4.1.4   Instrument performance verification

To ensure that the best possible measurement results are achieved, the DLS instrument performance should be checked regularly. ISO standard ISO 22412 [20] recommends that a verification procedure is performed by measuring 100 nm polystyrene latex spheres with a narrow size distribution. Intermittent checks should be performed as required using suitable reference materials. For more information about such reference materials, see Section 9.2.3.

9.4.2   Laser Diffraction

9.4.2.1   Principles

In a laser diffraction (LD) instrument, the basic set-up consists of a collimated laser beam, focusing lenses, sample, and an array of detectors located at different angles to the transmitted beam. As the laser beam passes through the particle suspension, the particles in the sample scatter the light at angles that are related to their size. The larger particles in the sample scatter light with strong intensity at small angles and smaller particles scatter light with lower intensity at larger angles. A set of detectors is used to monitor the scattering response, and the angular distribution of the scattered light intensity can be converted into a PSD using an appropriate model and a range of assumptions. Assuming that the particles are spherical, and that the refractive index (both real and imaginary) of the particles and the optical properties of the suspendant are known, Mie theory [4] can be used to convert the scattering pattern from all sizes of particles into a volume-weighted PSD. For large particles (diameter ≥ 50 μm) the Fraunhofer approximation [4] may be used, which assumes that the particles are opaque, that particles of all sizes scatter light with the same efficiency and that the particle size is much larger than the wavelength of the light. Particles with diameters much less than the laser wavelength will scatter light uniformly in all directions (Rayleigh scattering [4]) and LD systems are often equipped with backscattering detector to capture the signal from the smaller particles.

9.4.2.2 Advantages

LD is a well-established ensemble technique that allows measurements of particle suspensions over a wide range of particle sizes from 100 nm up to several millimetres. The technique is easy to use and fast (~1 min per measurement run).

The measurand in LD is the volume diameter, d_v, often presented as a volume-weighted distribution of diameters. Recent developments in instrumentation indicate that it is possible to lower the detection limits for LD down to around 30 nm by combining several different laser wavelengths and advanced detectors.

The detection range of the LD technique excludes size measurements of very small particles but can suffice as a very useful tool when assessing complex particle systems containing agglomerates and/or aggregates. In most cases it is possible to recover the samples after analysis.

9.4.2.3 Limitations

LD measurements require the sample concentration to be optimised to avoid multiple scattering effects or particle-particle interactions. Since LD is an ensemble technique, the sample is assumed to be of homogenous composition. One of the requirements for the application of Mie theory to model the results is that the optical properties of the particles and of the suspension medium must be known, and that assumptions about the particle shape have to be incorporated into the model. The technique is limited in its ability to discriminate between different particle populations with closely spaced mean diameters.

9.4.2.4 Instrument performance verification

To ensure that the best possible measurement results are achieved the instrument performance should be checked regularly. The ISO standard ISO 13320 [15] recommends performing measurements on traceable, spherical CRMs (see Section 9.2.3) to ensure accuracy.

9.4.3 Small Angle X-Ray Scattering

9.4.3.1 Principles

SAXS is an ensemble technique that can provide information on the size, shape and aggregation state of NPs. It differs from conventional XRD techniques primarily in that in a SAXS configuration the detector is measuring the scattering from the sample at very low angles (from a few degrees down to a few hundredths of a degree). The instruments use a collimated, monochromatic X-ray beam, and the sample is rotated through a range of angles generating a scan of scattering intensities versus angle. SAXS instruments use a series of beam conditioning optics to control the intensity of the X-ray beam in order to detect the scattering signal at these low angles.

9.4.3.2 Advantages

SAXS is an ensemble technique, which can be used to provide structural information from particles in suspension, powders or aerosols. The technique is non-destructive for most materials and does not require complex sample preparation. The resulting scattering signal is modelled and the results can provide specific information about the sample, such as the radius of gyration, x_{Gyr}, and particle shape, size and shape distributions. By using advanced data analysis software, it is possible to resolve complex multimodal size distributions.

9.4.3.3 Limitations

Traditional SAXS instruments are very expensive and even more so when considering the synchrotron sources that are now providing SAXS beam lines. Measurements and data analysis can be very complex. Sample preparation is important as this is a scattering technique and the presence of large particles or aggregates may suppress the signal from smaller particles. Both the experiments and the analysis can be time consuming.

9.4.3.4 Instrument performance verification

To ensure that the best possible measurement results are achieved the instrument performance should be checked, verified and calibrated as required. ISO standard ISO 13762 [16] recommends verification of instrument performance to be carried out at regular intervals using certified reference materials or particles with a known size distribution. There are NP reference materials available that can be used for instrument verification, as described in Section 9.2.3.

9.4.4 Transmission Electron Microscopy

TEMs are very useful tools in NP characterisation, as they can provide representative images of the samples, as well as measurements on a single-particle basis. For particle sizing, the measurand in TEM is a diameter such as a projected area diameter, x_a, and Feret's diameter, x_F. The number-weighted PSD can be constructed by analyzing a large number of particles.

9.4.4.1 Principles

The basic components of a TEM are an electron source, an electromagnetic lens system, a vacuum system, the necessary electronics, and control software. The principle is based on an electron beam, generated by a thermal filament (typically made from single crystal tungsten wire or LaB_6). For high precision beams, from a field emission gun made from a very sharp tipped tungsten needle where the electrons are expelled by applying a very powerful electric field in close proximity of the filament tip. In the TEM, the electron beam is focused and collimated through a multi-stage lens system before reaching the very thin, electron transparent sample (the thickness of the material is typically 80 nm or less). The beam then continues through another multi-stage lens system which projects an image of the sample onto a fluorescent screen (or camera image plane).

The TEM requires an ultra high vacuum and high voltage for operation. To be imaged, the sample needs to be transparent to the electron beam, and it is usually placed on a copper support grid (about 3 mm in diameter) and placed into a suitable holder.

9.4.4.2   Advantages

TEMs can have resolution limits as low as sub-nanometres and can be used to analyze both large numbers of NPs at lower magnifications and individual NPs at high magnifications. The projected image can be used to find information about the size and shape of the NPs.

The beam is quite wide in normal imaging mode, owing to the sophisticated system of lenses in a TEM, it can also be formed into a very fine (sub-nanometre) focused electron probe. Combined with the thin, electron transparent sample, this enables the generation of diffraction patterns for crystallographic information, and X-rays for sample composition information over very small volumes.

9.4.4.3   Limitations

TEMs are expensive and require expertise to operate and maintain properly. TEM sample preparation can be very time consuming and requires proficiency to achieve high-quality, reproducible results. The sample stage for a TEM is nearly always capable of holding only one sample grid at a time, which can make examining multiple samples very time consuming.

For NP suspensions, the sample has to be deposited on a TEM grid and thoroughly dried prior to imaging which may alter the properties of the particles. The drying process itself can generate artefacts such as apparent aggregation. In addition, for

NP suspensions, the particle concentration needs to be low which may lead to a poor sampling ratio. Particles with low electron density, such as polystyrene particles with diameters less than ~30 nm, are difficult to image using TEM. It can also be very time consuming to generate results that are statistically representative of the entire sample [28].

9.4.4.4 Instrument performance verification

To ensure that the best possible measurement results are achieved, the instrument performance should be checked, verified and calibrated as part of a routine maintenance schedule. Since a typical TEM measurement is usually based on image analysis, calibration of image magnification is crucial. ISO standard ISO 29301 [23] describes image/instrument calibration at a range of magnifications using an artefact that has been designed specifically for this purpose. There are also a range of particulate reference materials of different composition that can be used for image calibration, as described in Section 9.2.3.

9.4.5 Scanning Electron Microscopy

SEM also proves to be a useful tool in NP characterisation, since like TEMs it can provide representative images of the samples, as well as measurements on a single-particle basis. For particle sizing, the measurand in SEM is a diameter such as a projected area diameter, x_a, and Feret's diameter, x_F. The number-weighted PSD can be constructed by analyzing a large number of particles.

9.4.5.1 Principles

The electron-optics column in an SEM is shorter than in a TEM, as there are fewer lenses required to focus the beam. The column typically houses gun alignment coils, lenses that condition the beam into a fine spot onto the sample surface and the scan coils. In SEM, the focused electron beam is raster scanned across the sample. As the beam scans across the specimen, different interactions between the beam and the sample occur. In addition, the depth of interaction of the electron beam with the sample is proportional to the operating voltage of the electron gun. For greater surface sensitivity, low voltages (<5 kV) are often used.

The signals from these interactions can be detected by a range of detectors in the chamber above the specimen. The most commonly used detector picks up the signal from so called secondary electrons (SE); electrons which have been knocked out from their positions by the beam. Different interactions of the beam and sample give images based on topography, elemental composition (X-rays), density variation or crystalline structure of the sample.

9.4.5.2 Advantages

The resolution of an SEM is dependent on the size of the beam but ranges from a few nanometres up to several millimetres. The SEM can be used to study almost any kind of sample, as long as its surface is conductive. This makes sample preparation for SEM less complicated than for TEM, although drying effects and apparent aggregation can still be problematic.

The accelerating voltages used in SEM are much lower than in TEM (tens of kV compared with hundreds of kV) since the beam does not need to penetrate the specimen. As the imaging is not dependent on the density of the sample, it is possible to image samples with lower electron density. The sample chamber in a SEM is considerably larger than in a TEM and often capable of housing many different samples at the same time. The SEM has a large depth of field, which allows a large part of the sample to be in focus at one time and produces an image that is a good representation of the three-dimensional sample. The use of EDS allows for elemental detection, but as the beam-sample interaction volume in an SEM is quite large the element detection has a lower spatial resolution than for TEM.

9.4.5.3 Limitations

Like TEMs, SEMs are expensive and require expertise to operate and maintain. An SEM usually requires a conducting sample. If the sample is non-conducting, a conductive coating may be applied to prevent charge from being built up in the sample by sputtering a thin layer (a few nm thick) of a conductive metal, such as gold, onto the sample surface. However, if dimensional measurements are to be carried out, this has to be done carefully, as any measurement will now show both the feature of interest plus the applied coating. Particles suspended in a liquid have to be deposited onto a suitable substrate and thoroughly dried before inserting into the vacuum chamber. This may cause significant changes to the sample and generate artefacts such as apparent aggregation. The scanning motion of the electron beam in an SEM gives a topographical image, often with some edge distortion caused by strong beam interaction. The complex gray scale of the images can make image analysis challenging. As with TEM, it can be very time consuming to generate results that are statistically representative of the entire sample.

9.4.5.4   Instrument performance verification

To ensure that the best possible measurement results are achieved, the instrument performance should be checked, verified and calibrated as part of a routine maintenance schedule. For SEM calibration, there are two ISO standards, ISO   [19] and ISO/TS24597 [21]. There is also a range of reference particulate materials of different composition that can be used for instrument verification (see Section 9.2.3).

9.4.6   Atomic Force Microscopy

Scanning probe microscopes have facilitated the study of surfaces and surface structures at the nanoscale since their inception [3,29]. From this family of microscopy techniques, the most common is an AFM in which a solid tip probes the surface of a sample and measures the interactions between the tip and the surface.

9.4.6.1   Principles

In AFM the tip is attached to the free end of a cantilever and is brought very close to a surface. The tip is scanned in a raster pattern across the surface and its lateral (represented as the x- and y-directions of a coordinate system) and vertical movement (represented as the z-direction) is detected. The movement in the z-direction is usually detected by focusing a laser beam on the top side of the cantilever, and recording the movement of the beam as the sample surface is scanned. A topographic image is produced by mapping each recorded measurement of the tip-sample interaction and the position (x- and y-coordinates) in the scan where the measurement was made. Image analysis can be used to derive dimensions.

AFMs can be operated in three different modes: contact mode, non-contact mode and intermittent contact mode. Contact mode involves the tip staying in contact with the surface at all times. This has some drawbacks, including the possibility of the tip deforming and damaging the surface of very soft materials, and, on hard materials and rough surfaces, blunting of the tip resulting in reduced image resolution. Non-contact mode uses an oscillating cantilever with high stiffness. The tip is brought into close proximity with the surface such that attractive (predominantly Van der Waals) forces alter the amplitude and shift the frequency and phase of the cantilever oscillation. One limitation of this mode is the difficulty in detecting the weak forces involved which may not be capable of deflecting the tip sufficiently for signal generation. The third mode, intermittent contact mode, also uses an oscillating cantilever. In this mode, the tip is brought closer to the surface than in non-contact mode so that the tip touches (taps) the surface intermittently. At point of contact, the surface features are detected by the resultant change in the amplitude and phase of the cantilever oscillation.

9.4.6.2 Advantages

The resolution of an AFM can be very high. Features from 0.1 nm can easily be detected in the z-direction. The resolution in the x- and y-direction is directly linked to the sharpness and shape of the tip, but features as small as a few nanometres can be imaged. The AFM can generate three-dimensional maps of surfaces and image individual NPs. Some information about size and shape can be deducted from AFM images and it is possible to make particle size measurements based on height measurements (z-direction).

9.4.6.3 Limitations

For measuring the diameter of NPs with an AFM, it is necessary to deposit NPs onto a flat substrate in such a way that results in sub-monolayer islands of NPs or individual NPs attached to the substrate. Only the upward-facing surface of the NPs is imaged; so no information is provided about the downward-facing surface of the NPs.

The information generated using AFM is strongly dependent on the tip shape [39,41], and for imaging and measuring NPs, this presents a significant limitation. The illustration in Fig. 9.2 shows a NP on a substrate being scanned with (a) an infinitely thin tip which exactly tracks the upward-facing surface of the substrate and particle, (b) a more realistic AFM tip with a finite diameter (typically around 10 nm) which produces a broadened image of the particle which is a convolution of the tip and the particle shapes and (c) an AFM image of a polystyrene particle with a nominal diameter of 100 nm deposited on a flat silicon substrate scanned with a typical AFM tip. Figure 9.2c illustrates how only the upward-facing surface of the particle is imaged and how the image of the particle is broadened by the tip in the x- and y-directions. The evenly rounded upper surface of the particle image suggests that assuming the particle closely approximates a sphere and that the diameter can be derived from the height of the particle image is valid.

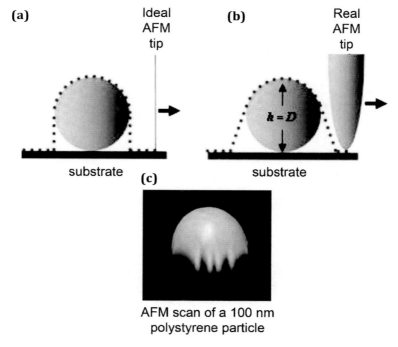

AFM scan of a 100 nm
polystyrene particle

Figure 9.2   A schematic NP on a substrate being scanned with (a) an ideally thin and sharp tip, (b) a more realistic AFM tip and (c) an AFM image of a polystyrene particle with a nominal diameter of 100 nm scanned with a typical AFM tip.

AFM measurements can be very time consuming. Sample preparation for AFM measurement can be complex and deposing particles by drop casting a particle suspension onto a substrate may change the properties of the particles as the suspendant evaporates. Furthermore, the typical range of an AFM scan is only a few micrometres up to less than 100 μm and a high-resolution image can take up to several hours to acquire. The scan range is often even more limited in the z-direction, where the maximum detection range is typically up to 10–15 μm. Like the other microscopy techniques, it can also be very time consuming to image a sufficient number of particles to obtain a measurement result which is statistically representative of the entire sample. It is often necessary to image several hundred particles to achieve this [27].

9.4.6.4 Instrument performance verification

To ensure that the best possible measurement results are achieved the instrument performance should be checked, verified and calibrated as required. To calibrate an AFM, the preferred method is to perform a set of measurements on an appropriate physical standard, which should be chosen according to the requirements. There is a range of suitable artefacts with regular periodic structures of well-known dimensions in one or two dimensions, a comprehensive list of which can be found at http://www.ptb. de/nanoscale/standards/ Standards_preliminary.pdf. In addition, a three-dimensional AFM artefact has recently been developed [33]. For suitable nanoparticle reference materials that can be used for performance evaluation, see Section 9.2.3.

9.4.7 Particle-Tracking Analysis

PTA is a recently established single-particle measurement technique. The detectable size range is ~20–1000 nm, the absolute limits of which depend on the scattering properties of the particles, their size, and the instrument configuration.

9.4.7.1 Principles

The technique is based on particles in suspension being illuminated by a laser beam. When particles undergo Brownian motion are exposed to the laser beam, their scattering cross section can be observed through a microscope and/or tracked using a video camera system. The instrument dynamically tracks individual particles in real time. The individual particle positions are monitored over a set time, and the resulting length of the particle path as well as the scattered intensity is recorded. The particle size is derived from the analysis of the track length and time, through determining the diffusion coefficient which can be related to the hydrodynamic diameter, d_h, via the Einstein–Stokes equation [20].

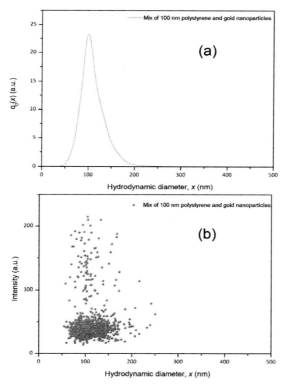

Figure 9.3 PTA measurements of a mixed sample of PSL and Au particles with nominal diameters of 100 nm. (a) Number-weighted PSD, (b) The same dataset, plotted as scattering intensity versus hydrodynamic diameter, indicating the difference in scattering intensity between the larger number of PSL particles with lower scattering intensity versus the smaller number of Au particles with a higher scattering intensity.

The PSD measured using PTA is a number weighted distribution of the hydrodynamic diameter. Different laser wavelengths allow studies of small-sized particles and/or fluorescent particles. By simultaneously measuring size and scattering intensity, it is possible to distinguish similarly sized particles of materials with different scattering properties. This is illustrated in Fig. 9.3, which shows PTA results for a mixed sample of polystyrene and gold particles with nominal diameters of 100 nm.

9.4.7.2 Advantages

The technique only requires a small amount of sample ($\sim$250–500 µL, depending on the sample cell) and the concentration range is 10^7 to 10^9 particles/mL. As the particle size measurement is based on the tracking of individual particles, the results are not greatly affected by size-dependent scattering intensity. It is possible to resolve particle sizes in multimodal mixes to a moderate degree. The technique can be used for *in-situ* studies of particle aggregation [31]. The technique is suitable for use with a range of suspendants, although the viscosity and temperature of the liquid must be known. Being a single-particle measuring technique, PTA can be used to determine particle number concentration.

9.4.7.3 Disadvantages

The PTA analysis model for hydrodynamic diameter assumes that the particles are spherical. The suspendant needs to be optically transparent and the viscosity and measurement temperature must be known. The particle motion is being tracked in a two-dimensional focal plane, and an approximation is made to fit the diffusion behaviour for a particle undergoing three-dimensional Brownian motion. Only a limited number of particles are analyzed, hence only limited statistical relevance can be gained. Even if analysis is carried out for extended periods of time to generate more data, there is a chance that the same particles are analyzed repeatedly. In addition, the data analysis is susceptible to user interpretation and requires a high level of understanding for correct interpretation.

9.4.7.4 Instrument performance verification

To ensure that the best possible measurement results are achieved the instrument performance should be checked and verified as required. PTA is a newly developed technique and thus no standards for this type of instrument currently exist. However, there is a range of reference materials available of different composition that can be used for instrument verification, as discussed in Section 9.2.3.

9.4.8 Differential Centrifugal Sedimentation

9.4.8.1 Principles

DCS is an ensemble technique based on the principle of particle sedimentation induced by centrifugal forces. The configuration described below covers particles that have a density greater than that of their suspending medium, although other configurations are available. The instrument consists of a hollow spinning disc into the centre of which a sample is injected. The disc is mounted on a drive shaft that rotates at a known speed (typically from ~600 rpm up to 24000 rpm). An experiment commences by setting the disc in rotation at the desired speed. The disc is then filled with a density gradient fluid mix, where the density of the fluid increases with the radial distance from the centre. The composition of the gradient can be tailored to suit the material and measurement system. For particles with higher density, a more concentrated gradient is desirable. A barrier layer may also be added to limit any suspendant evaporation.

Only a small amount of sample is required (less than 200 L) and the samples should have a concentration in the range ~50 μg L^{-1} to 1 mg L^{-1}. After injection, the particles sediment radially from the centre of the disc through the fluid following the centrifugal force. A detector beam (usually a light beam, but an X-ray beam configuration is also an option in some instruments) passes through the liquid near the outside edge of the disc, and as the particles pass through the beam, the intensity is reduced as the particles obscure the beam proportionally to the concentration and particle size.

The measurand is the sedimentation time, which can be converted to diameter, x_{Stk}, using Stokes' law and to a volume-weighted PSD using Mie theory, if the optical properties of the particles are known. A schematic illustration of the detection zone of a typical DCS instrument is shown in Fig. 9.4, illustrating the larger or higher density particles passing through the laser beam, followed by the smaller or lower density particles.

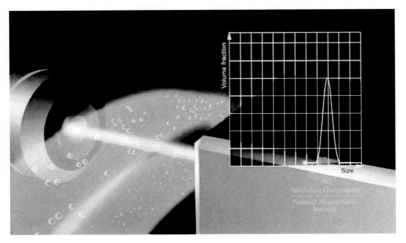

Figure 9.4   Schematic illustration of particles passing through the gradient fluid. The laser beam (left) is obscured by the passing particles, which is detected by the detector (right). Particles with larger sizes or higher densities will reach the detector beam first, followed by the smaller or less dense particles.

9.4.8.2 Advantages

The disc centrifuge is capable of measuring a wide range of particle sizes, from 5 nm to 30 μm, depending on the rotation speed and the density of the particles and density gradient fluid. It is possible to achieve very good size discrimination, as illustrated in Fig. 9.5, which shows the results from a typical DCS measurement of a six-modal gold NP suspension, with nominal particle diameters ranging from 5 to 50 nm.

Typical gradients can be created using sucrose and water, which can easily be adjusted to change the density of the gradient. Other suspendants, such as oil and cell culture media can be used to create the gradient, which means that particles can be measured in an environment that is representative of typical applications such as cosmetics or toxicological studies.

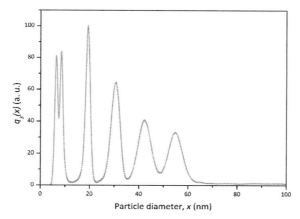

Figure 9.5   Volume-weighted PSD of a six-modal mix of Au nanoparticles, measured by DCS. The plot illustrates the ability of the technique to clearly separate each of the particle populations in the 6 modal Au suspension (nominal diameters: 5, 10, 20, 30, 40 and 50 nm).

9.4.8.3   Disadvantages

The requirement that the particles have to have a higher density than the gradient is the most limiting factor of the technique. The DCS analysis for the Stokes diameter assumes that the particles are spherical and that the sample is of homogeneous composition. The densities and optical properties of both the particle material and the gradient fluid have to be known. Only dilute samples, in the concentration range of 10^7 to 10^9 particles/mL, can be measured. Low-density particles require a low density gradient and high rotational speed. The measurement duration may be long for complex samples, with for example broad PSDs, and it may not be possible to keep measurement conditions constant during the experiment. For the case where the sedimentation velocity is calibrated for each sample, one of the other limitations of the technique is the lack of suitable reference materials available for calibration.

9.4.8.4   Instrument performance verification

To ensure that the best possible measurement results are achieved the instrument performance should be checked and verified as required. In DCS, the sedimentation velocity can be calibrated using reference particles of known diameter and density directly preceding each measurement run. There is a range of reference materials available of different composition that can be used (see Section 9.2.3).

Several ISO standards provide information about best practices in DCS measurements [12–14].

9.4.9 Field Flow Fractionation

FFF covers a group of high-resolution liquid separation methods that can be applied to a wide range of particle suspensions with sizes from approximately 1 nm to 100 μm. The technique provides sequential separation of particles based on a size-dependent interaction of the particles with a force field and is frequently interfaced with various detection systems, such as ultraviolet and visible (UV-Vis) detectors, DLS, MALS, differential refractive index (DRI) detectors and inductively coupled plasma mass spectrometry (ICP-MS).

9.4.9.1 Principles

Several different "fields" can be used to achieve separation in FFF. This includes flow field-flow fractionation (FFFF, often used with an asymmetric flow channel and then called asymmetric FFFF, AFFFF or AF4), sedimentation field-flow fractionation (SFFF), thermal field-flow fractionation (ThFFF) or electrical field-flow fractionation (EFFF). For NP size separation, FFFF or SFFF is the most commonly used methods.

FFF is based on evenly distributed particles flowing through a channel. The channel flow is then subjected to an external field applied perpendicular to the sample flow direction using for example a fluid cross flow in FFFF and centrifugal force or sedimentation in SFFF. Once the external field is applied, the particles in the suspension are forced into a narrow layer along one wall of the flow channel. The particles in this compressed layer interact with both the axial channel flow and the external field, and separation (dependant on the particle size, density, diffusion coefficient or thermal diffusion coefficient, depending on which type of field is applied) occurs.

The mode of fractionation depends on the size of the particles in the suspension, as shown in Fig. 9.6. For small particles undergoing Brownian motion, the particle diffusion coefficient determines the elution sequence: the smaller particles diffuse at a higher rate and are eluted first, followed by the larger sized particle fractions.

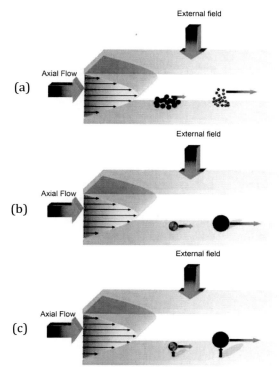

Figure 9.6   Schematic diagram illustrating the typical operation of FFF, showing the mechanism of separation for particles of different size. (a) shows the separation sequence based on particle diffusion coefficients, (b) shows particle separation in steric mode and (c) illustrates the particle separation in hyperlayer mode.

For larger particle systems, with diameters above ~1 μm a different principle occurs, often denoted the steric mode (see Fig. 9.6b). In this mode, the particle sizes are large enough that they no longer interact with the channel wall by diffusion but instead are forced by the external field into a thin layer close to the opposite channel wall. The parabolic channel flow now interacts more directly with the particles, and the elution sequence is reversed: larger particles elute earlier than smaller particles.

Another wall interaction mode is also possible, called the hyperlayer mode, where the particles are subjected to flow-induced hydrodynamic forces and form thin layers some distance away from the wall (see Fig. 9.6c).

9.4.9.2 Advantages

The dynamic range of FFF is very broad allowing particle sizes from nanometres up to several micrometres to be measured. FFFF is capable of continuously separating out PSDs with diameters spanning the range from 1 to 1000 nm The particle size discrimination is very good: narrow peaks in the PSD with as little as 5% difference in mean diameter can be resolved. The technique can be used stand-alone or integrated with auxiliary methods for further characterisation. It is possible to collect the eluted fractions for further analysis, for example by AFM or EM. Using FFF in combination with MALS and DLS can give information about particle shape by comparing the x_{Gyr} (or RMS radius) from MALS to the hydrodynamic diameter from DLS.

9.4.9.3 Disadvantages

The technique requires extensive method development. Without external detection systems such as MALS, DLS, UV-VIS and others, the algorithms for determining the size separation are very complicated. Some of the FFF techniques such as SFFF may require very long experimental run times to separate PSDs of very small particles or broad distributions.

The FFFF techniques uses a channel with a permeable membrane, and issues with the pore size of the membrane with respect to the size of the particles and/or suspendant may occur. The type of material used for the membrane may also affect the results, by causing the particle to interact with the membrane or even irreversibly attach to the membrane.

9.4.9.4 Instrument performance verification

To ensure that the best possible measurement results are achieved the instrument performance should be checked and verified as required. Again, there is a range of NP reference materials available (see Section 9.2.3).

9.5 Summary

This chapter presents some of the challenges associated with measurements at the nanoscale, particularly focusing on nano-particles in suspensions.

Some key concepts common in the metrology area have been defined and the international efforts in developing standards and protocols for use in measurements have been discussed. When first considered, to measure the properties of ENPs may not appear to be a major challenge but, as discussed here, there is a range of issues that present themselves. These challenges are often unique to the ENPs as the properties of nanoscale materials often differ from the properties of the same material in bulk form. Depending on the process used to produce the ENPs and the techniques used to characterise them, different parameters are being investigated.

The use of several different techniques, that complement each other as well as provide some redundancy, is a suitable approach to better understand the particle system. A sample of ENPs may have a very uniform distribution of shapes and sizes, but often they are more complex. The challenge is then to characterise an ensemble of particles by a small number of descriptors such as, for example, size. As indicated in Section 9.2.4.2, even this task is not always trivial. It is often necessary to use separation techniques. These present the particle ensemble to the measure-ment technique in such a way that sub-populations can be measured separately by (for example) separating close multimodal PSDs.

Other challenges involve the following:

- establishing the relevant properties and measurands for different applications
- addressing problems such as agglomeration/aggregation and associated suspensions dynamics
- resolving issues around representative sampling.

Currently, there is no single technique that covers all aspects of particle characterisation. Often the only solution is to use a range of complementary methods in order to ensure that any measurement performed on an unknown sample is representative of that sample. Another challenge is to be aware of the limitations of any technique that is being used. For most of the techniques summarised in this chapter, there are established protocols, such as the ISO standards, that can be used in calibrating the instrument and verifying the measurement technique. These documentary standards also provide a greater insight into the limitations of the applied method.

The use of RMs and CRMs in combination with test protocols is useful in ensuring that the instrumentation is working properly. The use of appropriate RMs can also reduce the uncertainty of a measurement result.

Acknowledgements

The authors wish to thank Dr Jan Herrmann, Mr Malcolm Lawn, Dr Victoria Coleman and Dr Heather Catchpoole for many invaluable discussions and assistance in finalising this manuscript.

Glossary

- AFFFF, asymmetric flow field-flow fractionation
- AFM, atomic force microscope
- AQSIQ, General Administration of Quality Supervision, Inspection and Quarantine of the People's Republic of China
- ASTM, American Society for Testing and Materials
- BAM, Bundesanstalt für Materialforschung und -prüfung (Federal Institute for Materials Research and Testing)
- BET, Brunauer, Emmett, and Teller
- BIPM, Bureau International des Poids et Measures (International Bureau of Weights and Measures)
- BSI, British Standards Institution
- CRM, certified reference material
- DCS, differential centrifugal sedimentation
- DIN, Deutsches Institut für Normung (German Institute for Standardization)
- DLS, dynamic light scattering
- DMA, differential mobility analysis
- DRI, differential refractive index

- ECOS, European Environmental Citizens Organisation for Standardization
- EDS, Energy dispersive X-ray spectroscopy
- EM, electron microscopy
- ENP, engineered nanoparticle
- ERM, European Reference Materials
- ETUI, European Trade Union Institute
- FFF, field-flow fractionation
- FFFF, flow field-flow fractionation
- ICP-MS, inductively coupled plasma mass spectrometry
- IEC, International Electrotechnical Committee
- IRMM, Institute for Reference Materials and Measurements
- ISO, International Organization for Standardization
- ITU, International Telecommunications Union
- IUPAC, International Union of Pure and Applied Chemistry
- JCGM, Joint Committee for Guides in Metrology
- JRC, European Commission Joint Research Centre
- JSR, JSR Corporation, Japan
- LD, laser diffraction
- LIBD, laser-induced breakdown spectroscopy
- MA(L)LS, Multi angle (laser) light scattering
- NIST, National Institute for Standards and Technology
- nm, nanometre (10^{-9} m)
- NMI, national metrology institute
- NP, nanoparticle
- OECD, Organisation for Economic Co-operation and Development
- PSD, particle size distribution
- PSL, polystyrene latex
- PTA, particle tracking analysis
- QCM, quality control material
- RM, Reference Material
- SA, Standards Australia
- SAED, selected area electron diffraction
- SAXS, small-angle X-ray scattering
- SC, subcommittee
- SCENHIR, Scientific Committee on Emerging and Newly Identified Health Risks

- SEM, scanning electron microscopy
- SFFF, sedimentation field-flow fractionation
- SI, Système international d'unités (international system of units)
- SPR, surface plasmon resonance
- TC, technical committee
- TEM, transmission electron microscopy
- ThFFF, thermal field-flow fractionation
- UV–Vis, ultraviolet and visible
- VAMAS, Versailles Project on Advanced Materials and Standards
- VIM, vocabulaire international de métrologie (international vocabulary of metrology)
- XRD, X-ray diffraction

References

1. Allen, T. (1997). *Particle size measurement*, 5th ed. (Chapman and Hall, London).

2. Berne, B. J., and Pecora, R. (2000). *Dynamic light scattering: with applications to chemistry, biology and physics* (Dover Press, New York).

3. Binnig, G., Quate, C. F., and Gerber, C. (1986). Atomic Force Microscopy, *Phys. Rev. Lett.*, **56**(9), pp. 930–933.

4. Bohren, C. F., and Huffman, D. R. (1983). *Adsorption and scattering of light by small particles* (John Wiley & Sons, New York).

5. Chu, B., and Liu, T. (2000). Characterization of nanoparticles by scattering techniques, *J. Nanopart. Res.*, **2**, pp. 29–41.

6. European Commission Scientific Committee (2010). *Scientific basis for the definition of the term "Nanomaterial"*, European Union. Available at: http://ec.europa.eu/health/scientific_committees/emerging/docs/scenihr_o_032.pdf/, last accessed on 12 April 2013.

7. Federal Institute for Materials Research and Testing, Nanoscaled Reference Materials, http://www.nano-refmat.bam.de/en/, last accessed on 15 April 2013.

8. Finsy, R. (1994). Particle sizing by quasi-elastic light scattering, *Adv. Coll. Int. Sci.*, **52**, pp. 79–143.

9. ISO 9276–1:1998: *Representation of results of particle size analysis— Part 1: graphical representation*, International Organization for Standardization, Geneva, Switzerland, 1998. Available at: http:// www.iso.org/iso/iso_catalogue/catalogue_tc/catalogue_detail.htm? csnumber = 25860/, last accessed on 15 April 2013.

10. ISO 9276–2:2001: *Representation of results of particle size analysis— Part 2: Calculation of average particle sizes/diameters and moments from particle size distributions*, International Organization for Standardization, Geneva, Switzerland, 2001. Available at: http:// www.iso.org/iso/iso_catalogue/catalogue_tc/catalogue_detail.htm ?csnumber = 33997/, last accessed on 03 April 2013.

11. ISO 9276–6:2008: *Representation of results of particle size analysis— Part 6: Descriptive and quantitative representation of particle shape and morphology*, International Organization for Standardization, Geneva, Switzerland, 2008. Available at:_http://www.iso.org/iso/iso_ catalogue/catalogue_tc/catalogue_detail.htm?csnumbe = 39389/, last accessed on 03 April 2013.

12. ISO 13318–1:2001, *Determination of particle size distribution by centrifugal liquid sedimentation methods—Part 1: General principles and guidelines*, International Organization for Standardization, Geneva, Switzerland, 2011. Available at: http:// www.iso.org/iso_ catalogue/catalogue_tc/catalogue_detail.htm?csnumber = 21704/, last accessed on 15 April 2013.

13. ISO 13318–2:2007, *Determination of particle size distribution by centrifugal liquid sedimentation methods—Part 2: Photocentrifuge method*, International Organization for Standardization, Geneva, Switzerland, 2011. Available at: http://www.iso.org/ iso/iso_ catalogue/catalogue_tc/catalogue_detail.htm?csnumber = 45771/, last accessed on 15 April 2013.

14. ISO 13318–3:2004, *Determination of particle size distribution by centrifugal liquid sedimentation methods—Part 3: Centrifugal X-ray method*, International Organization for Standardization, Geneva, Switzerland, 2011. Available at: http://www.iso.org/ iso/iso_ catalogue/catalogue_tc/catalogue_detail.htm?csnumber = 31503, last accessed on 15 April 2013.

15. ISO 13320:2009 *Particle size analysis—Laser diffraction methods*, International Organization for Standardization, Geneva, Switzerland, 2009. Available at: http://www.iso.org/iso/iso_catalogue/catalogue_ ics/catalogue_detail_ics.htm ?csnumber = 44929/, last accessed on 12 April 2013.

16. ISO/TS 13762:2001 *Particle size analysis—Small angle X-ray scattering method*, International Organization for Standardization, Geneva, Switzerland, 2001. Available at: http://www.iso.org/iso/iso_catalogue/catalogue_tc/catalogue_detail.htm?csnumber = 22376/, last accessed on 12 April 2013.

17. ISO 14488:2007, *Particulate materials—ampling and sample splitting for the determination of particulate properties*, International Organization for Standardization, Geneva, Switzerland, 2007. Available at: http://www.iso.org/iso/iso_catalogue/catalogue_tc/catalogue_detail.htm?csnumber = 39988/, last accessed on 15 April 2013.

18. ISO 14887:2000, *Sample preparation-Dispersing procedures for powders in liquids*, International Organization for Standardization, Geneva, Switzerland, 2000. Available at: http://www.iso.org/iso/iso_catalogue/catalogue_tc/catalogue_detail.htm?csnumber =25861/, last accessed on 15 April 2013.

19. ISO 16700:2004 *Microbeam analysis—Scanning electron microscopy—Guidelines for calibrating image magnification*, International Organization for Standardization, Geneva, Switzerland, 2004. Available at: http://www.iso.org/iso/iso_catalogue/catalogue_tc/catalogue_detail.htm?csnumber = 30420/, last accessed on 15 April 2013.

20. ISO 22412:2008 *Particle size analysis–Dynamic light scattering (DLS)*, International Organization for Standardization, Geneva, Switzerland, 2008. Available at: http://www.iso.org/iso/iso_catalogue/catalogue_tc/catalogue_detail.htm?csnumber = 40942/, last accessed on 12 April 2013.

21. ISO/TS 24597:2011 *Microbeam analysis—Scanning electron microscopy—Methods of evaluating image sharpness*, International Organization for Standardization, Geneva, Switzerland, 2011. Available at: http://www.iso.org/iso/iso_catalogue/catalogue_tc/catalogue_detail.htm?csnumber = 55760/, last accessed on 12 April 2013.

22. ISO/TS 27687:2008 *Nanotechnologies—Terminology and definitions for nano-objects—Nanoparticle, nanofibre and nanoplates*, International Organization for Standardization, Geneva, Switzerland, 2008. Available at: http://www.iso.org/iso /iso_catalogue/catalogue_tc/catalogue_detail.htm?csnumber = 44278/, last accessed on 15 April 2013.

23. ISO 29301:2010 *Microbeam analysis—Analytical transmission electron microscopy—Methods for calibrating image magnification by using reference materials having periodic structures*, International Organization for Standardization, Geneva, Switzerland, 2010. Available

at: http://www.iso.org/iso/iso_catalogue/catalogue_tc/catalogue _detail.htm?csnumber = 45399/, last accessed on 12 April 2013.

24. ISO technical committee TC229 Nanotechnologies, http://www. iso.org/iso /iso _technical_committee.html?commid = 381983/, last accessed on 12 April 2013.

25. ISO technical committee TC24/SC4, Particle characterisation, http:// www.iso.org/ iso/iso_technical_committee.html?commid = 47176/, last accessed on 12 April 2013.

26. JCGM 200:2008 *International vocabulary of metrology—Basic and general concepts and associated terms (VIM)*. International Organization for Standardization, Geneva, Switzerland, 2008. Available at: http:// www.bipm.org/utils/common/documents/jcgm/JCGM_200_2008. pdf/, last accessed on 12 April 2013.

27. Jillavenkatesa, A., Dapkunas, S. J., and Lum, L-S. H. (2001). *Particle Size Characterization*, National Institute of Standards and Technology, Special Publication 960–1, Available at: http://www.nist.gov/public_ affairs/practiceguides/SP960–1.pdf/, last accessed on 12 April 2013.

28. Masuda, H., and Gotoh, K. (1999). Study on the sample size required for the estimation of mean particle diameter, *Adv. Powder Technol.*, **10**, pp. 159–173.

29. Meli, F. (2005). *Nanoscale calibration standards and methods*, eds. Wilkening, G., and Koenders, L., Chapter 27 "Lateral and vertical diameter measurements of polymer particles with a metrology AFM" (Wiley-VCH, Weinheim) pp. 359–374.

30. Merkus, H. J. (2009). *Particle size measurements: fundamentals, practice, quality* (Springer Particle Technology Series, Springer).

31. Montes-Burgos, I., Walczyk, D., Hole, P., Smith, J., Lynch, I., and Dawson, K. (2010). Characterisation of nanoparticle size and state prior to nanotoxicological studies, *J. Nanopart. Res.*, **12**, pp. 47–53.

32. Morrison, I. D., Grabowski, E. F., and Herb, C. A. (1985). Improved techniques for particle size determination by quasi-elastic light scattering, *Langmuir*, **1**, pp. 496–501.

33. Ritter, M., Dziomba, T., Kranzmann, A., and Koenders, L. (2007). A landmark-based 3D calibration strategy for SPM, *Meas. Sci. Technol.*, **18**, pp. 404–414.

34. Roco, R. (2005). *The national nanotechnology initiative: research and development leading to a revolution in technology and industry*, National Science and Technology Council Report, Nanoscale Science, Engineering and Technology Subcommittee on Technology.

35. Royal Society and the Royal Academy of Engineering (RS-RAE) (2004). *Nanoscience and nanotechnologies: opportunities and uncertainties*, London: Royal Society & Royal Academy of Engineering Royal Society and Royal Academy of Engineering report, Section 47, p. 76.

36. Ruf, H. (1993). Data accuracy and resolution in particle sizing by dynamic light scattering, *Adv. Coll. Int. Sci.*, **46**, pp. 333–342.

37. Stintz, M., Babick, F., and Roebben, G. (2010). *Workshop report.* Available at http://www.euspen.eu/page1420/Resources/Eng-Nanoparticles/, last accessed on 12 April 2013.

38. Thomson, W. (1891). Electrical units of measurement, *Popular Lectures*, **1**, pp. 73–76.

39. Villarrubia, J. S. (2004). *Applied scanning probe methods*, eds. Bhushan, B., Fuchs, H., and Hosaka, S., Chapter 5 "Tip characterization for dimensional nanometrology" (Springer, Berlin Heidelberg) pp. 147–168.

40. Webb, P. A., and Orr, C. (1997). *Analytical methods in fine particle technology*, Chapter 3 "Surface area and pore structure by gas adsorption," (Micromeritics Instrument Corporation, Norcross, GA) pp. 53–152.

41. Yacoot, A., and Koenders, L. (2008). Aspects of scanning force microscope probes and their effects on dimensional measurement, *J. Phys. D: Appl. Phys.*, **41**, pp. 103001–103046.

Chapter 10

Safety of Engineered Nanomaterials and OH&S Issues for Commercial-Scale Production

Paul F. A. Wright[a,c] and Neale R. C. Jackson[b,c]

[a]*School of Medical Sciences, RMIT University, Bundoora VIC 3081, Australia*
[b]*School of Applied Sciences, RMIT University, Melbourne VIC 3000, Australia*
[c]*Nanosafe Australia*

paul.wright@rmit.edu.au

10.1 Introduction

The development of engineered nanomaterials (ENMs) and nano-technology applications has grown exponentially, along with the awareness of nanosafety issues in government, industry and public groups. A problem for the nanotechnology-related industries is that there is also a growing public concern from the negative perception in certain high-profile groups of an uncontrolled proliferation and release of nano-enabled products that have not been adequately tested for their safety to humans and the environment. There are indications that this may be impacting on nanotechnology development, with the rapid increase in the worldwide "nano" share of venture capital funding from 2001 to 2002, levelling off after 2002 along with the initial wave of nanotoxicology reports in the scientific literature [47].

Nanotechnology Commercialisation
Edited by Takuya Tsuzuki
Copyright © 2013 Pan Stanford Publishing Pte. Ltd.
ISBN 978-981-4303-28-6 (Hardcover), 978-981-4303-29-3 (eBook)
www.panstanford.com

It is important to first note that there are many types of ENMs—not all are alike or a potential hazard, and the possibility of exposure to nanoparticles (NPs) shed by ENMs is a major factor. The next step is to undertake a whole of life cycle analysis to determine if there are any "hotspots" of potential exposure to shed NPs. From this one can identify who or what may be exposed, as well as the eventual fate of these NPs, and whether there is the potential for adverse biological effects from the exposure scenarios. The physicochemical characteristics of NPs are very important in cellular uptake, which leads to their biological effects and their potential to cause toxicity.

The smart development and manufacture of nano-enabled products would therefore involve a safety-by-design approach (also termed by green chemistry specialists as the "benign-by-design" approach), where nanosafety information is used in the early development of nanomaterials to provide products that can be marketed as being safer to use, for consumers and the environment. Effective nanosafety research for those nanomaterials being scaled up to production will also help provide nano-manufacturers with a "licence to operate", as well as demonstrate their legal duty of care to provide a safe workplace for their employees, and also minimise environmental impact.

It is therefore important for nanomaterial developers and manufacturers to understand the major factors influencing the bioactivity of NPs in order to re-engineer nanomaterials to reduce their intrinsic hazard, or use appropriate workplace and environmental controls to reduce the potential risk of exposure to nanomaterials of concern.

This chapter provides an overview of the toxicology of ENMs and the occupational health and safety (OH&S) issues and workplace controls for using ENMs. In line with the special focus of this book, the majority of examples in this chapter concern inorganic engineered nanoparticles.

10.2   Overview of Nanotoxicology

Focus Point 1: Nanoparticles are most likely to have higher toxicity than bulk material if they are insoluble, penetrate biological membranes, persist in the body or are long and fibre-like.

Focus Point 2: The physicochemical characteristics of nanoparticles are very important in cellular uptake, which leads to their biological effects.

Focus Point 3: Incorporating the safety-by-design approach in nano-material development provides a marketing edge for products that have a reduced potential impact on health and the environment.

10.2.1 Toxic Potential of Nanoparticles

Epidemiology and toxicology studies have recently linked exposure to NP from anthropogenic sources, such as diesel exhaust, to increased mortality rates, impaired lung function and cardiovascular disease. Evidence is mounting that some nanomaterials may alter critical body processes, while further potential health risks are posed by the rapidly expanding applications of nanostructured materials, and the development of numerous engineered NP and their hybrid forms with modified surface characteristics to enhance their functionality and applications (reviewed in [46]).

Minimal knowledge of health effects for nanomaterial exposure is now seen as a major hurdle to the growth of this burgeoning industry. Information concerning the potential hazards, biomarkers and safe handling of these materials is urgently needed to generate evidence-based risk assessments, which will allow the scientific development and technical exploitation of nanomaterials [30,40].

To date, there have been few published systematic or comprehensive nanotoxicology studies undertaken that would allow adequate risk assessment of engineered NP, although the findings will soon be made available of the safety testing of representative manufactured nanomaterials by international research consortia, co-ordinated by the Organisation for Economic Co-operation and Development (OECD) (as detailed in Section 10.2.7 on latest initiatives in nanosafety research). The comprehensive database of nanotoxicology publications listed by the International Council on Nanotechnology (ICON) (http://icon.rice.edu/, based at Rice University, Texas, USA [23]) shows that the initial emphasis has been on inhalation studies and exotic NP, such as carbon nanotubes (CNTs) and fullerenes, with a growing number of studies now investigating well-established NPs in the marketplace, i.e. zinc oxide (ZnO) and titanium dioxide (TiO_2) NP in sunscreens, personal care products and coatings, and silver NP (nano silver) used for biocidal applications.

There does not appear to be a nano-specific toxicity that is novel compared with toxicity caused by the bulk form of the same material—only changes in biological potency and distribution within the body. Nanoparticles are most likely to have higher toxicity than the bulk material, if they are insoluble, penetrate biological membranes, can persist in the body, or are long and fibre-like (termed "nanoparticle of concern", NPOC) [26]. The insolubility and biological penetrance characteristics impart an increased uptake by the body (termed "bioavailability") compared with the bulk material, while persistence within the body may be due to either extensive tissue distribution and binding, or sequestration and slow remobilisation from such tissue deposition sites within the body [26]. The physical characteristic of being long, thin and fibre-like relates to the well-known toxicity of bio-persistent long fibres, including long fibre asbestos [9].

Previous research has suggested that a nanomaterial's toxicity is likely to be due to physicochemical properties that are not routinely monitored in traditional toxicity studies for substances, including particle size and size distribution, agglomeration state, shape, crystal structure, chemical composition, surface area, surface chemistry, surface charge and porosity [12,41].

The small size of NPs and their greatly increased surface area per mass, means that there is also a far greater proportion of more reactive surface atoms with elongated electron orbitals [40]. The enhanced chemical reactivity of NPs compared with their larger particle counterparts potentially results in increased production of reactive oxygen species (ROS), including free radicals, that may result in oxidative stress of exposed cells leading to mitochondrial and DNA damage [58]. Photoactivity is also important for some NPs, as TiO_2 and ZnO NPs can produce free radicals that cause *in vitro* DNA damage to human skin cells when exposed to ultraviolet light, even though they are considered to be relatively non-toxic metals [35]. However, results from such *in vitro* photogenotoxicity studies need to be evaluated as part of an overall "weight of evidence" approach, which assesses a range of both *in vivo* and *in vitro* studies, considers the dose ranges that are necessary to achieve the biological effects, and the likelihood of these effects occurring at the target site of toxicity. In the case of ZnO NPs, the photogenotoxic effect only occurred in a few *in vitro* genotoxicity tests, while all *in vivo* and most *in vitro* genotoxicity tests were negative for ZnO—indicating that ZnO is not likely to cause this effect during sunscreen use.

Therefore, to fully understand the potential toxicity of engineered NPs, their structural, physical and chemical properties need to be comprehensively elucidated, in both starting materials and within the actual physiological environments they encounter. This is easier to perform in simple *in vitro* test systems that involve biologically relevant fluids and/or cells, but characterising NP-cell interactions *in vivo* can be further complicated by difficulties in tracking NPs in an organism, as well as it's active/adaptive processes of bio-transformation, clearance and defensive/immune responses [45]. Fortunately, the potential for NPs to interact with several specific biological pathways may be screened for by using simple *in vitro* test systems with proteomic/genomic arrays and/or functional assays. But the relevance of such *in vitro* toxicity findings should always be confirmed where possible using *in vivo* studies via appropriate exposure routes, as this takes into account the complex interactions between organ systems that may modify toxic effects, and whether the NPs reach the relevant target organ site in sufficient amounts, and for long enough, to cause toxicity (known as the "toxicokinetic" characteristics).

It is also important to remember that humans are not necessarily the most sensitive species when exposed to substances, especially to dissolved metal ions, so ecotoxicology issues and the potential for accumulation of some NP types in certain ecosystems should also be considered.

10.2.2 Toxicokinetic Characteristics of Nanoparticles

To assess the potential for nanoparticles to be taken up the body and accumulate following exposure, it is important to understand the nanomaterial characteristics that influence the toxicokinetic parameters of absorption, distribution, metabolism and excretion (also known as the "ADME" parameters) (reviewed in [12]).

10.2.2.1 Absorption

Exposure to NPs can potentially occur via three routes: inhalation, oral ingestion and dermal exposure. Animal studies indicate that the inhalational route of exposure leads to absorption of NP into the body. The size of inhaled particles influences the region of the respiratory tract that receives the majority of particle deposition, with smaller particles finding their way into the alveolar sacs. Large inhaled particulates can be removed from the lung by muco-ciliary clearance, after which they can also be swallowed, adding to the amount of exposure via the oral route. Smaller particulates are engulfed by alveolar macrophages or translocate across the alveolar wall into the capillaries and, thereafter, mainly to the liver and spleen via the blood stream [41].

Inhaled metal oxide, gold or carbon nanoparticles can also deposit at the rear of the nasal airway and translocate across cell membranes into the olfactory bulb and along nerves to the brain [11]. Numerous inhalation studies in rodents have used very high doses of NPs by intratracheal instillation that were well in excess of the <0.4 mg/kg dose recommended to avoid pulmonary particle overload–the resultant overwhelming of the pulmonary clearance mechanisms is difficult to relate to low-dose exposures in humans but typically leads to macrophage activation, inflammation, lung fibrosis and cancer [9].

Oral ingestion of NPs can result in absorption across the intestinal wall, with smaller particles being absorbed better—these can pass through the portal blood stream to the liver and other organs. The mechanism of uptake from the gastro-intestinal tract is under investigation for certain NPs; however, smaller particles are taken up more readily, and it is presumed that specific cell types are involved (e.g. M-cells that sample the gut lumen contents). There is also the potential for some NPs (TiO_2 and SiO_2) to trigger immune responses in intestinal dendritic cells [12].

Skin penetration following NP exposure is minimal in humans. Comparative animal dermal studies have shown that human skin is less penetrable by NPs that pig or rodent skin [7]. Indications of negligible penetration of certain NPs through healthy human epidermis in the short term [61], are now being followed up studies involving NP exposure to damaged (e.g. sunburnt) human skin [4] and skin flexure, which may lead to increased transdermal penetrance of NPs. However, the issue of chronic dermal exposure over decades (e.g. with cosmetic use) will be more difficult and complex to assess.

10.2.2.2 Distribution

The exposure route influences the target sites of NP distribution in the body, but overall the main target organ for particulate accumulation is the liver, followed by the spleen, lungs and kidneys [12]. This is not surprising as the liver sinusoids are responsible for the capture and elimination of viruses and other particulates larger than 10–20 nm from the circulation. However, NPs can also directly penetrate across the blood–brain and testes barriers [41].

10.2.2.3 Metabolism

Unlike fullerenes (e.g. C_{60}), there is no evidence of metabolism of inorganic engineered nanoparticles, although partially oxidising and dissolving NPs (e.g. nano silver) can generate metal ions that are then complexed or otherwise processed within biochemical pathways that handle either mono- or di-valent metal ions [12].

10.2.2.4 Excretion

Particle size is very important in the mode of elimination from the body. Generally, particles with a hydrodynamic diameter (i.e. including the plasma proteins that bind to the NP surface) of greater than 5 nm cannot readily pass through the glomerular filter of the nephrons within the kidney in order to get into the urine and so do not undergo renal clearance [4].

10.2.3 Mechanisms of Nanoparticle Toxicity in Biological Systems

In order to understand the potential hazard of certain engineered nanomaterials once they have entered the body, it is important to consider the mechanisms by which nanoparticles enter cells and subsequently cause toxicity. Nature itself is the master nanotechnologist as cellular processes occur at the nanoscale. The most critical characteristics that affect nanoparticle uptake by cells are the particle size, surface charge or modification and the specific cell type involved (Table 10.1). Changes in one (or more) of these parameters can cause major differences in the efficiency and type of cellular uptake [52].

Table 10.1   Different factors which affect nanoparticle uptake by cells

NP factor	Effect on cellular uptake of nanoparticle
Particle size	• Smaller nanoparticles are taken up/internalised with greater efficiency than larger particles of the same chemical composition. • Small nanoparticles bypass cellular degradation pathways better than larger particles.
Particle surface charge	• Cellular membranes are generally negatively charged and therefore positively charged particles are taken up preferentially into live cells. • Nanoparticle surfaces that are positively charged at the low pH found in endosomes may undergo "endosome escape" into the cytosol.
Particle surface modification	• Receptor-mediated uptake mechanisms occurs for some nanoparticles with specific ligands conjugated to their surfaces—dependent on the expression of specific cell surface receptors; can also be used to target specific cell types. • Endosomes can be bypassed and rapid uptake achieved by conjugating protein transduction domains onto the surface of nanoparticles. • Oligodeoxynucleotide conjugation aids specific sub-cellular localisation due to the presence of complementary cellular deoxyribonucleic acid (DNA) sequences in organelles, such as the nucleus or mitochondria. • Nanoparticle surfaces bind proteins when they come in contact with biological fluids, forming a protein corona that influences how the particles are processed by the cell.

Source: Adapted from [52].

10.2.3.1   Particle uptake pathways in cells

Due to their small size, nanoparticles are easily taken up into cells by a range of uptake pathways used for macromolecules. Such macromolecules must be carried into the cell in membrane-bound vesicles derived by the invagination and pinching-off of pieces of the plasma membrane—a process termed "endocytosis". Endocytosis occurs by multiple mechanisms that involve either "phagocytosis" (cell eating), which takes up large particles, or "pinocytosis" (cell drinking) that involves the phagocytosis of fluid-filled vesicles by all cell types [5].

Receptor-mediated endocytic mechanisms involve receptors on the cell membrane that capture particles or substances via binding to specific ligands. These include the following, with the first three uptake pathways being part of pinocytosis [59]:

- clathrin-mediated endocytosis, the classical and well-described mechanism involving clathrin, the cage-forming protein that forms pits in the membranes of most nucleated cells (<150 nm);
- caveolin-mediated endocytosis, the clathrin-independent uptake mechanism via flask-shaped cholesterol-rich invaginations of cell membranes (50–80 nm);
- caveolin- and clathrin-independent, yet lipid raft-associated endocytosis, which is specific for the internalisation of certain cytokine receptors and proteins (~90 nm); and
- phagocytosis is typically restricted to specialised mammalian cells, i.e. predominantly by macrophages and Langerhans cells (immature dendritic cells in the skin), which are antigen-presenting cells that act as sentinels of the immune system (>500 nm). Langerhans cells, and stimulated macrophages and endothelial cells, also perform the receptor-independent uptake process of macropinocytosis (0.5–5 µm).

Consequently, endocytosis of virus-sized particles (20–200 nm) occurs in most nucleated eukaryotic cells, while the endocytosis of bacteria and larger particles (>0.5 µm) occurs in antigen-presenting immune cells. Based on the relative size of these invaginations in the cell membranes, individual nanoparticles will be taken up by receptor-mediated endocytosis, while the uptake of their larger agglomerates/aggregates is via phagocytosis and the receptor-independent endocytic pathway of macropinocytosis, whereupon the nanoparticles have essentially free access to all cellular compartments, depending on the intracellular processing that occurs [59].

10.2.3.2 Protein corona effects

Once nanoparticles have entered the body, the nanoparticle surface binds proteins present in biological fluids—this forms a protein corona, conferring a "biological identity" that greatly influences how the particles are "seen" and processed by the cell [31]. Thus, the exposure route is significant, as it affects the nature of the endogenous coating of protein and lipids absorbed *in situ* onto the NP surface. The high surface energies of pristine NPs means that they are rapidly coated with biomolecules, but these exchange over time until it stabilises to form a hard protein corona that survives entry to cell or early processing. However, lysosomes are able to digest off the hard protein corona via proteolytic activity in their acidic microenvironment of pH 4.8. This initial exchange of the soft corona before the NP enters the cell implies that there may be a time-scale to the surface appearance [32]. This also has important implications for environmental exposures with regard to the potential for NPs to act as nanovectors and also for some of the early short-term *in vitro* toxicity studies which may not have pre-incubated the test NPs sufficiently to obtain a stable corona.

So while entry into non-immune cells is faster for smaller particles, the surface characteristics matter because the bio-distribution of NPs is governed by the surface protein binding, as NPs travel into cells via existing energy-dependent pathways to sorting endosomes [8]. Thus, most NPs are processed as protein-coated particles and as there is no cellular export, there is a different rate of accumulation at different cell cycle phases. However, slowly oxidising and/or dissolving nanomaterials, such as Ag and ZnO, are different as they can be degraded and release high localised concentrations of ions. Similarly, phototoxic NPs may exhibit additional toxicity if they distribute to living cells that are also exposed to UV light. Larger-sized particles go into the phagosomes of macrophages, where cells can get burst and damage of lysosomes by different mechanism [8]. Unfolding of proteins in the corona caused by the tight binding and wrapping of proteins around the curvature of NPs is not common, but under some circumstances could potentially lead to inflammatory pathways [32]—as described in the section on the potential for immune effects (Section 10.2.4.6). Protein refolding by NPs may also cause protein fibrillation and the formation of protein aggregates and plaques; however, this uncommon effect and its implications are also under further investigation.

The effects of the protein corona highlights the differences between particles and chemicals in test systems, i.e. chemicals partition into cells, while nanoparticles are processed by cells.

Small particles that are not excreted by the kidneys or evade capture by the reticulo-endothelial system (including the liver and spleen), can circulate longer than larger particles and interact with non-specialised cells and cross blood–organ barriers. Recent studies have reported that different kinds of proteins can bind the NP surface at the different plasma/serum concentrations present for *in vitro* tests versus *in vivo* exposures. As many are signalling proteins, we cannot solely depend on *in vitro* assays that typically use 10% serum in cell culture media, and therefore need to also investigate the effect of pre-incubating in whole serum [32].

10.2.4 Summary of Nanoparticle Bioactivity

Certain NP characteristics are important in exerting their biological effects, in which several intracellular organelles are potentially involved–especially the endosomes that are involved in the processing of endocytosed particulates, and the mitochondria that are the "energy powerhouses" of the cell and very sensitive to redox imbalances within the cell, which can lead to oxidative stress. These NP characteristics include particle/agglomeration size, surface area, charge and chemistry, particle shape and crystalline form, target cell characteristics, and the presence of contaminants (reviewed in [9,12,25]).

10.2.4.1 Nanoparticle size, and surface area, charge and chemistry effects

The effects of these particle characteristics on cellular uptake are listed in Table 10.1. Thus, NP cytotoxicity can be progressively reduced by polymer coatings of increasing thickness, preventing cellular uptake and degradation. The formulation of the NP is also a very important determinant of their behaviour and penetrance characteristics in a biological system, as polymer (polyethylene glycol) coated NPs can diffuse far more rapidly through mucus than the uncoated versions.

Polarity (hydrophilicity) can influence toxicity; for instance, carbonaceous NPs (fullerenes) are less cytotoxic when derivatised to increase their polarity, although this will also alter their agglomeration state in aqueous systems and, therefore, will also change their delivered dose to humans and the environment. Surface chemistry affects the extent of serum protein binding to

NPs, which forms the dynamic coating of NPs when in the body that can alter the pathway and rate of clearance from the body depending on the final hydrodynamic diameter of the particulate in biological fluids (as detailed in the ADME parameters in Section 10.2.2).

10.2.4.2 Nanoparticle shape and form effects

Different forms of the nanomaterial can differ in toxicity despite having the same chemical composition, e.g. the many forms of carbonaceous NPs (bucky balls and CNTs) only consist of pure carbon, but their cytotoxicity varies widely. The aspect ratio of fibre-like nanomaterials is important, as not all CNTs are alike in their biological effects. Only straight CNTs that form "asbestos-like" aggregates longer than a macrophage may engulf (i.e. over 15 μm) can induce the asbestos-like effects of mesothelioma in animal models. Short or tangled CNTs (or indeed short fibre asbestos) do not appear to cause mesothelioma but can still generate other proinflammatory particle effects. The crystallinity state is also important for certain nanomaterials such as nano silica and TiO_2, for which NPs of different crystalline forms have different reactivities (as detailed in Sections 10.2.6.1 and 10.2.6.6 on specific inorganic nanomaterials).

10.2.4.3 Nanoparticle effects on target cells

Cell types can vary greatly in their susceptibility to cytotoxicity induced by identical NPs. Selection of cell types for *in vitro* exposure studies is critical as immortalised cells often lack components of apoptotic and cell cycle processes (and various transport, biotransformation and inflammatory pathways), that may render them less sensitive to cytotoxicity from NP exposures. Conversely, cell lines with altered tumour suppressor genes may also be more sensitive to genotoxicity caused by *in vitro* NP exposures. Cell death by either apoptosis (controlled cell death) or by necrosis (cell death that leads to inflammation) have profoundly different implications for the surrounding tissues, and determining the type of cell death is important for designing effective handling strategies for different NMs.

The immune system is very important in clearing particulates from the circulation but can result in their accumulation/ sequestration within cells making them crucial targets for potential toxicity from NP exposure (as detailed in Section 10.2.4.6 on immunotoxic potential).

Overall, when employing the weight of evidence approach concerning cell-specific effects of nanomaterials, a greater weight should be placed on data derived from *in vivo* rather than *in vitro* studies. With regard to *in vitro* studies, the best evidence will come from studies using human primary cells, whereas a lower level of evidence is provided by studies using immortalised/tumour cell lines and animal cells. The premise is that the closer a study mimics a human system, the more useful are the data that it provides.

10.2.4.4 Contaminant effects

NPs can act as nanovectors for toxins, or contain dopants/impurities that may exert their own cytotoxicity, especially if these agents produce ROS and oxidative stress. Impurities may be residual catalysts or precursors present in the nanomaterial from the synthetic process. For example, unlike "clean" acid-washed CNTs, those containing up to 5% iron after synthesis are as potent inflammatory and fibrogenic agents as quartz or asbestos, in part due to transitional metal catalysed Fenton chemistry producing ROS and oxidative stress.

10.2.4.5 Neurotoxicity potential of nanoparticles

There have been few investigations of the potential for neuronal effects *in vivo* from engineered NPs; however, it should be considered due to the ability of inhaled NPs to travel to the brain via the olfactory bulb. Translocation of particles along nerves is well known in neuroscience research, which has employed fluorescent NPs injected systemically to track peripheral neural pathways via retrograde transport back to the brain. Metal NPs injected *in vivo* into rats can bypass the blood–brain barrier and also disrupt its function with the leaking of blood factors to cause brain oedema, and neurotoxicity enhanced by hyperthermia (Ag, Cu > Al; 50–60 nm) [51]. *In vitro* exposure of brain cells (astrocytes) to anatase TiO_2 results in ROS production leading to mitochondrial disruption and oxidative stress. Interestingly, nanomedicine applications for nanoneuroprotection are under investigation, in which NPs are being developed as nanopharmaceuticals for drug delivery across the blood–brain barrier, as well as for therapeutic antioxidant effects in ROS-dependent neurological diseases [51].

10.2.4.6 Immunotoxic potential of nanoparticles

The immune system is a complex multi-organ system involving biomolecules and cells present throughout the body that have evolved to respond to biological particulates such as viruses and bacteria. Many NPs are similar in size to viruses or the components of larger pathogens and consequently have the potential to cause immune activation. Some nanoparticles can act as nanovectors for antigens to the immune system, and also have a strong stimulating ("adjuvant") effect, which is desired in vaccine technology. Therefore, the immune system is a crucial target for toxicity from exposure to NPs, with its unique responses to viral-sized particles and great importance in clearing particulates from the body [34].

The immune system is unable to eliminate insoluble nanoparticles because its arsenal of mechanisms for killing biological pathogens is ineffective with persistent nanoparticles, which can lead to "frustrated" macrophages that produce ROS and promote chronic inflammatory responses. Following exposure to certain non-biodegradable nanoparticulates, there is potential for direct immunosuppression of immune cell function, due to immune cells phagocytosing and accumulating NPs but being unable to degrade them; as well as inappropriate immunostimulation and/or cell activation (e.g. frustrated phagocytes producing ROS) causing cytokine release, which may result in nanoparticle-enhanced allergy and/or nanoparticle-induced autoimmunity.

Animal studies have indicated that allergen (ovalbumin)-coated polystyrene NPs <50 nm may invoke virus-like responses from the immune system (known as a T-helper 1 or "Th1" type response), while particles >90 nm may induce bacteria/parasite-like responses that are similar to that seen for allergic reactions (known as a T-helper 2 or "Th2" type response), includ dermatitis and asthma [34]. Metal and metal oxide NPs can elicit the release of a cytokine profile from exposed immune cells that is neither a typical Th1 nor Th2 type response but is instead a general pro-inflammatory response that for particulates typically involves cytokines such as interleukin-8 (IL-8) [18].

Depending on the type of cytokines secreted, we may observe responses that promote excessive inflammation and tissue injury or responses that support allergic reactions. Each response type produces a range of specific cytokines, including key cytokines that could be important in biomonitoring for the effects of nanoparticles on the immune system. However, there is currently little information available on the immune effects of industrial NPs.

Furthermore, workers with a genetic predisposition to allergy (e.g. "atopic" individuals who already have an immunoglobulin-E response to an inhaled allergen, which comprises 30–40% of the general population, and up to 50% of the workforce in certain workplaces) are potentially at greater risk from an interactive response between concurrent nanoparticle and allergen exposures. These are important factors that need to be considered when developing workplace controls and monitoring strategies to reduce the risk of allergy from nanomaterial exposure in the working environment. Consequently, the development of appropriate exposure assessment practices in the workplace is fundamental to evaluating the effectiveness of risk management processes [41].

The formation of a protein corona by adsorption of proteins onto NPs may also result in the exposure of amino acid residues on the surface of these proteins that are normally buried in the core of the native protein. These conformationally changed protein areas can act as "neo-allergens" and, if recognised by immune cells, are called "cryptic epitopes" that may trigger inappropriate cellular signalling events leading to autoimmune reactions (as opposed to being rejected by the cells as foreign bodies). However, the identification of such surface-exposed epitopes is very difficult, and models that account for the biophysical properties of nanoparticles, molecular protein structures and cell signalling are currently being developed [32].

Overall, nearly any influence of NPs on immunity may be regarded as having a potential negative impact on our health status, as the immune system is a finely tuned mechanism that may be unbalanced by NP exposure.

10.2.5 Potential Biomarkers of NP Exposure

The analysis of various biomarkers after NP exposure is of fundamental importance in developing future diagnostic tests of exposure and consequent risk assessment. Also, the investigation of potential biomarkers of the effect will help determine the toxic mechanisms involved, which can be used in the "green" design of safer nanomaterials (i.e. the "safety-by-design" approach).

The cellular responses to *in vitro* NP exposures generally indicate that oxidative stress caused by ROS and free radical generation initially upregulates stress responses (anti-oxidant defence), and can then lead to inflammatory signalling and finally cytotoxicity [37]. Bioindicators for each of these stress responses include:

(1) antioxidant defence proteins, e.g. glutathione transferases and hemoxygenase,

(2) inflammatory mediators leading to the production of cytokines and chemokines, and

(3) cytotoxic events involving cell membrane and mitochondrial disruption, leading to cell death via either apoptotic (caspase-dependent) or necrotic pathways [37].

Currently, comprehensive investigations of such pleiotropic effects from chemical exposure commonly involve the toxicogenomic approach of employing gene chips to compare the differential gene expression profiles of control and exposed cells/tissues. Using human and animal cells from different critical target organs enables the elucidation of the effects of chemical exposure on organ-specific gene expressions. These techniques are increasingly being used to investigate the effects of NP exposure in both *in vitro* and *in vivo* human and animal test systems, in order to confirm the most appropriate biomarkers for future diagnostic tests of NP exposure and risk assessment.

10.2.6 Toxicology of Specific Inorganic Engineered Nanomaterials

10.2.6.1 Nano titanium dioxide *(reviewed by [9])*

TiO_2 is regarded as an insoluble particle in biological systems, exhibiting low acute toxicity in a variety of traditional hazard/safety tests. Experimental animals exposed to high concentrations of nano TiO_2 through inhalation, may exhibit impairment of normal alveolar clearance mechanisms in the lung, resulting in particle build-up leading to excessive lung burdens that cause chronic alveolar inflammation. The International Agency for Research on Cancer (IARC) has classified TiO_2 as a Group 2B carcinogen–possibly carcinogenic to humans. The main data used for this classification concerned lung tumours in rats after high chronic exposures that would have caused the "particle overload phenomenon" (whereas mice and hamsters did not develop tumours). IARC considered the biological mechanisms occurring in the rat were still relevant to workers in situations of high exposures. However, with the mode of action involving a sustained inflammation, this implies that there is a threshold dose below which exposures will not be harmful. Some acute and intratracheal inhalation studies did not give the expected results of increased pulmonary toxicity due to aggregation and agglomeration of TiO_2.

Dermal absorption studies for nano TiO_2 have consistently concluded that TiO_2 is not absorbed through healthy human skin. Oral studies indicated that nano TiO_2 is absorbed from the gastrointestinal tract and distributed throughout the body. Intravenous and oral studies both resulted in TiO_2 retention in the lung, kidney, liver and spleen, but not in the lymph nodes, brain, plasma or blood cells.

Current research implies that the toxicity of TiO_2 is related to decreasing particle size and crystallinity. Anatase nano TiO_2 is more toxic/phototoxic than the rutile crystalline form, and anatase has also shown *in vitro* genotoxicity from the generation of oxyradicals via oxidative stress. However, both forms can induce a pulmonary inflammation response if the concentration of exposure is high enough in rat inhalation studies. Some of the toxic effects of nano TiO_2 can be scaled based on surface area considerations from the toxicity of micro-sized TiO_2 [12].

Concerns about potential biological effects from the photoactivation of nano TiO_2 were raised when BlueScope Steel researchers Barker and Branch (2008) reported that metal oxide nanoparticles in some sunscreens had photocatalytic activity capable of bleaching the painted surfaces of coated steel [2]. The relevance of photobleaching of pre-painted steel roofing materials to potential biological effects of human sunscreen is highly unlikely, as both the nature of the target site affected and the UV-sunscreen co-exposure circumstances are very different. Several studies have shown that there is negligible penetration of metal oxide nanoparticles past the outer dead cell layer (in the stratum corneum) of healthy or damaged skin. The skin is also constantly shedding its outer layer of dead cells.

In conclusion, when formulating a new nano TiO_2 product or use, its potential toxicity can be controlled by varying the crystalline form used. For example, the UV-photocatalytic activity of nano TiO_2 can be reduced by partial or full replacement of the anatase form with the less reactive rutile form, especially in sunscreen applications and personal care products. The use of dopants and coatings of nano TiO_2 to attenuate the effects of photocatalysis should also be considered in such applications where UV-absorption, but not reactivity, is required. In applications involving the use of an industrial-grade TiO_2 photocatalyst, such as the coatings on "self-cleaning" glass surfaces, then the potential for shedding of anatase nanoparticles into the environment needs to be assessed and minimised to reduce environmental impact.

10.2.6.2 Nano cerium dioxide *(reviewed by [25])*

Many of the unique properties the rare-earth oxide cerium dioxide (CeO_2, ceria) are enhanced in the nano form due to the increase in both the Ce^{3+} concentration and the lattice constant, including its oxygen storage capacity and redox activity. Nano ceria can be made in a range of shapes, including spheres, rods and cubes. Nano ceria could be used several biomedical applications within biological systems, including drug delivery, prevention of cell death or injury caused by oxidative stress and associated diseases.

Although nano ceria exhibits antioxidant (superoxide dismutase-mimetic activity) and biocompatible properties under specific conditions, its redox cycling ability may be pro-oxidant outside of this range of conditions and cause *in vitro* oxidative DNA damage. Nanoparticle-sized CeO_2 has exhibited higher photo-chemically induced toxicity than similar micron-sized material in a freshwater algae aquatic test system [48].

In conclusion, nano ceria has useful antioxidant and bio-compatible properties, but its redox cycling ability may also be pro-oxidant in certain conditions. These characteristics can be modified by doping the crystalline lattice with other elements to modify its antioxidant activity, while the minimal cytotoxicity of nano ceria can be further reduced by surface modification.

10.2.6.3 Nano zinc oxide *(reviewed by [9])*

Nano ZnO has low systemic toxicity and is partially soluble in biological systems, especially in acidic conditions (e.g. skin, gut and intracellular endosomes) and environments (e.g. acidic water and soils). Dissolving ZnO releases zinc ions that readily precipitate as salts, while zinc itself is the second most common essential metal in the body (after calcium) and is used as a cofactor in over 200 enzymes, interacting with other essential metal pools (i.e. iron and copper).

Concerns about ZnO toxicity have been raised by reports that both nano and micron-sized ZnO can be clastogenic (i.e. causing chromosomal damage, a form of genotoxicity) and induce DNA damage in cultured mammalian cells *in vitro* under the influence of UV light. Such often-mentioned *in vitro* studies showing that ZnO nanoparticles are slightly positive for causing DNA damage in only a few specific cell exposure systems, have been difficult to relate to the whole body situation, because of an identified experimental artefact known as "psuedo-photoclastogenicity" [10]. This pseudo effect may occur because of an increased sensitivity of these cells after UV exposure.

Importantly, all *in vivo* and most *in vitro* genotoxicity tests were negative for ZnO and no evidence has been found of phototoxicity on intact skin from studies involving human volunteers. Furthermore, there are issues with the high concentrations used in many *in vitro* studies that report the cytotoxicity of nano ZnO, and many studies do not report a direct comparison with the cytotoxicity profile of soluble zinc in the test system. Interestingly, the report by Barker and Branch (2008) about the photocatalytic activity of metal oxide nanoparticles in some sunscreens also included a sample containing nano ZnO [2]. However, the relevance of these findings to human sunscreen usage is highly unlikely, as ZnO is much more water soluble than TiO_2 and far less likely to persist on the skin as intact nanoparticles following sunscreen application.

Therefore, the main issue is whether intact ZnO nanoparticles are able to penetrate the skin–the weight of evidence indicates that nano ZnO when applied to the skin remains on the surface and in the outer layer, and does not penetrate through the stratum cornea. This has been confirmed in several studies and most recently in an animal study of UVB-damaged skin [33], which showed only a slight enhancement of penetration within the stratum corneum itself, but no absorption through the skin, as well as nano ZnO being even less penetrating than nano TiO_2. Therefore, it is extremely unlikely that sufficient ZnO nanoparticles penetrate and remain in the important target zone of the basal epidermal layer for long enough to generate a significant amount of free radicals by photocatalysis, compared with the known large amounts of free radicals already being produced by UV exposure itself.

Finally, the concerns about nano-metal oxides easily penetrating human skin specifically have not been supported by the latest research. A landmark Australian study recently used a rare stable zinc isotope to directly measure the absorption of zinc from sunscreen formulations in humans with concurrent UV exposure under typical beach conditions [21]. It showed minimal absorption of zinc from both nano and bulk ZnO, following twice-daily dermal application over 5 days. Zinc ions found in the blood from sunscreen during the researcher's five-day beach trial were only 1/1000th of the total blood zinc levels. As zinc is an essential metal for maintaining good health, the body would easily handle this very small amount of zinc absorbed from the sunscreen.

Environmental toxicity seen in ecotoxicity test systems for nano ZnO appears to be related to the release of zinc ions and therefore should be scalable to the toxicity of bulk ZnO with its less efficient ion leaching from its smaller surface area per mass [12].

In conclusion, nano (and micro) ZnO used in sunscreen type products and for other similar applications exhibits negligible toxicity and penetration into the human body. There are surface modification options available that can make ZnO more bio-compatible by reducing particle dissolution and the release of zinc ions, which would further reduce the *in vitro* cytotoxicity seen at relatively high concentrations. The crystalline lattice of ZnO particles can also be doped with other elements to modify photoactivity.

10.2.6.4   Nano gold *(reviewed by [28])*

Colloidal gold has been used in the treatment of rheumatoid arthritis since the 1920s, while nano gold has been used in biotechnology for nuclear transfection and targeting in non-viral gene delivery applications because they easily enter cells. Gold nanoparticles and tunable nanoshells have most recently been employed in the field of nanomedicine as multipurpose contrast agents for tumour imaging (due to the increased "leakiness" of tumour blood vessels) and killing via their photothermal properties. Their *in vitro* cytotoxicity in certain cell types at high concentrations has limited the use of uncoated nano gold and prompted the search for more biocompatible coatings that also aid in cell uptake. There is also evidence of other adverse effects, such as the potential for affecting immunological responses, and particle shape-dependent effects for the *in vitro* cytotoxic potential of nano gold particles, shells and rods.

In conclusion, it is possible to use surface coatings or encapsulation with biocompatible polymers to reduce the cytotoxic potential of nano gold, whilst retaining its functionality and usability. The particle surface can also be capped (e.g. with alkanethiol groups) to increase biocompatibility, and also functionalise the nano gold for a range of biomedical applications.

10.2.6.5 Nano silver *(reviewed by [57])*

Of the more than 1300 nano products on the market in 2011, over a quarter contain nano silver and include clothing and medical applications such as antibacterial wound dressings [47]. As one of the main functions of nano silver is its antimicrobial toxicity, it should therefore be regarded as a toxic agent. This antimicrobial activity is due to the oxidation of surface silver atoms to release Ag^+ ions, which is dependent on both the size and dispersal of nano silver particles.

Rodent inhalation studies have shown toxicity in the lungs (with mixed inflammatory cell infiltrate, chronic alveolar inflammation, and small granulomatous lesions) and liver (bile-duct hyperplasia) following sub-chronic exposure to nano silver, but shorter oral exposure studies did not show toxicity, despite a dose-dependent accumulation of silver content in all the tissues examined, especially the kidneys. Dermal exposure in damaged skin, as in the case of wound dressings, can result in systemic exposure with the liver being a target organ. Silver ions are often considered not to be toxic to humans, as some people have intentionally ingested considerable quantities of nano silver as a dietary supplement over long periods of time, without overt adverse effects except for skin discolouration (argyria) [12].

The main concerns have been the potential of nano silver to adversely impact the environment due to the high sensitivity of aquatic organisms to Ag^+, and also the possibility for bacterial resistance to arise. However, recent studies indicate that the environmental fate of nano silver is to generally form insoluble precipitates (e.g. silver sulphide) that are unlikely to be sufficiently bioavailable to interfere with nitrogen-fixing bacteria that are important to effective agriculture.

In conclusion, nano silver can be surface-modified with hydrophilic groups to increase biocompatibility, but such modifications would also decrease its antibacterial activity and potential usefulness in many current applications. These modifications may also slow oxidation and dissolution of nano silver particles to make them more biopersistent. However, further functionalisation of biocompatible forms of nano silver may also provide potential new applications, such as in biomedical diagnostics and biosensors.

10.2.6.6 Nano silica *(reviewed by [25])*

The toxicity of silica is dependent on crystallinity, as amorphous silica is far less hazardous than crystalline silica. Excessive and prolonged exposure to fine particulate crystalline silica (silica dust) is well known for causing silicosis in miners, a non-cancerous lung disease. In 1996, the IARC categorised crystalline silica as a Group 1 carcinogen—carcinogenic to humans, based on occupational lung cancer incidence after a review of the available epidemiological data. Exposure standards for amorphous silica are set on the basis of having less than 1% crystalline silica content, e.g. fumed silica (respirable dust) has a National Exposure Standard of 2 mg/m^3 in Australia, and a number of other amorphous silica types have a National Exposure Standard of 10 mg/m^3.

Nano silica exhibits concentration dependent toxicity in several cell types, and the mechanism of toxicity is thought to be due to its highly adhesive properties to cellular surfaces, which can lead to changed membrane structures and cellular surfaces. The particle shape of nano silica is also an important factor in its toxic potential as silica nanowires, but not nanospheres, are highly toxic to embryonic zebrafish—causing toxicity, deformities ("teratogenic") and interfered with nerve development [38].

In conclusion, the surface of nano silica can be modified to increase its hydrophobic character, causing increased particle aggregation and reduced direct membrane effects, and thereby improving its biocompatibility. Due to potential toxicity of silica nanomaterials with high aspect ratios, consideration should also be made as to whether nanowires may be substituted with nano-spheres, while retaining functionality for a particular application.

10.2.6.7 Quantum dots *(reviewed by [9,28])*

Many of the core elements in quantum dots (QDs) are known to be toxic at low concentrations, e.g. cadmium, selenium, lead and arsenic. Therefore, stability is a critical factor in the cytotoxic potential of QDs, as toxicity will eventuate from the leaching of toxic metal ions under conditions that promote QD degradation, such as an oxidative environment.

Bare QD cores, and some QDs with biodegradable shells, are readily endocytosed by cells into endosomes, which are then trafficked to various cellular compartments. These include the acidic (pH 4–5) and oxidative environments of lysosomes and peroxisomes, which can degrade QDs and result in the leaching of toxic metals from the core.

Studies of the cytotoxic potential of QDs have mainly concerned CdSe/ZnS cores, as these are considered the most versatile for biological applications. Cytotoxicity of quantum dots (QD) is progressively reduced by polymer coatings of increasing thickness, as this prevents cellular uptake QDs—intravenous injection of these modified QD do not appear to cause acute toxicity. Depending on the surface coating, QDs may be recognised as being foreign to the body and be sequestered by the reticuloendothelial systems in the major organs. Zwitterionic or neutral organic QD coatings can prevent the adsorption of serum proteins that would otherwise increase QD hydrodynamic diameter >5 nm and prevent rapid renal excretion through the glomerular filter of the nephron [4].

In conclusion, it is possible to encapsulate QD cores with stable shell coatings made from biocompatible polymers to significantly reduce their cellular uptake and degradation, and consequently their cytotoxicity, whilst retaining functionality and usability.

10.2.7 Latest Initiatives in Nanosafety Research

As part of the international effort to co-ordinate safety testing of common nanomaterials, the Organisation for Economic Co-operation and Development (OECD, www.oecd.org), through its Working Party on Manufactured Nanomaterials (WPMN) launched the "Sponsorship Programme for the Testing of Manufactured Nanomaterials" at the end of 2007, with a priority list of 13 nanomaterial types and specific endpoints relevant for human health and environmental safety (www.oecd.org/env/nanosafety) [42]. A range of national research consortia were formed to become lead sponsors, co-sponsors or contributors for conducting the safety testing of a representative set of manufactured nanomaterials, including fullerenes (e.g. "bucky balls"), single-wall (SWCNTs) and multi-wall (MWCNTs) carbon nanotubes, silver, iron, titanium dioxide, aluminium oxide, cerium oxide, zinc oxide, silicon dioxide, dendrimers, nanoclays and gold. Direct comparisons were also made in most test systems between the nanoparticulate and bulk forms of these materials. A total of 59 selected endpoints have been investigated, which included specific endpoints for nanomaterial information/identification, physico-chemical properties and material characterisation, environmental fate, environmental and mammalian toxicology, and material safety.

At the time of the publication of this book, the various national consortia were completing the collation of their research findings for Phase 1 of the testing programme. This initial exploratory phase necessarily involved the development of methods both for preparing nanoparticle samples and for testing them. This work has identified that in order to correlate physicochemical parameters with specific bioactivities, it is essential for nanoparticle characterisation, including agglomeration state, to be performed by appropriate metrology methods, both before and after nanoparticle addition to the biological test systems.

These experimental findings will be uploaded onto the NANO-hub database hosted by the European Commission Joint Research Centre's Institute for Health and Consumer Protection (EU JRC-IHCP, http://ihcp.jrc.ec.europa.eu/our_databases/nanohub/) [17]. This is in preparation for Phase 2 of the programme, which will consider those cross-cutting issues or tests that were identified by analysis of the Phase 1 data. The outcomes of this important nanosafety research programme will be progressively made available to the public via the OECD and related websites and scientific publications.

This co-ordinated international research effort is of immense benefit to the developers and manufacturers of nanomaterials, as it will provide crucial hazard-specific information that will enable the correlation of important physico-chemical properties with biological activities—whether they are desirable bio-activities to be exploited, or unwanted bioactivities to be avoided through re-engineering of the nanomaterial.

10.3 Overview of Occupational Health and Safety Issues and Workplace Controls

Focus Point 4: It is important to conduct a whole of life cycle analysis of potential "hotspots" for exposure to nanomaterials, especially where there is a likelihood of the nanomaterial shedding particulates.

Focus Point 5: When implementing workplace control measures, the processes need to move up the hierarchy of controls for scale-up from R&D processes to the manufacturing phases.

Focus Point 6: Adopt the Precautionary Principle to limit work-place exposure when the exposure standard is unknown; benchmark exposure levels and the "control banding" approach should be used where there is limited hazard and risk information.

Focus Point 7: Adopt the "As Low As Reasonably Practicable (ALARP)" approach when the exposure standard is known.

10.3.1 Occupational Health and Safety Issues Relating to Engineered Nanomaterials

Over the last decade, nanotechnology-enabled productivity has been promoted as the solution to solve a broad range of current problems, including energy production, conversion and storage, water purification, agricultural productivity, and more effective medicines and diagnostics. To harness the benefits of nanotechno-logies and successfully deliver these promised benefits to industry and society, nano-related industries must ensure the smart development and manufacture of such nano-enabled solutions, by using a safety-by-design approach for nano-products that are safer for workers, consumers and the environment. Nano-manufacturers must also ensure that they provide a safe workplace for their employees, by using appropriate workplace and environmental controls to reduce the potential risk of exposure to nanomaterials of concern.

In many countries, the public attitude towards nanotechnology is seeing the perceived benefits moving to strongly to outweigh the perceived risks (details in Chapter 12). However, this current positive public opinion towards nanotechnology is heavily dependent on the participants in nano-related industries acting responsibly and proactively, as well as being aware of the potential risks versus benefits for the whole life cycle of nano-products").

The risk assessment process for a whole life cycle analysis involves the identification of potential exposure risks to a nanoparticle of concern during the life cycle of the nano-product [26]. It includes the following stages:

- starting/precursor materials
- process of manufacture
- method of isolation/purification/separation
- storage and transport
- product applications
- potential for release in use/abuse of products
- disposal of product
- removal/clean-up of waste streams
- additional components added to life cycle as necessary.

At each step through the life cycle there are multiple owners of the safety problem. As most companies are involved in only part of the life cycle, some European countries have instituted measures to ensure disclosure in business-to-business interactions of the "nano-ness" of industrial feedstock. The growing concern about end-of-life and safe disposal issues is also prompting the incorporation of built-in recycling options at the development phase.

Nanowaste management is a growing issue for which some treatment options have been proposed, but their overall effective-ness is unknown in many cases (detailed in Chapter 11). These options include the following: treatment as hazardous waste; treatment based on known physicochemical properties; containment, e.g. in solid matrices; recycling; solubilising, e.g. high alkalinity for ZnO particulates; and incineration (e.g. for CNTs).

There are a multitude of ENM types and hybrids being developed with a wide range of physicochemical and biological properties—this means that risk assessment and management processes need to be appropriately adapted to each workplace setting. In the absence of comprehensive hazard and toxicity data for each individual nanomaterial in development or production, it is appropriate to employ existing workplace control measures where necessary that have been successfully applied in the handling of ultrafine particles (such as in mining) or in maintaining a "clean room" environment (e.g. in biotechnology processes).

There are currently few ENMs for which exposure standards have previously been established (i.e. fumed silica and carbon black)—this indicates that there is evidence of safe levels of exposure to some ENMs, but such evidence is lacking for most engineered NMs. However, several recommended Occupational Exposure Limits have been published [26].

There are also issues associated with limited amount of nano-specific information currently provided on product labels and (material) safety data sheets or "(M)SDSs" that impacts on how well ENMs are controlled in the workplace. Changes in the regulation of nanomaterials are ongoing in several countries and regions, especially for nanolabelling, with European Union countries implementing mandatory labelling from mid 2013 by the inclusion of "(nano)" on the packaging of products containing nanomaterials. The information on SDSs for nanomaterials will also include nano-specific information as part of E.U. REACH program (Registration, Evaluation, Authorisation and Restriction of Chemical substances) and the Global Harmonization Scheme (GHS) for chemical regulation.

A number of public interest groups have called for specific regulation of nanomaterials, including for nanomaterials to be assessed as new chemicals. In some cases this would require new safety testing to determine whether a nanomaterial presents a different hazard potential to its bulk counterpart, before being permitted for commercial use. Due to the long time period needed to adopt new regulatory legislation in many countries and regions, the generally accepted approach has been to adapt existing regulatory frameworks to bridge perceived gaps in the regulation of nanomaterials (addressed in Chapter 8).

10.3.2 Nanomaterial Health Risk Assessment

The process of health risk assessment is the analysis of possible negative health effects from current or future processes that may be caused by a hazardous material and/or process, taking into account the actions taken to mitigate or control exposure, i.e. "risk = probability × severity" [22]. The goal of risk assessment is to evaluate whether or not the exposure in a specific workplace is above an acceptable level of risk defined by the specific legislation or by decision-makers, in order to inform decision-makers about the need for the further strengthening of risk management processes. This process is sufficiently flexible that it can be adapted to the risk assessment of nanomaterials [26]. The four components of the process of risk assessment are [22]:

(a) hazard identification—hazards (intrinsic toxicities) are identified that contribute significantly to risk and exposure

(b) hazard (dose-response) characterisation—potential adverse health effects that are related to identified hazards are determined

(c) exposure assessment—pathways by which individuals can be exposed to the hazard in the workplace, and the level of this exposure, are evaluated

(d) risk characterisation—this incorporates the information from (a), (b) and (c) in order to evaluate the potential risk of exposed individuals in the workplace.

Aspects of hazard identification and dose-response characterisation for nanomaterials have been discussed in the first part of this chapter. Additional information for both will soon to become available through the OECD safety testing programme for ENMs, especially with regard to the generation of quantitative (surface) structure activity relationships (Q(S)SAR) from the growing number of mechanistic toxicology studies of related ENM, as biological properties can vary greatly within ENM classes.

The key OH&S issues required to be addressed concerning potential exposure to ENMs in nanotechnology workplaces have been identified by the U.S. National Institute for Occupational Safety and Health (NIOSH) on their nanosafety website (www.cdc.gov/niosh/topics/nanotech/safenano/) [54], which has a wide range of resources for nanomanufacturers. These issues include the following [54]:

- Are the nanoparticles hazardous to workers?
- How can workers be exposed?
- Can nanoparticles be measured?
- Can worker exposures be controlled?

The potential for exposure of workers during nanomaterial production depends on the method employed, i.e. whether it is a "bottom-up" or "top-down" approach (detailed in Chapter 4), that is, whether NPs are built from the bottom up using atoms and small molecules, as in the case of precipitation, microemulsion and vapour phase reactions, or NPs are derived from bulk material by attrition milling or reactive grinding (Table 10.2). Each of these NP production routes have their relative cost, and own set of production process, occupational health, safety and environmental issues.

Table 10.2 Nanoparticle production routes and their relative cost, process issues and occupational health, safety and environmental issues

Method	Production costs ($/kg)	Process issues	OH&S issues
Precipitation	1–20	Poor dispersity, Separation	Liquid disposal, redispersion
Microemulsion	15–50	Separation, recycle	Liquid disposal, redispersion
Vapour phase reactions	20–200	High capital cost, scale-up, poor dispersity	Free particulates, redispersion
Attrition milling	2–20	Energy costs, contamination	Free particulates, redispersion
Reactive grinding	5–40	Separation, contamination	Free particulates, redispersion

Source: Adapted from [29].

Major challenges still exist for the nanomaterial health risk assessment in the area of workplace exposure assessment and monitoring. Workplace monitoring can involve the following:

- personal monitoring—using active sampling of breathing zones by personal pumps
- area monitoring—usually of work stations by air sampling and surface wipes
- biological monitoring—measuring industrial contaminants in blood urine or exhaled breath.

Although standard industrial hygiene methods can be used, there is still the need for the following: well-validated detection methods for exposure assessment monitoring in workplaces; benchmarking dosimetry measurements with Standard Reference Materials (SRMs); and suitable biomarkers of exposure and effect in workers. However, none are currently available as they are presently complicated by the need to fully understand *in vivo* "particokinetics" following exposure to nanoparticles.

While traditional monitoring instruments assess mass per volume (i.e. mg/m^3), other NP parameters are also relevant, such as the particle number, size distribution, surface area and chemistry. The measurement of multiple NP parameters is desirable as no single metric will completely characterise exposure. NIOSH's Nanoparticle Emission assessment Technique (NEAT) recommends a graded approach to nanomaterial workplace monitoring, i.e. [19]:

- Step 1: screen area and process, using particle counters and simple size analysers
- Step 2: collect samples at source, obtaining filter based samples for electron microscopy (EM) and elemental analysis
- Step 3: collect personal samples, also of filter based samples for EM and elemental analysis
- Step 4: use less portable equipment, such as more sensitive aerosol sizing equipment.

Although a greater range of multipurpose transportable detection instruments are starting to become available for use in the workplace, most standard instruments are expensive and still require specialist expertise to operate. The issues of nanoparticle measurement and standardisation are discussed in detail elsewhere (see Chapter 9).

Given the current issues with nanomaterial exposure assessment in the workplace, the remainder of this chapter details the workplace controls that are effective in reducing the exposure of workers to ENMs.

10.3.3 Hierarchy of Workplace Controls for Handling Nanomaterials

The control of worker exposure is paramount in the workplace and is achieved using the widely recognised "hierarchy of controls", i.e. elimination, substitution/modification, engineering controls, administrative controls and personal protective equipment.

There are a range of existing workplace control methods that are known to be effective to protect or minimise exposure of workers to engineered NMs during their life cycle of manufacture, handling, use and disposal (reviewed by [26], with several control measure recommendations adopted specifically for laboratory guidelines by the OECD [43]). These control methods are mainly based around the lower levels of the "hierarchy of controls", i.e. engineering controls (enclosure, ventilation/extraction), administrative controls and personal protective equipment (PPE).

In order to promote a safe workplace rather than a safe person strategy, it is necessary to move up the hierarchy of controls and consider options for the elimination, substitution and/or modification of engineered NMs that have the highest hazard potential. Further testing and data are needed in specific workplace situations to better understand the levels of protection afforded, and ensure effectiveness. The minimal amount of health effects data currently available for many ENMs means that it is not possible to fully inform risk management processes for working with ENMs. A further implication of the limited amount of toxicological data available is that a precautionary approach to the prevention and control of workplace exposures should be used.

A summary of the evidence for the effectiveness of workplace controls from each strategy in the hierarchy of control are shown below (reviewed by [26]).

10.3.3.1   Elimination controls

Since the specific properties of engineered nanomaterials are usually required for manufacturing a novel product, it is unlikely that this option will often be feasible.

10.3.3.2   Substitution and modification controls

The control options for substitution of nanomaterials and modification of nanomaterials and/or processes have not yet been widely used in the workplace for nanomaterials. It should be noted that the likelihood of NP exposure in the workplace is highest when handling free particles (e.g. during equipment maintenance, clean-up of spilled NMs, transfer of NMs in open systems, and cleaning "dust" collection systems), and lowest when working with embedded NPs and during manufacturing in enclosed systems. Therefore, simple generic modification options can be considered, e.g. using wet pastes or pelleted nanomaterials instead of dry powders wherever possible.

A range of nanomaterial modifications have been shown to reduce the *in vitro* cytotoxicity of various nanomaterials, including metal/metal oxide nanoparticles, quantum dots, fullerenes and CNTs (reviewed by [25]). The methods of surface modification, encapsulation, particle size control, functional group addition and crystalline phase type control can each be employed for different engineered nanomaterials to decrease their potential toxicity (see Table 10.3). Carbonaceous NPs are included in Table 10.3 to highlight the issue concerning fibre-like NPs (i.e. CNTs), and that while reduced hydrophilicity may decrease the cytotoxicity of some inorganic NPs like nano silica, the opposite is the case for the hydrophobic CNTs and fullerenes.

Table 10.3 Summary of substitution/modification options to reduce potential hazards from different engineered nanomaterial

NP type	Suggested strategies	Substitution/Modification options
Metal oxide NPs		
Titanium dioxide	Utilise the known differences in reactivity and phototoxicity between the two main crystalline phases of titania.	Substitute the less toxic and reactive rutile form for the more toxic and reactive anatase form in applications where it does not remove functionality.
Cerium Dioxide	Nano ceria is biocompatible in several applications, but can be pro-oxidant and is a potential issue for aquatic environments.	Not possible to suggest modifications at this time.
Zinc oxide	Already of low toxicity, and can reduce potential toxicity further through surface or crystal lattice modification.	Surface modification with silane (APTES) or oleic acid/ SiO_2 or polystyrene; crystal modification by lattice doping with other mineral elements.

	Metal NPs	
Gold	Encapsulation with a biocompatible polymer to reduce toxicity, whilst retaining functionality.	Encapsulation with chitosan or polyethylene glycol; surface coating with phosphatidylcholine; functionalisation by alkanethiol capping.
Silver	Surface modification by conjugation with hydrophilic moieties to reduce toxicity of nano silver products.	Surface functionalisation with phosphorylethanolamine, phosphorylcholine or alkanethiol capping, in applications where it does not remove functionality.
	Other inorganic NP	
Silica	Reduce hydrophilicity to increase aggregation and reduce direct membrane effects.	Partial hydrophobisation by polymer or protein coating, or alkylsilylation.
Quantum dots	Decrease cellular uptake and intracellular degradation to reduce toxicity from leaching metals, whilst retaining functionality in most cases.	Encapsulation of core with stable biocompatible shell, e.g. with chitosan, silica or PEG, that decreases endocytosis and endosome degradation.
	Carbonaceous NPs	
Carbon nanotubes	Increase hydrophilicity by surface modification to decrease toxicity and increase biocompatibility; modify biodistribution to reduce persistence; Keep CNTs shorter than 5 μm in length.	Sidewall functionalisation with hydrophilic functional groups; surface modification with grafted polyetherimides or low molecular weight chitosan.
Fullerenes	Increase hydrophilicity by surface modification in order to decrease toxicity and increase biocompatibility.	Functional group modification by attaching water soluble groups, such as alcohols or carboxyl groups.

Source: Adapted from [25].

When undertaking surface modification to change the efficiency and uptake of an engineered nanomaterial there is also the consideration of biocompatibility, which is important in negating or reducing potential toxic effects. However, in some cases, such modifications may affect the functionality of nanomaterials in relation to intended end-uses.

Recent surveys of nanotechnology-related activities show that substitution/modification processes are most often used to change the functional properties of the product by altering particle size, physical properties, agglomeration properties, chemical properties and conductive properties. However, the use of substitution/modification options to change the health or toxicological properties is minimal at present. Nevertheless, suitable substitution/modification methodologies are well known and used and thus there is an existing capability that might be applied more broadly to work health and safety related purposes [25].

If researchers, developers and manufacturers of ENMs adopt these methods, then it is possible to re-engineer nanomaterials in the early stages of development to reduce the potential toxicity of manufactured nanomaterials. The downstream effect of this will be to reduce the risk posed by the use of these nanomaterials, not only in the workplace but also in the general community [25].

10.3.3.3 Enclosure controls

Current evidence indicates that worker exposure is significantly reduced or negated if a process involving engineered nano-materials, which would otherwise result in the release of airborne particles, is performed in a properly designed enclosure/containment (reviewed by [26]). The method of containment or enclosure is designed for the specific processes but is usually implemented in combination with other control measures when enclosures are opened, e.g. administrative controls and/or personal protective equipment (PPE).

10.3.3.4 Extraction controls

Evidence indicates that worker exposure is significantly reduced or negated through the use of correctly designed and implemented local exhaust ventilation (LEV) and filtration for processes involving engineered nanomaterials that would normally result in the release of airborne particles (reviewed by [26]). This control measure is usually implemented in combination with other control measures, e.g. administrative controls and/or PPE. The better extraction methods have involved the use of high-efficiency particulate air (HEPA) filtration and electrostatic precipitation.

The EU Nanosafe2 program recently reported that HEPA filters and fibrous respirator and mask filters are efficient in clearing nanoparticulates, thus confirming the conventional filtration theory and disproving the "skimmer model" that only particles larger than the pore size should be stopped [15]. The capture efficiency depends on flow rate and type of filter material, with the maximum penetrating particle size (MPPS) of HEPA filters is 150–300 nm, but can be between 20–80 nm for other types of filters (i.e. N95). The primary mechanism of capture is a function of particle diameter, as larger particles are blocked by interception (i.e. collision with fibre) and inertia (so they cannot deviate with air flow around the fibre), while smaller particles are trapped by diffusion and collision enhanced by Brownian motion, with no thermal rebound as NPs lack sufficient kinetic energy [15].

Geraci (2008) has presented a number of important points about process enclosure and LEV, which needs to be very close to the emission source, including the relationship between the effectiveness of LEV in the capture of particles of different sizes [19]. For example, nano silver particles up to 500 nm are easily captured by LEV, whereas particles above this size show a lower capture efficiency, i.e. a reduction to 96% and 93% for particle sizes of 1 and 10 µm, respectively. Geraci (2008) also indicated that a reduction of 74–96% in air particulate mass concentration can result if efficient and well-maintained LEV is effectively utilised during a reactor cleanout operation [19].

Most traditional engineering controls are expected to be adequate, i.e. enclosure, fume hoods and biohazard cabinets, although it is important to emphasise that laminar flow hoods are not protective owing to outward air flow blowing material onto the user. HEPA and personal cartridge filters can be appropriate measures for NP risk management due to historical use with vehicle exhaust fumes and viruses [27].

In controlling exposure to ENMs, the same principles as those applied to the management of fine powders, dusts and dusty materials should be considered, i.e. [24]:

- preventing dust from becoming airborne
- handling combustible nanopowders in liquid form, when possible
- design of machinery to prevent ignitions and sparks
- operating temperature of the electrical equipment
- use of controlled-atmosphere production and storage processes
- risk of asphyxiation.

As the minimum ignition energy drops steeply with particle size certain highly reactive metallic nanomaterials (e.g. aluminium), the physical hazard of explosivity and flammability of nanopowders also requires workplace controls during handling [14]. When working with potentially explosive nanomaterials, the following additional steps may be necessary [16]:

- anti-static shoes and mats used in areas where materials handled
- distillation system for evaporating solvent from a colloidal dispersion housed within explosion proof enclosure.

Engineering control options can also be implemented to reduce the possibility of dermal exposure by the re-engineering of work processes to avoid immersion, splashes or spillage.

10.3.3.5 Administrative controls

There are a range of administrative workplace controls that may be implemented for workers involved in using engineered nanomaterials (reviewed by [26]). These are usually implemented in combination with other control measures, e.g. enclosure, extraction and PPE; however, their application should be based on a risk assessment of a specific process or situation and may in certain cases, but not usually, be sufficient on their own (e.g. for nanomaterials embedded in matrices that do not shed nanoparticles during specific processes).

The following are procedural controls are usually used in conjunction with engineering controls [24]:

- limiting the process to specified areas
- limiting access to areas
- reducing time spent in possible exposure area
- reducing the number of personnel that may be exposed

- specific personnel training, i.e. information provision about special measures and the potential for negative health effects from exposure to ENM particulates.

Administrative controls can also be used to supplement engineering controls, because even if these are very efficient, there could still be potential exposure to NPs during clean-up and maintenance. Also, normal good hygiene practices are applicable. Some nanomaterial-specific practices include [24] the following:

(i) work practices
- sticky mats at room entrances to prevent transfer
(ii) routine maintenance and clean-up of work areas and spills
- wet wiping and vacuum cleaning (with HEPA-filter)
- dry wipe only for liquid spills
- use of respirators and dermal protection
- frequent hand washing
(iii) waste disposal (of nanomaterials and used PPE, wipes and equipment)
- separate disposal containers
- recycling nanomaterials
- incinerating waste nanomaterials on-site (for carbonaceous ENMs)
- returning nanomaterials to suppliers.

10.3.3.6 Personal protective equipment

Evidence indicates that there are a range of PPE which may be used for engineered nanomaterials and can provide a good level of protection (reviewed by [26]):

- *Respiratory protective equipment (RPE):* there is good evidence that a range of facemasks and their filters (e.g. N95 cartridge filter pieces or better) effectively remove engineered NM, including N95, N100 and P100 face mask and N95 and N100 filters. Particulate filters are classified by U.S. NIOSH based their resistance to oil, i.e. "N" not resistant, "R" resistant (for one shift in oil mist) and "P" (oil proof), and also efficiency in particulate removal, i.e. "95" (95%), "99" (99%) and "100" (99.97%, HEPA filter) [55]. The US N95 filter classification corresponds approximately to the Australian type P2 filters, while P100 and N100 approximately correspond to P3 filters.

The main risk is from poor fitting, i.e. a lack of tightness between the face and mask. Otherwise, self-contained breathing apparatus (SCBA) is appropriate for respiratory protection in certain ENM handling situations with greater exposure potential.

- *Protection from dermal exposure:* There is good evidence that a glove protection plan (e.g. double-gloving using nitrile-type glove materials) and other garments made of non-woven air-tight (e.g. Tyvek polymer) materials are effective against NP penetration if applied correctly [15].
- *Eye protection:* There is evidence that a range of eye protection devices (e.g. goggles) have been successfully used in order to reduce ocular exposure [15].

The use of PPE should be considered as the last line of defence in the hierarchy of workplace exposure mitigation approaches, after all other available measures have been implemented. PPE should also be worn on a precautionary basis whenever the failure of a single control, including an engineering control, could entail a significant risk of exposure to workers. PPE will also be needed in situations where the use of engineering controls is impractical. PPE is usually implemented in combination with other control measures, e.g. process enclosure, extraction and administrative controls.

10.3.4 Risk Management and Control Banding

Risk management frameworks are now available, including the highly regarded Nano Risk Framework developed by the Environmental Defence Dupont Nano Partnership (EDDNP, http://nanoriskframework.org/) for use in projects involving engineered nanomaterials [13]. The EDDNP framework covers research, development, production, use, disposal/recycling stages of a nanomaterial's life cycle and also includes a communication strategy in the last of its six step framework (along with case studies for TiO_2, CNTs and nano FeO), i.e. [13]:

Step 1: Describe material and application.

Step 2: Profile life cycle(s), including properties, hazards and exposure characteristics.

Step 3: Evaluate risks.

Step 4: Assess risk management.

Step 5: "Decide, document and act".

Step 6: Review and adapt.

The risk management process that is generally proposed now for research and early development activities involving nano-materials, is that of "control banding", where similar control approaches are used within categories of nanomaterials that have been grouped ("banded") according to their exposure potential and hazardous properties, i.e. grouped according to risk [50]. Control banding is considered to be an appropriate method because of the current lack of data available for the risk assessment of many individual nanomaterials, but there is some understanding of hazards posed by different groups of nanomaterials, e.g. CNTs.

Control banding is possible because there are few basically different approaches to control (so risks can be banded), and many similar problems have previously been solved by these control methods. It is a qualitative risk management process used in pharmaceuticals industry that was developed by the UK Health & Safety Executive (HSE) in the Control of Substances Hazardous to Health (COSHH) Essentials model [53]. Control banding is defined as a strategy (or process) using a single control technology (e.g. general ventilation or containment) applied to a defined range or band of exposure to a chemical (e.g. 1–10 mg/m^3) that falls within a given hazard group (e.g. harmful by inhalation, or irritating to respiratory system) [53]. It is used to facilitate the control of chemicals in the workplace, especially used in the control of chemical hazards with limited toxicological information and when workplace exposure limits are absent [50].

The British Standards Institute (BSI) published a hallmark guide in December 2007, designed to specify protocols for the safe handling and disposal of engineered nanomaterials [3]. The document considered the different types of nanomaterials that may need to be handled, defined exposure and risk in the framework of nanomaterial types, and provided a general approach based on previous practice with respect to the handling of nanomaterials. Protocols were also defined for the handling of spills and accidental releases, based on the nanomaterial spilt or released. Disposal procedures were discussed based on the types of nanomaterials that were being handled, together with the requirements for preventing fire and explosion concerned with the handling of nanomaterials. The most notable aspect of the BSI document was the suggestion of benchmark exposure levels for proposed classes of nanomaterials based on hazardous properties that used safety factors compared with the bulk materials. It also suggested control approaches which are intended to keep potential exposures below the benchmark exposure levels using a control banding scheme. Following were the four categories (bands) of nanomaterials in the BSI guide [3]:

- fibrous
- carcinogenic/mutagenic/asthmagenic/reproductive toxin (CMAR)
- insoluble or partially soluble
- soluble

For this exercise, the CMAR nanomaterial was assumed to have increased bioavailability and thereby required a 10-fold safety margin over the bulk material, while fibrous nanomaterials received the most rigorous U.K. limit for fibres in the air (0.01 fibres/mL). The insoluble nanomaterial benchmark exposure level was based on the NIOSH time-weighted average proposed limit for ultrafine TiO_2 particulates, which is 15-fold lower compared with fine TiO_2 particulates. Soluble nanomaterials received a minor safety margin of two-fold over the bulk material, despite indicating that their nanoparticulate forms are unlikely to have greater bioavailability [3].

Several regulatory authorities internationally have since evaluated the BSI control banding method and have adopted most of the guidelines with minor modifications. A control banding nanotool has been developed by Paik *et al.* (2008) [44], and there are a range of online assessment tools to assist with risk management, with three being especially useful:

- Control Banding for Nanotechnology Applications (http://controlbanding.net/Services.html/) [6]
- Safe Work Australia–Work health and safety assessment tool for handling engineered nanomaterials (http://www.safeworkaustralia.gov.au/)[49]
- Workplace Health & Safety Queensland (WHSQ), Australia—Nanomaterial control banding tool worksheet, based upon the nanomaterial control banding approach described by" Paik *et al.* (2008) and incorporating changes suggested

by Zalk *et al.* (2009) [3,60], and containing content on identifying flammability of nanomaterials (http://www.deir. qld.gov.au/workplace/subjects/nanotechnology/control-banding/index.htm/) [56].

The control measures indicated by the control banding include similar workplace controls detailed above, including the engineering controls of enclosure, HEPA filtration and LEV for the nanomaterial process; administrative controls; and using face piece masks or self-contained breathing apparatus (SCBA) as appropriate for respiratory protection, double-gloving and non-woven fabrics for dermal/general protection. The same control measures that are expected to be able to protect workers in operations associated with research, including maintenance and cleaning of the work space, should also be effective in downstream operations such as manufacturing or construction activities, although their usage would have to be determined by an appropriate risk management process. However, if nanoscale materials are classified as potential carcinogens on the macroscale (i.e. Risk Phrase R45), then specialist advice should be sought when handling these materials.

Then in later development/production activities, and once the toxicological and other relevant properties of the nanomaterial have been determined, the control measures should be reviewed through a thorough process-specific risk assessment and, if warranted, modified accordingly. It is always recommended that a complete life cycle analysis of the nanomaterial should be made to identify potential "hotspots" of worker exposure, including construction, packaging, manufacturing, handling, maintenance or cleaning work, and end-of-life and safe disposal issues. There are a whole range of jobs and tasks that need to be considered. Existing ventilation systems that are effective for extracting ultrafine dusts in other industries should also be employed and optimally maintained where appropriate, in order to reduce exposure to ENMs.

A specific issue identified is the limited amount of data on the effectiveness of controls for nanomaterial types that are more commonly produced and used by nanotechnology industries, such as silicon, metal/metal oxide and CNT-based nanomaterials. Assessment requires research studies that involve actual workplace measurements taken before/after a nanomaterial process commences, and before/after control measures have been employed, thereby providing accurate comparisons of the levels of both engineered and incidental particulates between each situation. As indicated above, there are a whole range of jobs and tasks that need to be examined in the assessment process.

Several work safety authorities (such as the U.S. NIOSH [54], EU Nanosafe [15], Safe Work Australia [49] and the German programs of AP4:Tox NanoCare [1] and NanoGEM [36]) have ongoing specialised nanosafety research programs to progress the development of NMs emissions and exposure measurement capability in the workplace. There are also very useful Web information resources with training for safe handling of nanomaterials, such as the GoodNanoGuide (http://GoodNanoGuide.org/) [20] and the NIEHSnano webpage of the U.S. National Institutes of Environmental Health Sciences (NIEHS, http://is.gd/NIEHSnano/) [39]. Currently, the capability of measuring exposures to ENMs is limited. The development of easily operable, transportable, inexpensive and accurate real-time monitoring techniques to determine airborne concentrations of nanomaterials in the workplace is an urgent goal that would greatly assist in environmental monitoring, detection of airborne nanomaterials and validation of workplace controls.

The following are some initial questions to assist nanomaterial producers in considering potential OH&S issues for nanoenabled processes (from the nanosafety short course by Wright, Jackson and Allan of Nanosafe Australia, at the 3rd International Conference of Nanosciences and Nanotechnology, Sydney, 2010):

Q1. List the main characteristics of your nanomaterials that are most likely to increase their toxic potential to cause biological effects greater than those seen in the bulk material.

Q2. Briefly describe the potential toxic effects of the different nanoparticle types that you use.

Q3. What are the main hazards that concern you when fabricating engineered nanomaterials?

Q4. What part(s) of your nanomaterial's life cycle do you consider to be your highest potential risk?

Q5. Briefly outline the workplace controls that you use (or may want to use) to reduce exposure to nanomaterials in your workplace.

Q6. What strategies do you use for modifying/substituting your nanomaterials to reduce their hazard potential?

Q7. Where could you use "control banding" to help select appropriate workplace controls to reduce exposure to nanomaterials in your workplace?

Q8. What nanowastes do you generate in your workplace, how do you dispose of them, and what changes would you make to your nanowaste disposal practices?

Q9. What are your perceptions of how the risk and benefits of nanotechnology are being communicated to the public?

Q10. How do you think the public perceives your nano-technology-related activities?

In conclusion, the uncertainties about health and safety risks, limited measurement capability, and the possibility of a long latency period before any symptoms of disease develop, suggest a precautionary approach is required to control workplace exposures during the manufacture, use, storage, handling and disposal of nanoparticles. However, the principle of "As Low As Reasonably Practicable" (ALARP, also known as "As Low As Reasonably Achievable", ALARA) can be adopted once specific ENM data about the health and safety risks have been determined and defined, the accurate and detailed methods developed for their monitoring in workplace environment, and further validated evidence provided of the effectiveness of workplace controls to protect workers from exposure in the actual workplace.

References

1. AP4:Tox NanoCare http://www.nanopartikel.info/cms/lang/en/ Projekte/NanoCare/ NanoCare-Publikationen/.

2. Barker, P. J., and Branch, A. (2008). The interaction of modern sunscreen formulations with surface coatings, *Prog. Org. Coat.*, **62**, pp. 313–320.

3. British Standards Institute (BSI) (2007). Nanotechnologies—Part 2: A guide to safe handling and disposal of manufactured nanoparticles. BSI PD 6699-2:2007. ISBN 978-0-580-60832-2, http://www.bsi-global.com/.

4. Choi, H. S., Liu, W., Misra, P., Tanaka, E., Zimmer, J. P., Itty Ipe, B., Bawendi, M. G., and Frangioni, J. V. (2007). Renal clearance of quantum dots, *Nat. Biotechnol.*, **25** (10), pp. 1165–1170.

5. Conner, S. D., and Schmid, S. L. (2003). Regulated portals of entry into the cell, *Nature*, **422**(6927), pp. 37–44.

6. Control Banding for Nanotechnology Applications. Paik, S., and Zaik, D. M. at the Lawrence Livermore National Laboratory (LLNL). http:// controlbanding.net/ Services.html/.

7. Cross, S. E., Innes, B., Roberts, M. S., Tsuzuki, T., Robertson, T. A., and McCormick, P. G. (2007). Human skin penetration of sunscreen nanoparticles: *in-vitro* assessment of a novel micronized zinc oxide formulation, *Skin Pharmacol. Physiol.*, **20**, pp. 148–154.

8. dos Santos, T., Varela, J., Lynch, I., Salvati, A., and Dawson, K. A. (2011). Quantitative assessment of the comparative nanoparticle-uptake efficiency of a range of cell lines, *Small*, **7**(23), pp. 3341–3349.

9. Drew, R. (2009). Engineered nanomaterials: a review of the toxicology and health hazards. Safe Work Australia, Commonwealth of Australia publication. Online publication date: 1 Nov. 2009. ISBN 978-0-642-32921-9. 182 pages. http://www.safeworkaustralia.gov.au/.

10. Dufour, E. K., Kumaravel, T., Nohynek, G. J., Kirkland, D., and Toutain, H. (2006). Clastogenicity, photo-clastogenicity or pseudo-photo-clastogenicity: genotoxic effects of zinc oxide in the dark, in pre-irradiated or simultaneously irradiated Chinese hamster ovary cells, *Toxicol. Lett.*, **164**(Suppl. 1), pp. S290–S291.

11. Elder, A., Gelein, R., Silva, V., Feikert, T., Opanashuk, L., Carter, J., Potter, R., Maynard, A., Ito, Y., Finkelstein, J., and Oberdörster G. (2006). Translocation of inhaled ultrafine manganese oxide particles to the central nervous system, *Environ. Health. Perspect.*, **114**, pp. 1172–1178.

12. ENRHES (2010). Engineered Nanoparticles-Review of Health and Environmental Safety (ENRHES) project. Jan. 2010. 426 pages. http:// ihcp.jrc.ec.europa.eu/whats-new/enhres-final-report/.

13. Environmental Defence Dupont Nano Partnership (EDDNP) (2007). Nano Risk Framework, http://nanoriskframework.org/.

14. EU Nanosafe2 (2008). Explosivity and flammability of nanopowders. EU Nanosafe dissemination report. Feb 2008. DR-152-200802-2, http://www.nanosafe.org/.

15. EU Nanosafe2 (2008). European Strategy for Nanosafety: Safe production and use of nanomaterials. EU Nanosafe dissemination report. Jan. 2008. DR-325/326-200801-1. http://www.nanosafe.org/.

16. European Agency for Safety and Health at Work (EU-OSHA): Workplace exposure to nanoparticles (2009). European Risk Observatory Literature Review. Online publication date June 3, 2009. https://osha.europa.eu/en/publications/literature_reviews/workplace_exposure_to_nanoparticles/view.

17. European Commission Joint Research Centre's Institute for Health and Consumer Protection (EU JRC-IHCP) http://ihcp.jrc.ec.europa.eu/our_databases/nanohub/.

18. Feltis, B. N., O'Keefe, S. J., Harford, A. J., Piva, T. J., Turney, T. W., and Wright, P. F. (2012). Independent cytotoxic and inflammatory responses to zinc oxide nanoparticles in human monocytes and macrophages, *Nanotoxicology*, **6**, pp. 757–765.

19. Geraci, C. (2008). Nanotechnology: real world challenges for the industrial hygienist. Professional Conference on Industrial Hygiene, 11 Nov. 2008. http://www.americanceramicsociety.org/downloads/EHS_Nano/10_EHS_Geraci.pdf/.

20. GoodNanoGuide. http://www.goodnanoguide.org/Short+Courses/.

21. Gulson, B., McCall, M., Korsch, M., Gomez, L., Casey, P., Oytam, Y., Taylor, A., McCulloch, M., Trotter, J., Kinsley, L., and Greenoak, G. (2010). Small amounts of zinc from zinc oxide particles in sunscreens applied outdoors are absorbed through human skin, *Toxicol. Sci.* **118**, pp. 140–149.

22. Herber, R. F. M., Duffus, J. H., Christensen, J. M., Olsen, E., and Park, M. V. (2001). Risk assessment for occupational exposure to chemicals. A Review of Current Methodology, *Pure Appl. Chem.*, **73**(6), pp. 993–1031.

23. International Council on Nanotechnology (ICON) (based at Rice University, Texas, USA). http://icon.rice.edu/.

24. International Organization for Standardization (ISO) (2008). Nanotechnologies-Health and safety practices in occupational settings relevant to nanotechnologies. Document No. ISO/TR12885:2008.

25. Jackson, N., Tepe, S., and Wright, P. (2010). Engineered nanomaterials: Investigating substitution and modification options to reduce potential hazards. Safe Work Australia, Commonwealth of Australia publication. Online publication date: 1 Aug. 2010. ISBN 978-0-642-33100-7. 81 pages. http://www.safeworkaustralia.gov.au/.

26. Jackson, N., Lopata, A., Elms, T., and Wright, P. (2009). Engineered nanomaterials: evidence of the effectiveness of workplace controls to prevent exposure. Safe Work Australia, Commonwealth of Australia

publication. Online publication date: 1 Nov. 2009. ISBN 978-0-642-32884-7. 82 pages. http: //www.safeworkaustralia.gov.au/.

27. Kim, S. C., Harrington, M. S., and Pui, D. Y. H. (2007). Experimental study of nanoparticles penetration through commercial filter media. *J. Nanopart. Res.*, **9**, pp. 117–125.

28. Lewinski, N., Colvin, V., and Drezek, R. (2008). Cytotoxicity of nanoparticles, *Small*, **4**(1), pp. 26–49.

29. Lines, M., and Turney, T. W. (2008). Emerging markets for nano-materials, *Ind. Miner.*, **485**, pp. 60–65.

30. Maynard, A. D., Aitken, R. J., Butz, T., Colvin, V., Donaldson, K., Oberdörster, G., Philbert, M. A., Ryan, J., Seaton A., Stone, V., Tinkle, S. S., Tran L., Walker N. J., and Warheit, D. B. (2006). Safe handling of nanotechnology, *Nature*, **444,** pp. 267–269.

31. Monopoli, M. P., Bombelli, F. B., and Dawson, K. A. (2011). Nanobiotechnology: nanoparticle coronas take shape, *Nat. Nanotechnol.,* **6**(1), pp. 11–12.

32. Monopoli, M. P., Walczyk, D., Campbell, A., Elia, G., Lynch, I., Bombelli, F. B., and Dawson, K. A. (2011). Physical-chemical aspects of protein corona: relevance to *in vitro* and *in vivo* biological impacts of nanoparticles, *J. Am. Chem. Soc.*, **133**(8), pp. 2525–2534.

33. Monteiro-Riviere, N. A., Wiench, K., Landsiedel, R., Schulte, S., Inman, A. O., and Riviere, J. E. (2011). Safety evaluation of sunscreen formulations containing titanium dioxide and zinc oxide nanoparticles in UVB sunburned skin: and *in vitro* and *in vivo* study, *Toxicol. Sci.*, **123**(1), pp. 264–280.

34. Mottram, P. L., Leong, D., Crimeen-Irwin, B., Gloster, S., Xiang, S. D., Meanger, J., Ghildyal, R., Vardaxis, N., and Plebanski, M. (2007). Type 1 and 2 immunity following vaccination is influenced by nanoparticle size: formulation of a model vaccine for respiratory syncytial virus, *Mol. Pharm.*, **4**(1), pp. 73–84.

35. Nakagawa, Y., Wakuri, S., Sakamoto, K., and Tanaka, N. (1997). The photogenotoxicity of titanium dioxide particles, *Mutat. Res.*, **394**(1–3), pp. 125–132.

36. NanoGEM: Nanostructured Materials—health, exposure and material properties. http://www.nanopartikel.info/cms/lang/en/Projekte/nanogem/.

37. Nel A, Xia, T., Madler, L., and Li, N. (2006). Toxic potential of materials at the nanolevel, *Science,* **311**, pp. 622–627.

38. Nelson, S. M., Mahmoud, T., Beaux, M., Shapiro, P., McIlroy, D. N., and Stenkamp D. L. (2010). Toxic and teratogenic silica nanowires in developing vertebrate embryos, *Nanomedicine*, **6**, pp. 93–102.

39. NIEHSnano webpage of the U.S. National Institutes of Environmental Health Sciences (NIEHS). http://is.gd/NIEHSnano/.

40. Oberdörster, G., Oberdörster, E., and Oberdörster, J. (2005). Nanotoxicology: an emerging discipline evolving from studies of ultrafine particles, *Environ. Health Perspect.*, **113**, pp. 823–839.

41. Oberdörster, G., Maynard, A., Donaldson, K., Castranova, V., Fitzpatrick, J., Ausman, K., Carter, J., Karn, B., Kreyling, W., Lai, D., Olin, S., Monteiro-Riviere, N., Warheit, D., and Yang, H. (2005). Principles for characterizing the potential human health effects from exposure to nanomaterials: elements of a screening strategy, *Part. Fibre Toxicol.*, **2**, 8 (35 pages).

42. Organisation for Economic Co-operation and Development (OECD) (2011). Sponsorship programme for the testing of manufactured nanomaterials. http://www.oecd.org/document/47/0,3746,en_2649 _37015404_41197295_1_1_1,00.html/.

43. Organisation for Economic Co-operation and Development (OECD) (2010). Compilation and comparison of guidelines related to exposure to nanomaterials in laboratories. Series on the Safety of Manufactured Nanomaterials, No. 28. OECD Environment, Health and Safety Publications. NV/JM/MONO(2010)47. 81 pages. http://www.oecd. org/document/53/0,3746,en_2649_37015404_37760309_1_1_1_ 1,00.html/.

44. Paik, S. Y., Zalk, D. M., and Swuste, P. (2008). Application of a pilot control banding tool for risk level assessment and control of nanoparticle exposures, *Ann. Occup. Hyg.*, **52**(6), pp. 419–428.

45. Powers, K. W., Brown, S. C., Krishna, V. B., Wasdo, S. C., Moudgil, B. M., and Roberts, S. M. (2006). Research strategies for safety evaluation of nanomaterials. Part VI. Characterization of nanoscale particles for toxicological evaluation, *Toxicol. Sci.*, **90**(2), pp. 296–303.

46. Priestly, B. G., Harford, A. J., and Sim, M. (2007). Nanotechnology: a promising new technology–but how safe? *Med. J. Aust.*, **186**(4), pp. 187–188.

47. Project on Emerging Nanotechnologies, *Analysis-Consumer Products-Nanotechnology Project*, August 2009, http: //www.nanotechproject. org/.

48. Rogers, N. J., Franklin, N. M., Apte, S. C., Batley, G. E., Angel, B. M., Lead, J. R., and Baalousha, M. (2010). Physico-chemical behaviour and algal toxicity of nanoparticle CeO_2 in freshwater, *Environ. Chem.*, **7**, pp. 50–60.

49. Safe Work Australia (2010). Work health and safety assessment tool for handling engineered nanomaterials (Publication Date: 1/08/2010), http:// www.safeworkaustralia.gov.au/.

50. Schulte, P., Geraci, C., Zumwalde, R., Hoover, M., and Kuempel, E. (2008). Occupational risk management of engineered nanoparticles, *J. Occ. Env. Hyg.*, **5**, pp. 239–249.

51. Sharma, H. S. (2007). Nanoneuroscience: emerging concepts on nanoneurotoxicity and nanoneuroprotection, *Nanomedicine*, **2**(6), pp. 753–758.

52. Thurn, K. D., Brown, E. M. B., Wu, A., Vogt, S., Lai, B., Maser, J., Paunesku, T., and Woloschak, G. E. (2007). Nanoparticles for applications in cellular imaging, *Nanoscale Res. Lett.*, **2**, pp. 430–441.

53. UK Health & Safety Executive (HSE) (1989). Control of Substances Hazardous to Health (COSHH) Essentials, http://www.coshh-essentials.org.uk/.

54. U.S. National Institute for Occupational Safety and Health (NIOSH) Nanosafety website http://www.cdc.gov/niosh/topics/nanotech/safenano/.

55. U.S. National Institute for Occupational Safety and Health (NIOSH) (2004). NIOSH Respirator Selection Logic, http://www.cdc.gov/niosh/docs/2005-100/.

56. Workplace Health & Safety Queensland (WHSQ), Australia. Nanomaterial control banding tool worksheet (Publication Date: 13/07/2010). http://www.deir.qld.gov.au/workplace/subjects/nanotechnology/controlbanding/index.htm/.

57. Wijnhoven, S. W. P., Peijnenburg, J. G. M., Herberts, C. A., and Hagens, W. I. (2009). Nano-silver-a review of available data and knowledge gaps in human and environmental risk assessment, *Nanotoxicology*, **1** pp. 30–40.

58. Xia, T., Kovochich, M., Brant, J., Hotze, M., Sempf, J., Oberley, T., Sioutas, C., Yeh, J. I., Wiesner, M. R., and Nel, A. E. (2006). Comparison of the abilities of ambient and manufactured nanoparticles to induce cellular toxicity according to an oxidative stress paradigm, *Nano Lett.*, **6**(8), pp. 1794–1807.

59. Xiang, S. D., Scholzen, A., Minigo, G., David, C., Apostolopoulos, V., Mottram, P. L., and Plebanski, M. (2006). Pathogen recognition and development of particulate vaccines: does size matter? *Methods,* **40**, pp. 1–9.

60. Zalk, D. M., Paik, S. Y., and Swuste, P. (2009). Evaluating the control banding nanotool: a qualitative risk assessment method for controlling nanoparticle exposures, *J. Nanopart. Res.,* **11**, pp. 1685–1704.

61. Zvyagin, A. V., Zhao, X., Gierden, A., Sanchez, W., Ross, J. A., and Roberts, M. S. (2008). Imaging of zinc oxide nanoparticle penetration in human skin *in vitro* and *in vivo, J. Biomed. Opt.,* **13**(6), article number 064031.

Chapter 11

Managing Nanowaste: Concepts and Challenges for Nanomanufacturers

Jeremy Allan

School of Natural Science, University of Western Sydney, Richmond, NSW 2753, Australia

Jeremy.Allan@workcover.nsw.gov.au

This chapter explores a range of issues surrounding the management of nanowaste among nanotechnology research and manufacturing organisations ("nanowaste generators"). Principles are explored in terms of characterising, measuring and handling nanowaste. These efforts are essential if communities are to (a) derive benefit from nanotechnologies over the longer term and (b) be assured that human health and ecosystems are protected from the potential adverse effects of uncontrolled and increasing volumes of nanowaste entering the environment.

11.1 Introduction

With the rapid emergence of materials and products designed using nanotechnology, concerns have escalated regarding the potential

Nanotechnology Commercialisation
Edited by Takuya Tsuzuki
Copyright © 2013 Pan Stanford Publishing Pte. Ltd.
ISBN 978-981-4303-28-6 (Hardcover), 978-981-4303-29-3 (eBook)
www.panstanford.com

for adverse biological and environmental effects. Active promotion of social responsibility principles has required re-assessing the parameters associated with nanoproduct lifecycle assessments (LCAs). An increasingly diverse range of LCA parameters is emerging, from energy utilisation throughout the value chain, through to long-term measures of the impacts of conventional product displacement. With advances in nano-ecotoxicology enabling the detection of trace levels of nanowaste, the application of traditional waste management methods to treat some types of nanowaste is facing increased scrutiny.

One trend emerging among some nanomanufacturers is the identification of cleaner or greener fabrication methods. The application of nanotechnology to many industrial processes is often portrayed as potentially revolutionary in terms of improving the health of the environment and biological species. However, few studies have evaluated the relative efficacy of various approaches to managing wastes generated by nanomanufacturing facilities.

The extent to which nanomanufacturers are taking genuine efforts to detect, characterise, quantify, control and monitor nanowaste is under-researched. An examination of emerging trends in nanowaste management is essential to promote consistency among nanomanufacturers in order to effectively control nanowaste emissions. Developments in nanotechnology are progressing against the background of increasing anthropogenic pollution, unresolved environmental contamination from past decades and diminishing natural resources. Therefore, the challenge facing nanomanufacturers is not only to produce nanomaterials that are themselves ecologically benign but also to use fabrication processes that counter existing industry trends of increasing anthropogenic impact.

11.2 Conceptualising Nanowaste

There are many existing and proposed nanotechnology applications involving intentional, targeted interactions with biological and environmental systems. Information is scarce concerning the quantities, constituents and characteristics of nanowaste that are generated by such applications. Using probabilistic forecasting approaches, some studies have attempted to model the consequences of nanowaste released from commercial nanoproducts [4,14,31].

These assessments require information about bioavailability, persistence and other physicochemical properties of nanowaste constituents. To facilitate this process, a vocabulary is required which articulates the meaning of intentional versus incidental exposure, open and closed systems, system boundaries and the concepts of unintentional exposures and non-target receptors.

The concept of "nanowaste" has emerged alongside developments in nanometrology, value chain analysis and product lifecycle assessment. Terms such as "nanowaste", "nanopollution" and "nanometric debris", while not standardised, are increasingly identified in the literature. The term "incidental nanomaterial[1]" is often used to describe the background nanoparticulates detected in atmospheric sampling. In environmental and occupational hygiene studies, sampling requires assessing the contribution of incidental material to distinguish the relative contribution from various "external" sources unrelated to the processing, handling or application of engineered nanomaterials.

11.2.1 Nanowaste Streams

Nanowaste generated from manufacturing processes may enter the environment via emissions to the atmosphere, effluent, or as solid material. To facilitate the development of a dialogue on nanowaste, the British Standards Institute (BSI) put forward several descriptions of nanowaste streams:

- *pure nanomaterials* (e.g. discarded or fugitive nanodispersions, pristine nano-objects, contaminated, off-specification or impure nano-objects)
- *items contaminated with nanomaterials*, such as containers, wipes, disposable personal protective equipment (PPE)
- *liquid suspensions* containing nanomaterials (e.g. acid containing nano-objects)
- *solid matrices* with nanomaterials that are friable or have a nanostructure loosely attached to the surface such that they can reasonably be expected to break free or leach out when in contact with air or water, or when subjected to reasonably foreseeable mechanical forces [6]

---

[1]ISO/TS 80004-1:2010 (definition 2.10) defines *incidental nanomaterial* as "nanomaterial generated as an unintentional by-product of a process".

Direct occupational and consumer exposure to nanowaste is an increasing concern. One study, for example, identified nanowaste comprising exposed nano-objects embedded in larger nano-composite particles in an analysis of the debris generated by abrasive processes such as sanding (Fig. 11.1) [9].

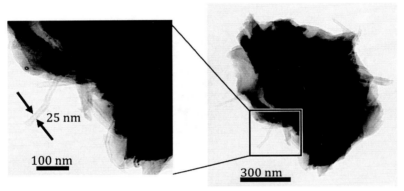

Figure 11.1 Sanding particle with detail of protruding fibres. Reproduced with permission from Ref. [9]. Copyright (2011) Taylor & Francis.

At nanomanufacturing facilities, nanowaste is generated by a diverse range of activities such as research and development, re-packaging of raw materials, plant commissioning, decommissioning and the maintenance or servicing of installations. It is also generated during the production of impure or contaminated product, product testing and routine nanomaterial production and handling. A study of the possible impacts of nanowaste exposures from point source emissions from such facilities was conducted using the test species drosophila [28]. The effect of carbon nanotubes and carbon black on functions such as locomotion, grooming behaviours and respiratory function were examined. The physical effects of exposure overwhelmed the capacity of drosophila to maintain grooming, blocked spherocytes and impaired motility, causing secondary metabolic impairment which eventually proved fatal (Fig. 11.2).

The authors of the study concluded that nanomaterial deposits might be expected in hot spots near manufacturing point sources or during intentional application of pest control agents. Transport and re-deposition (via grooming behaviours and direct contact) is also possible, bringing nano-objects into contact with human or environmental receptors that would not otherwise be exposed [28].

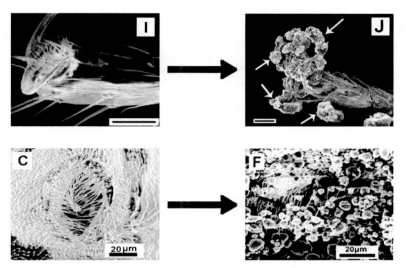

Figure 11.2 Physical impairment from nano-objects deposited on drosophila. Photographs I and J show the change in a drosophila's leg after nano-object exposure, while C and F show the change in a drosophila's spiracle. All scale bars are 20 μm. Reprinted with permission from Ref [28]. Copyright 2009 American Chemical Society.

Industrial and natural processes produce an infinite range of nanoscale waste forms. Industrial processes include non-renewable energy consumption, raw material extraction and various production processes. The vocabulary for describing nanoscale wastes, nanowaste and related wastes is under development. In one study, emissions from diesel forklifts as well as welding and grinding activities associated with the installation of new nano-manufacturing equipment were described as "incidental" nanomaterial [34]. Clarification is required concerning whether a definition of "nanowaste" should apply to such material.

By-products that contain nano-objects and/or nanostructured material associated with nanomanufacturing or end-user applications of nanoproducts would most likely satisfy a definition for "nanowaste". However, deliberations surrounding terms and definitions pertaining to nanowaste are at an early stage. Open for consideration is the extent to which such waste material contains engineered nanomaterials and whether such material may be recoverable or offset by nanotechnology-derived benefits.

11.3   Measuring Nanowaste

The uptake of information technology throughout industry is enabling vast amounts of data to be captured concerning resource utilisation and wastage. Over the past three decades, enormous volumes of data have been gathered from value chain activities in order to improve manufacturers' capacity to respond to changing patterns of consumption [2]. Such data are critical to identifying changes in consumer preferences, values and demands. Data are also necessary for meeting regulatory requirements associated with waste characterisation, quantification and control. Data are usually represented in terms of grams or kilograms of nanomaterial produced, or per nanoproduct or nanoproduct component produced.

Defining the various forms and quantities of nanowaste is essential, in order to unambiguously convey the potential impacts of nanoproducts throughout their lifecycle. The contribution from end-user applications is increasingly the subject of forecasting and modelling studies. For example, it was recently calculated that the amount of nanoscale titanium dioxide ($nTiO_2$) deposited in the vicinity of tropical reefs due to consumer use of sunscreens could be in the order of thousands of tonnes per year [4]. Another study examined the capacity of $nTiO_2$ to detach from paint applied to buildings, with significant concentrations detected near the source (building facade) and in distant natural waters [20].

In the major rivers in Europe, it has been estimated that up to 15% of the total concentration of silver could comprise nano-silver (nAg) due to release from consumer products alone (i.e. excluding emissions from nanomanufacturing facilities and other sources) [3]. Also on a large scale, concentrations of various nanomaterials (e.g. $nTiO_2$, nZnO, nAg) in sewage sludge are expected to rise by at least two orders of magnitude over a time frame of just four years [15]. As sewage sludge is applied to agricultural soil, the potential impact on soil flora, plants and live stock requires urgent assessment.

Contemporary approaches to LCAs provide a framework to ensure all potential sources of nanowaste are identified and measured, from the design phase through to product obsolescence. As far back as 2004, The Royal Society and The Royal Academy of Engineering advocated an LCA approach to nanomaterial risk assessment:

> We recommend that a series of lifecycle assessments be under taken for the applications and product groups arising from existing and expected developments in nanotechnologies, to ensure that savings in resource consumption during the use of the product are not offset by increased consumption during manufacture and disposal [39].

LCAs require examining both upstream and downstream processes. Upstream processes, for example, include precursor chemical manufacturing, component manufacturing, equipment manufacturing, upgrades and maintenance. Downstream processes include nanowaste capture, containment, recovery, recycling, treatment and disposal processes.

11.3.1 Indicators and Parameters

Product LCAs are increasingly incorporating a broad range of parameters throughout the value chain. Water and energy utilisation, particularly greenhouse gas (GHG) emissions [23], have become critical parameters of LCAs for industry with the advent of carbon trading [1][2]. Recent studies indicate that volumes of GHG emissions generated by some nanomanufacturing processes can be substantial, with regulatory environmental thresholds easily exceeded should current trends of consumer demand continue [35].

LCA parameters also encompass measures of energy and waste not only associated with the nanomaterial produced, but also with the construction of facilities, equipment and infrastructure utilised in nanofabrication [29]. Increasingly, research efforts are aimed at utilising renewable energy in nanomanufacturing as well as alternative methods of nanofabrication that use less energy and produce less GHG emissions.

11.3.2 Monitoring and Reporting

Monitoring the nanoparticulate constituents of nanowaste can be carried out using various occupational hygiene methods such as

---

[2]To distinguish between these parameters, energy consumption is represented as $MJ \cdot kg^{-1}$ and global warming potential as $kg\ CO_2\ equiv \cdot kg^{-1}$.

surface sampling, air sampling, wastewater sampling and numerous characterisation techniques. Occupational hygiene and environmental surveys can take place during any phase of the nanoproduct lifecycle. Surveys are possible during nano-manufacturing processes and value chain activities (e.g. importing, retailing, prototype testing, waste treatment, product use). As the levels of nanowaste are likely to vary widely in these scenarios, it is important that monitoring is carried out in a manner that minimises exposure via inhalation, ingestion, dermal or ocular routes of exposure.

11.4 Managing Nanowaste

The recognition of various forms of nanomaterials as being possibly "hazardous" or "toxic" has resulted in numerous calls for a precautionary approach to regulation. In 2004, The Royal Society and The Royal Academy of Engineering concluded:

> Specifically, we recommend as a precautionary measure that factories and research laboratories treat manufactured nanoparticles and nanotubes as if they were hazardous and reduce them from waste streams and that the use of free nanoparticles in environmental applications such as remediation of groundwater be prohibited [39].

This advice contrasts starkly with practices identified in two studies that examined nanowaste management practices among research and development firms, as well as nanomanufacturers of consumer nanoproducts. In one study, approximately 60% of nanotechnology firms classified nanowaste as hazardous [10]. However, only 30% had specific nanowaste practices in place. Where nanowaste practices were in place, nanowaste practices included (1) transferring nanowaste to contractors without labelling; (2) disposal of nanowaste down sink into general sewer; (3) sending nanowaste to landfill on the assumption that no regulation means nanowaste is not hazardous; (4) on-site incineration; and (5) storage on site until regulated.

Another study involving 40 companies producing nanoproducts across Germany and Switzerland examined the potential for release of nano-objects throughout the nanoproduct lifecycle [18]. It was found that the majority (60%) of companies believed that no release of nano-objects was possible throughout the nanoproduct lifecycle. Twenty percent of companies determined there was the potential for release of nano-objects throughout the nanoproduct lifecycle. However, only four firms conducted any investigation into the potential for uptake of nano-objects released from the production site or nanoproduct into biological systems. For the disposal stage, no investigations or analyses concerning potential uptake were reportedly undertaken.

The inconsistency of nanowaste management practices suggests greater efforts are required to promote the advice of credible authorities and nanotoxicologists. The intense competition among nanotechnology companies may be limiting transfer of knowledge concerning nanowaste management, while the lack of toxicity data for nano-objects and lack of specific industry codes and standards may be generating confusion [5].

To facilitate decision-making surrounding appropriate nanowaste management options, information concerning volumes and categories of nanowaste needs to be collected from duty holders throughout the value chain. Collected data should clearly identify the average and maximum volumes of residual nanowaste, end products and by-products involved in the processes of containment, storage, treatment or transformation.

Application-specific and scenario-modelling approaches to risk assessment are essential to evaluate the potential impacts of nanowaste. For example, the volumetric and environmental distribution profile of nanowaste generated by systems associated with food-related nanomanufacturing (including packaging) may be completely different from profiles associated with electronic component nanomanufacturing.

11.4.1   Corporate Nanowaste Management Policy

The development of a nanowaste management policy at corporate and site levels should specifically address legal and ethical obliga-tions. To guide policy development, principles derived from material flow accounting provide a useful starting point for defining the parameters to account for all inputs (i.e. material, energy and water) [8]. The framework in Table 11.1 outlines several principles to guide nanowaste management policy and program development [36]:

Table 11.1   Policy principles associated with nanowaste management

Policy principle	Purpose
Auditable	Nanowaste and its handling can be tracked
Transparency	Understanding among value chain entities, contract waste management operators, neighboring facilities, and the local community are promoted
Clarity	Confusion and misinterpretation of nanowaste characteristics are avoided
Strategically planned	Nanowaste generation is considered both locally and as part of an integrated site, or national, strategy
Managed	Facility conditions such as plant and product deterioration are incorporated
Optimised	Leading edge practice, innovation and review are promoted
Integrated	Interdependencies are identified among other facilities and other waste categories
Implemented	Operational compliance with procedural requirements is demonstrable

11.4.2   Disclosure and Transparency

Effective surveillance of generated nanowaste and its subsequent fate requires the identification of an appropriate range of measurement parameters. Voluntary schemes for capturing data have sought information on the following types of parameters:

- waste generated in the extraction and production of raw materials
- waste related to transport and storage throughout the supply chain
- ageing of nano-objects—diminution in quality of nanomaterial over time
- types or categories of nanowaste created during synthesis, fabrication or processing
- recipients of the nanowaste
- current and projected presence of nanowaste in the environment that results from manufacturing, distribution and end-of-life treatment of nanomaterials

- quantity of discarded off-specification nanomaterials and methods of handling and treatment of that material
- average storage time of nanowaste before disposal or transfer off-site
- storage container specifications and packaging methods for nanowaste
- annual generation of nanowaste
- nanowaste generated during different nanoproduct processes (consolidated)
- records of all nanowaste generated and shipped each year and
- on-site treatment of nanowaste generated by nanomanu-facturing

11.4.3 Contract Nanowaste Management Services

Contract waste management services fulfil a specialised role in terms of risk management, involving the transfer of responsibility for waste handling and treatment to facilities that have the necessary capability. Contract waste management services are often portrayed as a legitimate means of risk transfer, with legal obligations transferred from the waste generator to the contractor or service provider. Additional contractors may be involved, such as transport, storage, recycling and logistics service providers.

Contract firms provide numerous other services involving nanowaste, such as clean room maintenance, exhaust stack servicing, equipment calibration, routine cleaning, prototype testing of nanoproducts and evaluation of nanoproducts. A wide variety of legal entities are therefore required to demonstrate due diligence concerning processes for either eliminating or minimising the exposure of people and the environment to nanowaste.

Across industry, methods of waste handling, transport, storage and treatment vary widely, resulting in highly variable waste flow-allocation patterns [27]. Recent applications of legal jurisprudence surrounding waste management suggest that knowledge of the methods of waste handling, containment, transport, storage, transformation and treatment, should be shared not only among nanowaste generators and contractors, but also civil society [11]. Such an approach would seem compatible with LCA principles, necessitating the exchange of risk-related information beyond the traditional supply chain.

11.5 Nanowaste Risk Management

Increasingly sophisticated methods of resource recovery, material transformation and energy utilisation are challenging many preconceptions associated with "waste". In the manufacturing sector, discourse on waste management is traditionally associated with maximising productivity and minimising downtime. Critical path methodology is increasingly integrating objectives associated with waste management, energy efficiency and resource recovery. Philosophies such as *total production manufacturing* and *lean* have emerged, as well as ideals such as *benign-by-design* and *zero waste manufacturing*.

11.5.1 Legal and Regulatory Drivers

At the international level, legal and regulatory influences on methods of nanowaste management include policies associated with various conventions and trade agreements [12]. Legal jurisprudence surrounding the "polluter pays" principle continues to evolve, requiring liable organisations to pay not only for remediation but also for preventative measures to avert further pollution and environmental harm [32].

International instruments relevant to nanowaste include the Basel, Rotterdam and Stockholm Conventions. The complexity and duplication of these instruments, however, has given rise to calls to introduce a revised system of international conventions concerning waste governance [22].

A fundamental means of communicating risk information about substances is the Safety Data Sheet (SDS). Downstream users are required to consider the hazards associated with substances based on information from suppliers and to apply the appropriate risk management measures. Therefore, duty holders throughout the value chain must communicate to obtain the required information. The *intended* use of the substance is communicated to the manufacturer in order to facilitate exposure scenario development. If the SDS does not cover such use, downstream users are responsible for assessing the risks from substances. The relevant information to be communicated includes the following:

- composition and information on ingredients
- handling and storage
- exposure controls/personal protection
- physical and chemical properties
- toxicological information and
- ecotoxicological information

Regardless of the type of nanomanufacturing process, the use of the SDS is obligatory for substances that meet the criteria for classification as hazardous or are identified as substances of very high concern [13].

11.5.2 Risk Assessment Techniques

Numerous principles for assessing wastes generated in manufacturing environments are described in the literature [7]. As examples, four methods are set out here:

(i) *Mass and energy balance analysis*—identifying and quantifying system process losses to the atmosphere (dusts, liquids, gases, vapours), to effluent, to solids and to an accumulation in equipment; investigating reactant and utility requirements for different reactor modes (with and without recycling).

(ii) *Monitoring and targeting analysis*—monitoring the consumption of all raw materials and utilities, setting realistic targets below actual consumption levels and identifying waste minimisation options.

(iii) *Scoping audit*—identifying and quantifying parameters that include materials (raw materials, cleaning agents and packaging), utilities (electricity, heat and water) and waste streams (airborne, effluent and solid).

(iv) *True cost waste assessment*—based on waste streams and activities, calculating and prioritising direct costs (e.g. treatment and disposal costs) in addition to indirect costs (e.g. costs of unconverted raw materials, rework costs, storage costs, management time and monitoring costs) [7].

Nanowaste risk assessment also requires understanding the potential for system failures at all levels. This includes equipment malfunction, lack of conformity to procurement specifications, reliability of existing controls and likelihood of human error. Such principles are outlined by management system frameworks associated with various industry sectors and provide useful templates for nanowaste management. For example, one research group [21] adapted the *Hazard Analysis and Critical Control Point* (HACCP) framework (primarily a food safety management system) and applied this approach to managing risks associated with infectious wastes.

Others [36] have developed an approach coined *Waste and Source Material Operability Study* (WASOP), based on systems theory and *Hazard Operability Study* (HAZOP) principles. While developed for nuclear waste management, WASOP can also be applied to other waste forms and processes. In nanowaste management, the following activities could be considered when developing corporate protocols:

- system decomposition—identifying transportation routes, including vector transport, and cascading organisational triggers (upstream/downstream) for specific system failures
- exploration of the widest range of possible consequences for the facility
- selection of experts (with knowledge of constituents) to inform the WASOP, under various operating conditions among upstream/downstream facilities
- facilitator to guide the WASOP process, provide content support, encourage rigorous consideration of issues/actions and accurately record details

- identification of the major wastes generated during nanomanufacturing processes and during upstream/downstream value-adding activities
- for each risk issue, identifying actions to reduce likelihood of producing avoidable waste and
- documenting all issues and actions to form an audit trail.

The authors more recently put forward an alternative framework described as *Waste and Source-matter Analyses* (WASAN), for application to waste management scenarios [37].

11.5.3 Common Exposure Scenarios

In addition to fully examining all possible exposure scenarios and consequences, it is essential that those considered *common exposure scenarios* are documented. Common exposure scenarios include fugitive emissions and potential zones of settled nanowaste. Potential zones could extend beyond the manufacturing area to neighbouring facilities, rooms, offices and surrounding soil. Typical indicators of settled nanowaste include the percentage of surfaces with positive findings; servicing of filters, roof exhaust stacks, air conditioning units and other controls associated with ventilation.

11.5.4 Corporate Social Responsibility

Ethical and corporate social responsibility values are increasingly important drivers of organisational practices. One recent study into international supply networks identified waste management metrics extending to both upstream and downstream suppliers [38]. The notion of "suppliers" included suppliers of alternative energy. Some sectors (e.g. pharmaceutical) were identified as engaging rigorously with both upstream and downstream suppliers, conducting activities such as random audits and inspections. The value chain, rather than individual links, was considered crucial to the performance of organisations as "synchronised systems". This holistic approach harnesses the collectivity and diversity of expertise throughout the value chain to improve corporate social responsibility outcomes [38].

Guidance such as *ISO 26000:2010 Social Responsibility* calls for increased transparency surrounding the disclosure of information, such as emissions produced during manufacturing [19]. Manufacturing components, extracting raw materials and other unit processes, are often carried out by subsidiaries and intermediaries. For energy-intensive processes, the trend of *off-shoring* such operations for cost-cutting purposes has given rise to the phenomenon of *off-shoring pollution* and the creation of pollution havens [26]. Communities are increasingly demanding information associated with such entities be included in the data associated with end products, including disclosure of wastes and other parameters.

11.5.5 Extended Producer Responsibility

Notions of *extended producer responsibility* have evolved over recent years, in response to downstream users continuing to bear the burden of harmful exposures. Regulatory instruments, e.g. the European Union's *Extended Polluter Responsibility Policy*, and several test cases have focused attention on the term "producer" and whether the meaning differs from "manufacturer" or "supplier" [25]. At the same time, purchasers, importers, distributors and retailers are increasingly demanding information about associated by-products, emissions and compliance with performance specifications.

Consumers are also demanding more information about products (e.g. state of origin) and generally examining the degree of transparency concerning corporate decision-making. Efforts to reduce the amount and/or ecotoxicity of nanowaste have spurred many organisations to explore new nanomanufacturing methods, principles and processes, such as "organic", "green manufacturing", "benign-by-design" and "zero waste manufacturing". Certification schemes for products are emerging, where manufacturing processes adhere to principles and conditions associated with low ecotoxicity methods. Certifiers of organic produce in Canada and the United States, for example, have amended certification criteria to specifically exclude the use of nanomaterials being used in livestock and crops. In Australia, there are at least two cases currently in the courts regarding the cross-pollination of organic crops with genetically modified crops growing in adjoining fields.

Efforts taken by some nanomanufacturers to minimise risk include conducting nanoecotoxicological characterisation at the design stage; clearly stating product disassembly and recycling instructions; embedding sensors in products to give warnings or prompt certain actions; and use of barcodes to enable user access to more detailed information. This latter emphasises the two-way responsibility inherent in such transactions, necessitating engagement with the public from the early design stages, supporting local government infrastructure, through to commercialisation phase initiatives (e.g. policies to incentivise recycling).

11.6 Nanowaste Handling

There are numerous pathways whereby nanowaste may be emitted to the surrounding environment. Critical objectives associated with controlling nanowaste include containment and immobilisation. Adherence to control implementation in line with the hierarchy of controls is fundamental. These principles should be considered from the design stage, when specific types of nanofabrication processes are being considered.

Various methods of preventing nanowaste release to the environment are available to nanomanufacturers. One example involves filtering exhaust material through a cellulose acetate filter and water bath. Further research is required to investigate more efficient methods of nanowaste control for furnace exhaust [40].

11.6.1 Routine Nanowaste Handling

Routine efforts to handle nanowaste include the collection of materials and equipment such as wipes, PPE and other contaminated items, in labelled and enclosed containers with secure caps or covers. The container should be stored in a fume hood until full, then double-bagged, labelled and securely sealed. Procedural controls should be determined in consultation with occupational hygienists, including frequency of filter changes, frequency of cleaning work surfaces and equipment, vacuum flow rates during debris accumulation, filter cleaning and installation of static pressure gauges in vacuum cleaners [17].

11.6.2 Contingency Response

Incidents resulting in spills of nanomaterials or materials contaminated with nanowaste require a contingency response. Early planning and consultation with occupational hygienists should be undertaken to determine the types and levels of protection. A number of protocols should be developed depending on the extent[3] of the spill and whether the spill involves dry materials, liquids, or a combination of both. Notification protocols and access control are always likely to be important considerations in the event of a spill or uncontrolled release. Nanowaste spill kits should be readily available and contain as a minimum the following types of equipment in Table 11.2.

---

[3]A "significant" spill, for example, may be a spill involving more than a few grams of nano-objects.

Table 11.2   Recommended contents of nanowaste spill kit

• Barricade tape	• Certified HEPA-filtered vacuum cleaners
• Gloves	• Barriers to minimise air currents
• Respirators (e.g. powered)	• Sealable plastic bags
• Adsorbent material	• Walk-off mats
• Wet wipes	

11.6.3   Nanowaste Containment, Storage and Treatment

The efficacy of nanowaste containment, storage and treatment is determined by many factors. Decisions made at the design phase, such as raw material selection, choice of synthesis process, facility location and process layout (e.g. proximity to enclosures and trafficable areas), will in turn determine the additional nanowaste controls required.

The primary risk associated with immobilisation methods is the potential for leaching of nanomaterials and other chemical contaminants, from the matrix and subsequent release into the environment. Continuous improvement in the design of equipment and containment methods should therefore be an objective for nanowaste generators.

The incineration of regulated wastes is under increasing scrutiny due to lack of disclosure concerning efficacy, with reports of incompletely transformed materials and unknown quantities of hazardous emissions produced [16]. Further, from a societal point of view, incineration is often viewed as aesthetically unappealing and unsustainable. Consequently, efforts to introduce alternative forms of treatment, such as plasma gasification are reported in the literature as possible alternatives to incineration [24].

Methods for recycling and recovering nanomaterials are increasingly described in the literature. Researchers recently described an approach for recycling valuable SnO_2 nanowires from tinplate electroplating sludge, a process that could be applied to other types of nanowaste [41].

The use of biological systems in nanomanufacturing presents an additional challenge regarding the types of nanowaste controls required. Understanding and controlling cell processes to ensure productive synthesis does not result in unforeseen nanowaste production will be a key challenge in advancing such manufacturing techniques.

11.7   Future Directions

Trends in nanomanufacturing towards greener, waste-free processes are likely to intensify. Future nanomanufacturing processes will most probably utilise traditional waste streams to produce nanomaterials (e.g. carbon nanotubes derived from waste plastic) [42]. Processes such as *phytomining* are also emerging with application to both nanomaterial synthesis and nanowaste management.

Efforts to minimise energy consumption and contributions to GHGs are also intensifying. With new types of catalysts, solar reactors may be capable of housing synthesis processes for cleaner production of industrial grade carbon black and carbon nanotubes [33].

Nanosensors embedded in nanomanufacturing equipment and nanoproducts may help guide processes associated with decommissioning, including plant disassembly and product recycling. Nanoporous membranes and other solutions utilising nanotechnology, are also being developed for the decontamination of contaminated materials.

The zoning of precincts or clusters to support and promote collaboration among nanotechnology organisations has obvious benefits in terms of nanowaste management [30]. For instance, collective waste management strategies and onsite wastewater treatment may be implemented in a precinct zone in order to prevent uncontrolled release [33].

In conclusion, nanotechnology itself may be used to treat existing nanowaste or prevent potential nanowaste problems. Collaboration among duty holders throughout the value chain is essential to minimise nanowaste output and to prevent the "waste haven" effect, where certain entities generate nanowaste disproportionately. The quest for more inherently cleaner modes of nano-manufacturing suggests zero release of nanowaste is a feasible objective. Key questions facing stakeholders include whether to tolerate increasing rates of nanowaste output and to determine quantitative measures of potential impacts.

References

1. Anctil, A., Babbitt, C. W., Raffaelle, R. P., and Landi, B. J. (2011). Material and energy intensity of fullerene production, *Environ. Sci. Technol.*, **45**, pp. 2353–2359.

2. Attaran, M., and Attaran, S. (2007). Collaborative supply chain management: the most promising practice for building efficient and sustainable supply chains, *Bus. Proc. Manage. J.*, **13**, pp. 390–404.

3. Blaser, S. A., Scheringer, M., Macleod, M., and Hungerbühler, K. (2008). Estimation of cumulative aquatic exposure and risk due to silver: contribution of nano-functionalized plastics and textiles, *Sci. Total Environ.*, **390**, pp. 396–409.

4. Botta, C., Labille, J., Auffan, M., Borschneck, D., Miche, H., Cabié, M., Masion, A., Rose, J., and Bottero, J.-Y. (2011). TiO_2-based nanoparticles released in water from commercialized sunscreens in a life-cycle perspective: structures and quantities, *Environ. Pollut.*, **159**, pp. 1543–1550.

5. Breggin, L. K., and Pendergrass, J. (2007). *Where does the Nano go?, End-of-life Regulation of Nanotechnologies* (Woodrow Wilson International Center for Scholars Project on Emerging Nanotechnologies, USA).

6. British Standards Institute (2008). Nanotechnologies—Part 2: Guide to Safe Handling and Disposal of Manufactured Nanomaterials. PD 6699–2:200(BSI Standards, UK).

7. Brown, N. J., and Dempster, H. J. (2004). Waste minimization techniques and options for the wet and pretreatment sections of coil coating plants, *Environ. Prog.*, **23**, pp. 185–193.

8. Browne, D., O'Regan, B., and Moles, R. (2009). Assessment of total urban metabolism and metabolic inefficiency in an Irish city-region, *Waste Manag.*, **29**, pp. 2765–2771.

9. Cena, L. G., and Peters, T. M. (2011). Characterization and control of airborne particles emitted during production of epoxy/carbon nanotube nanocomposites, *J. Occup. Environ. Hyg.*, **8**, pp. 86–92.

10. Conti, J. A., Killpack, K., Gerritzen, G., Huang, L., Mircheva, M., Delmas, M., Harthorn, B. H., Appelbaum, R. P., and Holden, P. A. (2008). Health and safety practices in the nanomaterials workplace: results from an international survey, *Environ. Sci. Technol.*, **42**, pp. 3155–3162.

11. Davies, A. (2007). A wasted opportunity?, Civil society and waste management in Ireland, *Env. Pol.*, **16**, pp. 52–72.

12. Ederington, J. (2009). Should trade agreements include environmental policy?, *Rev. Env. Econ. Policy*, **4**, pp. 84–102.

13. European Commission (2008). Follow-up to the 6th Meeting of the REACH Competent Authorities for the Implementation of Regulation (EC) 1907/2006 (European Commission, Belgium).

14. Gottschalk, F., Scholz, R. W., and Nowack, B. (2010). Probabilistic material flow modeling for assessing the environmental exposure to compounds: methodology and an application to engineered nano-TiO_2 particles, *Environ. Modell. Softw.*, **25**, pp. 320–332.

15. Gottschalk, F., Sonderer, T., Scholz, R. W., and Nowack, B. (2009). Modeled environmental concentrations of engineered nanomaterials (TiO_2, ZnO, Ag, CNT, Fullerenes) for Different Regions, *Environ. Sci. Technol.*, **43**, pp. 9216–9222.

16. Health Protection Scotland, S. E. P. A. (2009). *Incineration of Waste and Reported Human Health Effects*, (Health Protection Scotland, Glasgow), pp. 129.

17. Heitbrink, W. A., and Santalla-Elias, J. (2009). The effect of debris accumulation on and filter resistance to airflow for four commercially available vacuum cleaners, *J. Occ. Environ. Hyg.*, **6**, pp. 374–384.

18. Helland, A., Scheringer, M., Siegrist, M., Kastenholz, H. G., Wiek, A., and Scholz, R. W. (2008). Risk assessment of engineered nanomaterials: a survey of industrial approaches, *Environ. Sci. Technol.*, **42**, pp. 640–646.

19. International Organization for Standardization (2010). *ISO 26000:2010 Guidance on Social Responsibility*, (International Organization for Standardization, Switzerland).

20. Kaegi, R., Ulrich, A., Sinnet, B., Vonbank, R., Wichser, A., Zuleeg, S., Simmler, H., Brunner, S., Vonmont, H., Burkhardt, M., and Boller, M. (2008). Synthetic TiO_2 nanoparticle emission from exterior facades into the aquatic environment, *Environ. Pollut.*, **156**, pp. 233–239.

21. Kojima, S., Kato, M., Wang, D., Sakano, N., Fujii, M., and Ogino, K. (2008). Implementation of HACCP in the risk management of medical waste generated from endoscopy, *J. Risk Res.*, **11**, pp. 925–936.

22. Koloutsou-Vakakis, S., and Chinta, I. (2011). Multilateral environmental agreements for wastes and chemicals: 40 years of global negotiations, *Environ. Sci. Technol.*, **45**, pp. 10–15.

23. Krishnan, N., Boyd, S., Somani, A., Raoux, S., Clark, D., and Dornfeld, D. (2008). A hybrid life cycle inventory of nano-scale semiconductor manufacturing, *Environ. Sci. Technol.*, **42**, pp. 3069–3075.

24. Kwak, T. H., Maken, S., Lee, S., Park, J. W., Min, B., and Yoo, Y. D. (2006). Environmental aspects of gasification of Korean municipal solid waste in a pilot plant, *Fuel*, **85**, pp. 2012–2017.

25. Lakovou, E., Moussiopoulos, N., Xanthopoulos, A., Achillas, C., Michailidis, N., Chatzipanagioti, M., Koroneos, C., Bouzakis, K. D., and Kikis, V. (2009). A methodological framework for end-of-life management of electronic products, *Resour. Conserv. Recy.*, **53**, pp. 329–339.

26. Levinson, A. (2010). Offshoring pollution: Is the United States increasingly importing polluting goods?, *Rev. Environ. Econ. Policy*, **4**, pp. 63–83.

27. Li, Y. P., Huang, G. H., Qin, X. S., and Nie, S. L. (2008). IFTCP: an integrated method for petroleum waste management under uncertainty, *Pet. Sci. Technol.*, **26**, pp. 912–936.

28. Liu, X., Vinson, D., Abt, D., Hurt, R. H., and R., and D. M. (2009). Differential toxicity of carbon nanomaterials in Drosophila: larval dietary uptake is benign, but adult exposure causes locomotor impairment and mortality, *Environ. Sci. Technol.*, **43**, pp. 6357–6363.

29. Meyer, D. E., Curran, M. A., and Gonzalez, M. A. (2010). An examination of silver nanoparticles in socks using screening-level life cycle assessment, *J. Nanopart. Res.*, **13**, pp. 1–10.

30. Morose, G., Shina, S., and Farrell, R. (2011). Supply chain collaboration to achieve toxics use reduction, *J. Clean. Prod.*, **19**, pp. 397–407.

31. Musee, N. (2011). Simulated environmental risk estimation of engineered nanomaterials: a case of cosmetics in Johannesburg City, *Hum. Exp.Toxicol.*, **30**, pp. 1181–1195.

32. Oelofse, S. H. H. (2008). Protecting a vulnerable groundwater resource from the impacts of waste disposal: a South African waste governance perspective, *Int. J. Water. Resour. Dev.*, **24**, pp. 477–489.

33. Ozalp, N., Epstein, M., and Kogan, A. (2010). Cleaner pathways of hydrogen, carbon nano-materials and metals production via solar thermal processing, *J. Clean. Prod.*, **18**, pp. 900–907.

34. Peters, T. M., Elzey, S., Johnson, R., Park, H., Grassian, V. H., Maher, T., and O'Shaughnessy, P. (2009). Airborne monitoring to distinguish engineered nanomaterials from incidental particles for environmental health and safety, *J. Occup. Environ. Hyg.*, **6**, pp. 73–81.

35. Plata, D. L., Hart, A. J., Reddy, C. M., and Gschwend, P. M. (2009). Early evaluation of potential environmental impacts of carbon nanotube synthesis by chemical vapor deposition, *Environ. Sci. Technol.*, **43**, pp. 8367–8373.

36. Shaw, D., and Blundell, N. (2008). WASOP, a qualitative methodology for waste minimization: Systems thinking, HAZOP principles and nuclear waste, *Int. J. Energy Sect. Manage.*, **2**, pp. 231–251.

37. Shaw, D., and Blundell, N. (2010). WASAN: The development of a facilitated methodology for structuring a waste minimisation problem, *Eur. J. Oper. Res.*, **207**, pp. 350–362.

38. Tate, W., Ellram, L., and Kirchoff, J. F. (2010). Corporate social responsibility reports: a thematic analysis related to supply chain management, *J. Supply Chain Manage.*, **46**, pp. 19–44.

39. The Royal Society and The Royal Academy of Engineering (2004). *Nanoscience and Nanotechnologies: Opportunities and Uncertainties* (The Royal Society, UK).

40. Tsai, S. J., Hofmann, M., Hallock, M., Ada, E., Kong, J., and Ellenbecker, M. (2009). Characterization and evaluation of nanoparticle release during the synthesis of single-walled and multiwalled carbon nanotubes by chemical vapor deposition, *Environ. Sci. Technol.*, **43**, pp. 6017–23.

41. Zhuang, Z., Xu, X., Wang, Y., Wang, Y., Huang, F., and Lin, Z. (2011). Treatment of nanowaste via fast crystal growth: With recycling of nano-SnO_2 from electroplating sludge as a study case, *J. Hazard. Mater.*, (DOI:10.1016/j.jhazmat.2011.09.036).

42. Zhuo, C. (2009). *Synthesis of Carbon Nanotubes from Waste Polyethylene Plastics*, (North eastern University, ProQuest Dissertations and Theses).

Chapter 12

Public Engagement

Craig Cormick

Department of Industry, Innovation, Science, Research and Tertiary Education, Canberra, Australia

craig.cormick@csiro.au

12.1 Introduction

If you are serious about understanding or undertaking public engagement, there are two vital questions that you need to answer:

First,

(1) Why should you engage with the pubic?

Once you have answered that,

(2) What does good engagement look like?

We will address each in turn. Public engagement can be undertaken for many reasons, varying from seeking better market intelligence, obtaining better policy inputs, gaining a better understanding of public concerns and aspirations, having a preference for

Nanotechnology Commercialisation
Edited by Takuya Tsuzuki
Copyright © 2013 Pan Stanford Publishing Pte. Ltd.
ISBN 978-981-4303-28-6 (Hardcover), 978-981-4303-29-3 (eBook)
www.panstanford.com

receiving information from the public rather than those who seek to represent the public, or for seeking guidance on technology futures for research and development that are most likely to be accepted by the public.

Also, as new technologies are increasingly becoming the focus of public concerns, unless the causes of these concerns and the factors driving them are better understood, new and contentious technologies can run the risk of public rejection. This can be diminished by good public engagement processes, however, which have been shown to lead to improved public input to policy, research and product development, as well as to diminishing concerns about products and processes using new technologies, when those products and processes meet community needs [3].

Underpinning this is the fundamental belief that in a democracy, citizens should have a say in decisions about technological developments that will significantly affect their lives [26]. Without over-repeating the types of impacts that nanotechnologies are predicted to have, they are widespread and impact most fields, from new materials to medical advances to renewable energies and information technology. Indeed, it might seem at times that there are very few fields of human endeavour that have not been promised improvement through some developments in nanotechnology.

And amongst the plethora of promises sit the consumers (or citizens, or members of the public), sometimes bemused at the breadth of promises, and generally not well aware of, nor understanding of, what nanotechnology is and is not. And yet, the crucial point is that it will be the members of the public that ultimately decide upon the fate of many applications of nanotechnologies in the market place.

This is a key issue about public engagement: what do the public make of it all, and is that an accurate reflection of things? In a broader sense this question tends to be answered not just in terms of a particular nanotechnology application, but in terms of what type of a society we want to live in, and what impacts will new technologies have on that (as well as who should be deciding what technologies we adopt or reject?) So it can be very useful to step back and look at the processes of engagement, and how they are framed, before the public is even involved in them, as how different stakeholders view nanotechnology and view public engagement is integral to the types of public engagement that occur.

But before looking at different perspectives on public engagement, we should also consider the contra question: What might happen if we do not engage with the public?

The spectre of GM foods and crops looms over most discussions on public engagement with new technologies as a case study of too little too late. GM foods and crops can be typified as a technology that was developed before being presented to the public—who it turned out did not especially want the technology. The reasons for this are many and often more complex than the descriptions given but can be summed up as, the public were being given a technological solution to a problem that they did not see as their problem. In addition, they were being asked to take whatever risks related to GM foods might arise, but all the benefits were going to others: predominantly the crop companies and farmers.

Imagine how different the GM debate might have played out, and the types of GM crops we would have seen developed, if researchers had held early discussions with members of the public over what might be the best applications of gene technology. It is highly likely that we would have seen small niche crops with high value-add, such as pharmaceuticals being grown in plants in greenhouses, rather than GM broad acre crops with herbicide resistance.

Even if we held those debates now, we will never really know how it might have played out in actuality, as any discussions of GM foods will be forever framed around the way that GM foods were introduced into society. For nanotechnology, however, there is still time to get it right. Towards this, it is perhaps more relevant to look at the ways in which the nanotechnology debate is different from the GM debate, rather than to look at the ways that it might be similar.

The situation at present could be summarised as follows: The majority of the public have little awareness of what nanotechnology is but tend to be favourable towards it [21], and media reports still tend to concentrate on the potential of nanotechnologies [5]. But things could change rapidly, especially following a real or perceived public health scare involving nanotechnologies.

Nanotechnology is a technology that is still emerging, as are public attitudes towards it. And attitudes will continue to form as more sections of the public become more aware of nanotechnology, and its risks and potentials, and are then able to articulate their thoughts and feelings about the impacts of nanotechnology (both good and bad).

12.2 Nanotechnology in Society

Mirroring the development of nanotechnologies has been the development of social research into nanotechnology in society, looking at the impacts of nanotechnology on individuals and society. Three major drivers of this have been the following:

- rapid internationalisation and globalisation of science
- changes in the role of science and society
- increasing interdependence of science with society [17].

As a result, most developed countries have made the decision that there are strong benefits to be gained from engaging with the public on nanotechnology developments, being driven by a variety of sources, including governments, researchers and NGOs, with little industry input to date.

In an ideal world, good engagement goes something like this: A scientist develops a new process or innovation, and before applying it, he or she has a discussion with the community that will be most affected by it, as to how they would like the technology developed and used. They discuss, in clear and reasoned ways, what types of applications should have resources put into them, and what types of products should be developed. Then, with a firm understanding of public support or rejection, or preferred direction of further research, capital for development is easier to acquire, and products are developed, and the public, the scientists, and developers are all happy with the outcomes.

In reality, it tends to go a little like this though: A scientist develops a good idea and then hunts around for a use for that idea, focusing on areas most likely to attract funding. When the idea is developed into an application, it is taken to the market— where it succeeds or fails, for a variety of reasons. If, there is community backlash at that point, then engagement is undertaken to try and sell the benefits of the product and process, and minimise the risks, and/or better determine how the members of the public became so misguided as to reject the product or process.

There are not many examples of the first model that spring to mind—and too many of the second, mostly based on the assumption that if an idea gets capital funding then it must be a sound idea. This comes from the traditional triple-helix model of technology development, where the key players are government–researchers–industry. With some technologies, such as mobile phones or iPads, it works well. But with many other technologies, particularly those are socially disruptive in any way, such as biotechnologies and nanotechnologies, it is not such a suitable model, and a "quintuple helix" (government–researchers–industry–NGOs/community groups–the public/s) provides a more inclusive approach.

Overall, the trend amongst public engagement practices has generally been towards better engagement, with strong attempts to get it right, and increased demonstration of commitment to good engagement. A study of community engagement in the United Kingdom, undertaken by the Ipsos-Mori Social Research Institute in 2006, found that experiences of public engagement (although not specifically about nanotechnology) were increasingly under-pinned by the following core values:

- People have the right to participate in decisions that affect their lives.
- Beneficiaries of public policy can add value to its development and implementation.
- Participation should lead to change for the better [16].

There has been an explosive growth in applications of public engagement in recent years, tied to more open government and policies of social inclusion, and a desire to reach beyond professional lobby groups who tend to dominate the mediascape and public debates, and those working in the nanospace cannot afford to get left behind. And while in the public sector democratic deliberation is an important driver for engagement processes it is not such a key driver for industry. But the same models can provide tremendous insights into public values and thinking. And in addition, as industry are largely influenced by regulation, legislation and public attitudes, then industry necessarily has a vested interest in the best outcomes of good wider engagement.

Jose Manuel de Cozar-Escalante, of the University of La Laguna in Spain, summarised this well, as:

> Over the past few years, the mistrust of traditional methods of political representation has driven Western countries to experiment with a multitude of new deliberative mechanisms aimed at fostering wider representation in the political decision making process. As these experiments have proliferated rapidly, the temptation to naively ignore their problems has become more and more difficult to overcome [8].

Likewise with nanotechnology, once public engagement has begun, it cannot be turned back as public expectation grows with it, and it will become more and more important to improve on engagement practices and processes to meet growing expectations. But it is worth emphasising that the uptake of nanotechnologies will be enhanced by public engagement, as the types of nano-technologies that are developed will better align with public needs and values if the public are engaged in discussions about nanotechnologies and their applications in the early stages of development.

12.3   So What Does Good Engagement Look Like?

> They (the best public engagement) have opened up science-governance processes to public scrutiny and debate, and have demonstrated that public deliberations can generate important messages for scientists and decision-makers about the concerns and aspirations held by members of the general public for their work. They have also demonstrated how public engagement can generate mutual learning, build new skills, and overcome preconceptions and social barriers between different groups [12].

This second point can be a little harder to answer well, as public engagement can look very different to different interest groups, and can be defined in many different ways. It can be about simply educating and informing the public about the needs of researchers or industry or government, but it is better engagement when it is also about researchers, industry and government adapting to the needs of the public.

It is perhaps more instructive to being by looking at what good engagement does not look like, and unfortunately many public engagements, while not necessarily bad, are only "almost good", which can be like having a bridge that will almost get you across a wide chasm—but not quite—dumping you into the river of public criticism flowing rapidly beneath you.

Poor engagement is usually developed in isolation from the stakeholders needing to be engaged with, is more about proselytising or converting a stakeholder group to another's way of thinking, and a common outcome, even amongst many good forms of engagement, is that it makes no impact on policy or technology development. This is an outcome of some otherwise very good engagement exercises, such as the United Kingdom's 2005 Nanojury and the 2009–11 Dutch Societal Dialogues on Nanotechnology, which ultimately had little impact upon policy formulation. Also, as many models of engagement only include two key groups, such as researchers and the public, or government and the public, they fail to be inclusive of key participants who are integral to any outcomes being widely adopted. This is sometimes due to deeply rooted ways of thinking and doing within an organisation that need to be overcome, lest they become an obstacle to good engagement.

12.4   Obstacles to Good Engagement

One of the key obstacles is based on the two fundamental principles of good engagement that have a habit of working against each other. These are how to include all key stakeholders into a process while accommodating the very different and competing perspectives of different stakeholders. The best that can sometimes be achieved is to find some way for stakeholders to come to an accord, rather than expecting them to see the issues from a perspective beyond their own self-interest point of view.

Carolyn M. Hendriks has observed that most stakeholders, whether they are a part of civil society groups or not, use the public sphere for their own purposes. She states [14]:

> When we take this phenomenon seriously, we see how easily discursive models of public deliberation might collapse into the very kind of interest group pluralism that deliberative democracy as sought to reject.

The fact that many actors approach public deliberation strategically reinforces the important of designing moments of collective reasoning that encourage 'we' rather than 'I' thinking.

The problem can be best depicted through the analogy of a reverse panopticon prison. To explain: a panopticon is a large prison where each cell is open on one side and facing towards a single point, where a jailor may observe the prisoners freely, but the prisoners cannot observe the jailer, who may or may not be observing them at any given time, leading to perpetual unease. The panopticon was designed by the English philosopher and social theorist, Jeremy Bentham, in 1785, and the concept of the prison as a new mode of obtaining power of mind over mind was later popularised by the French philosopher Michel Foucault [10].

The concept was taken further by George Orwell in his novel 1984, with people being viewed through their televisions at all times, which is interesting given that one of the concerns being raised about nanotechnologies is the ability to contribute to miniature surveillance technologies that might be able to monitor citizens much more widely than even now occurs through closed circuit TV monitoring.

The reverse panopticon prison, however, is a model where there is a single point of focus that all the prisoners in their cells can view, through a peep hole in their cells. And like the analogy of the different people grabbing hold of the different limb of an elephant in a darkened room and proclaiming it to be a different creature, the prisoners can only ever see one façade of a complex shape before them.

This is played out time and time again in planning, consultations and examples of public engagements. Participants are trapped in their cells with their singular view of the complex object before them, and cannot imagine the perspective of another prisoner in another cell. So single issue solutions to complex problems are dogmatically placed on the table. "It just needs better education!"—"It just needs better regulation!"—"It just needs people to better understand the risks!"

If it does not yet seem the case for nanotechnology, consider global climate change and it is easy to see how many single perspectives compete with each other without attempting to assemble the disparate views into a new way of seeing a complex issue. Again to quote Jose Manuel de Cozar-Escalante [8]:

In short, we should seek a broader conception of representation for the politics of science and technology, a representation that is better suited to the intricacies of our increasingly technological and globalised world.

For nanotechnology, perspectives on the problems and their solutions tend to be framed differently by all the key stakeholders: researchers, government, industry, non government organisations and the public, and they tend to have great reluctance to leave their cells to consider alternative perspectives. While there is increasing agreement amongst key stakeholders that public engagement on nanotechnology is a good thing—the reasons for doing it can differ greatly, and attempts to bring different stake-holders together to try and share perspectives can lead to a tendency to competing perspectives. In Australia, workshops on public engagement on nanotechnologies and biotechnologies have tended to accentuate the differences between key stake-holders [3]. Also, in a recent workshop held in Canberra, stake-holders representing researchers, social scientists and engagement practitioners tended to have homogenous attitudes towards the key engagement issues within their own groups but differed significantly to other groups. For the researchers, better education was the key issue. For the social scientists, better processes was the key issue. And for the practitioners, having a better under-standing of audiences was the key issue [33].

12.5   A Short History of Public Engagement

Historically the scientific/industrial view point—that if the public only understood the science better they'd accept it better—has tended to dominate much early engagement, based on educating the public about nanotechnology. This now discredited argument, known commonly as the Deficit Model, still emerges in discussions on the impact of science on society, though with decreasing frequency.

Many engagement processes undertaken under this model tended to see awareness raising as the end game, but in practice only had an impact on those who agreed with the facts or arguments being put forward. Of key interest here is a study by Druckman and Bolsen that found that factual information was actually of limited value in influencing opinions, as it did not have any greater impact than information that lacked factual basis [9]. The key finding of this study was that an individual's pre-existing opinions will bias what information they are willing to accept. In addition, providing people with different points of view tends to make them become more polarised or extreme towards the position they already held [9]. This effect has been summarised as: "People think in frames... to be accepted, the truth must fit people's frames. If the facts do not fit a frame, the frame stays and the facts bounce off" [19].

Added to this is the finding of Binder that when people talk about risks associated with unfamiliar science and technologies, most people will adopt an initial position of support or opposition, and the more they talk to people with different positions the more likely they are to defend their initial stance, regardless of the factual basis upon which it was made [4].

The significance of such findings is quite important to under-stand, as it shows that engagement campaigns that are based on informing and educating an audience with existing views will have very little impact. This suggests that new ways of thinking about how to inform people are needed, and rather than seeking to challenge people's frames, we need to provide different ways of thinking that allow members of the public to experiment with different frames in their own thinking.

12.6   Deficit 2.0

The Deficit model of communication or engagement has now been replaced by something I term the Deficit 2.0 model. It goes like this: If only the public had a more science-centric view of the world they would understand things as well as scientists do. And flowing on from that, they would see risk in terms of scientific or mathematic likelihoods, and discuss new technologies in terms of facts and not emotions, and so on.

This perspective, or course, fails to accept that there is a public view of risk that while different to a scientific view of risk, is no less valid, whether it is underpinned by emotions or not.

By contrast, an industry, or private sector view of the public, is one whereby they are often described as being primarily consumers, and needing to be engaged through traditional consumer models. They are the "market" that can be influenced by sophisticated advertising and marketing. But it can be surprising how often technology-driven products do not well address the first principle of marketing: find a consumer need and meet it.

Again, GM food is the poster child for technology developers falling in love with the technology and attempting to sell it to an unresponsive public, largely based on their own enthusiasm.

Government agencies tend to define the public increasingly as stakeholders, which has a connotation of being shareholders, or having a vested interest. But this does not hold up well to scrutiny of the different motives and different levels of engagement amongst the public. Something that is not often acknowledged amongst those involved in discussions on community engagement with new technologies, is that many members of the public really do not give a damn about science and technology issues. In fact, according to figures from the Victorian Department of Innovation, Industry, and Regional Development it might be as high as 35% [6].

Turning to NGOs and civil society groups, many view the publics as their members, concentrating on those who align with their perspectives or ideologies or are actively engaged in social issues. Europeans have a preference for calling these active members of the public "citizens", as articulated by Wickson *et al.* [32], who examine how the public are categorised as laity, consumers or stakeholders. Citizens, however, tend to have a strong relationship with the "state" and are actively engaged in inputs to policy formation. Unfortunately this does not account very well for those who do not know they are citizens, or could not care if they were.

We could go on and on, analysing the way that different interest groups define the public, and argue how they should best be reached, but it would only serve to continue to demonstrate the analogy of the reverse panopticon prison. Rather we might now look at the public themselves, and ask who really represents them and their interests? For the public are consumers and citizens and public/s and stakeholders and the unengaged and engaged and need to be represented by ways of thinking that understands this huge diversity, not just to better understand the public but also to better understand all the interest groups and stakeholders who are seeking to engage with the public on new technologies.

So, returning to the question of what good engagement is, we can look at some models and principles that have been put forward by different organisations, and see what a comparison of their definitions tells us. The first model was developed by Involve, a UK "think-do tank" committed to "exploring how new forms of public participation can strengthen democracy" [12]. Their principles were

(1) **Makes a difference**: The purpose of participation is to achieve change in relation to the purpose identified; it may also make a difference to all those involved in terms of learning, confidence and sense of active citizenship. This requires active commitment to change by all parties.

(2) **Transparency, honesty and clarity** about the purpose, the limits of what can and cannot be changed, who can be involved and how, and what happens as a result.

(3) **Accessibility** so that no participant is excluded because of lack of physical access to meeting places, timing, appropriate support (e.g. child care), etc.

(4) **Learning and development**: Participatory processes should seek to support a climate of mutual learning and development among all those involved.

(5) **Power**: Participatory processes should have sufficient power to achieve the agreed objectives. This may require a change in the existing power sharing arrangements.

(6) **Appropriate participants**: Representative and/or inclusive, depending on the purpose of the exercise, with traditionally excluded groups given special support and encouragement when their involvement is appropriate.

(7) **Adequate resources**: To manage the process well and to deliver on the results.

(8) **Voluntary**: People may be encouraged to be involved, and even paid for involvement, but effective participation requires them to choose to be involved. Participation cannot be compulsory [12].

By comparison, the Australian National Toxics Network (NTN) supports a list of 10 protocols for community engagement:

(1) Shared commitment to the process

(2) Accountability

(3) Clarity of roles and responsibilities

(4) Openness, procedural fairness and equity

(5) Timeliness of decision-making and information delivery

(6) Access to information, expertise and personnel

(7) Easily comprehended information and flexible processes

(8) Continuity

(9) Commitment to consensus

(10) Feedback mechanisms [23]

And the Dutch Rathenau Institute, one of the leading NGOs in this field, has developed 10 lessons for good engagement on nanotechnology:

(1) Differentiate between the risk issue and the broader debate about social impacts of nanotechnology.

(2) Actively address the risk issue, as a failure to do so can undermine the legitimacy of broader societal dialogue.

(3) Involve NGOs in developing policy, particularly in relation to risk.

(4) Provide clear information about nanotechnology products, risk governance strategies and any uncertainties about that exist.

(5) Create a public agenda which enjoys wide support with the input of all participants respected.

(6) Build upon ongoing discussions wherever possible, and establish what is to be discussed under the heading of "nanotechnology", and what issues can be more appropriately examined within other, already existing discussions. This will also promote participation from existing institutions and societal organisations.

(7) Facilitate the involvement of smaller NGOs, assisting them in capacity building.

(8) Remain open to the agendas of interest groups.

(9) Inform the wider public, who have little awareness of nanotechnology, about the societal impacts of nanotechnology to assist them to learn more.

(10) Give a wide section of the community a voice by means of small-scale engagement activities [30].

An inclusive multi-stakeholder process, known as STEP (Science and Technology Engagement Pathway), run over 2010–2011 to develop common principles amongst key stakeholders, undertaken by the Australian Department of Innovation's National Enabling Technology Strategy, came up with

(1) **Commitment and Integrity**: A high level of commitment and integrity, including mechanisms for transparency and accountability;

(2) **Clarity of objectives and scope**: Clarity about the purpose, objectives and scope of the engagement;

(3) **Inclusiveness**: Inclusiveness of the diversity of people and views;

(4) **Good process**: Including an appropriate and structured method, communication and consultation with participants throughout, and appropriate, independent oversight and evaluation;

(5) **Quality information/ Knowledge sharing**: Relevant, accurate and balanced information and knowledge sharing;

(6) **Dialogue and open discussion**: Genuine, interactive deliberative dialogue; opening up discussion rather than closing it down;

(7) **Impact on decision making**: Demonstrated influence on decision making [22].

12.7   Understanding NGOs and Affected Publics

These different models show both that even amongst NGOs there can be differing perspectives, and that not all NGOs can be broad-banded together as having the same perspectives. And the role of NGOs deserves some further comment, as while they have been integral to the adoption of much public engagement, some interest groups find them very difficult to work with.

It is important to understand the different types of NGOs and their differing agendas, to assist in finding ways to work with them. A useful way of considering NGOs is to understand whether they are primarily representing a public constituency or membership, whether they do not primarily represent a public constituency but represent the public's interests, or whether they primarily represent their own interests.

need to be discussed application by application. Medical health and environmental applications were the areas most strongly supported and miniature surveillance devices, and integrating computers into clothing or consumer goods were least supported (see Fig. 12.3).

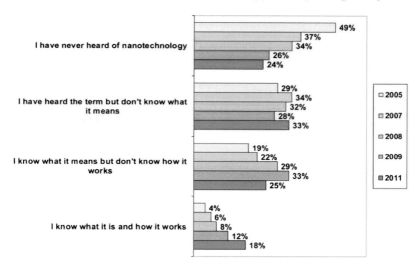

Figure 12.2 Attitudes to risks and benefits of nanotechnology 2005–2009.

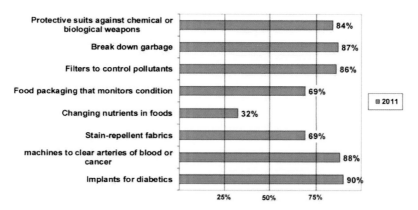

Figure 12.3 Attitudes towards different applications of nanotechnology.

Using nanotechnologies in foods to change their nutrients was also not highly supported, but using nanotechnology in food packaging received moderate support [21], showing again that the public are able to differentiate between different applications.

It is worth adding, however, that on most topics, the public do not fall into simple for or against categories, and seeking opinions against a Likert scale over a scale of 10 increments shows that there are minorities who are extremely for or extremely against contentious technologies, with the bulk of the population in the middle. Yet public forums, as a popular form of engagement, tend to only attract the extreme fors and againsts, and become ideological blood sports that then to lead audience members polarising into for or against camps, when their own prejudices, values or attitudes are reinforced.

12.9   Engaging the Unengaged

And then there are the unengaged publics, who do not as yet care too much about nanotechnology and its impacts, and do not show up at most engagement activities. In an effort to better understand these members of the public, the Australian Government held a series of "nanodialogues" on various topics such as water, bionics and new materials, recruiting members of the public who were generally disinterested in science and technology. Participants were recruited by a market research company and were paid for taking part.

The parameters of the dialogues were that the participants led the discussions more than would happen in a focus group that technologies were framed in terms of applications, and that the discussions should lead to what type of a world we want to live in. The key finding was that disengaged and unengaged members of the public have different values, interests and levels of awareness in science and technology issues to those sections of the public who tend to self-select to attend most information or engagement activities.

The unengaged also tend to have had poor experiences with science at school that has turned them off science. They also tend to seek information on science and technology issues primarily from friends and family, and they respond to S&T discussions overwhelmingly in terms of their applications only, and as such need to be engaged in different ways to the highly engaged or affected members of the public.

So how does this align with any proscriptive list of good engagement principles, such as those put forward by three organisations cited? It is clear that principles need to be broad enough to encompass different audiences' needs and preferences, and their differing attitudes towards different applications of nanotechnology. Any process must also find a way to reduce mistrust between competing stakeholders, and allow different individuals and groups to genuinely learn about the perspectives and opinions of others. In practice, however, this can sometimes prove a little harder than it might seem.

As challenging as the unengaged are those who are "alternatively engaged" and who have set beliefs based on their values, are not receptive to counter messages even if they are based on scientific evidence. The growth of anti-science ideologies has come to the fore during public and political debate on climate change, attracting enough social research to outline some key principles that make some people immune to science-based messages. These include

(1) When information is complex, people make decisions based on their values and beliefs.

(2) People seek affirmation of their attitudes (or beliefs)—no matter how fringe—and will reject any information or evidence that are counter to their attitudes (or beliefs).

(3) Attitudes that were not formed by logic are not influenced by logical arguments.

(4) Public concerns about contentious science or technologies are almost never about the science—and scientific information therefore does little to influence those concerns.

(5) People most trust those whose values mirror their own.

Case Study: Social Inclusion and Community Engagement Workshop. Canberra 2008.

The Australian Office of Nanotechnology (AON) sought to develop a new framework for public engagement by inviting all key stakeholders to take part in a workshop held over one day in December 2008. Participants represented the five key stakeholder groups of Government, Industry, Researchers, Community Groups and Change Agents, and the General Public (change agents consisted of NGOs, social scientists and unions, who generally advocated change to the status quo).

Following was the definition of social inclusion proposed by the AON prior to the workshop: "A socially inclusive approach allows for genuine inputs into the development on products and policies and ensures equitable outcomes from them."[3] And public engagement was defined as "one of the processes that allows for increased social inclusion. Engagement seeks to achieve increased two-way information flow and knowledge exchange as well as increase overall technological literacy." [3]

A more holistic approach that incorporates the full spectrum of players, from industry, government, research, community and activist groups was expected to

- enable social inclusion to be incorporated across the full range of policy, research and product development,
- establish the relationships and networks needed for ongoing broader engagement,
- create meaningful mechanisms for key players to engage with each other in equal exchanges and
- provide the critical mass needed to make a difference.

Yet despite agreeing on the importance of social inclusion and engagement, the workshop demonstrated that stakeholder groups do view the world from their own perspectives and are rarely likely to change these positions even when exposed to different perspectives. For instance, when each group was asked to identify key points for their sector, it was not a surprise that they nominated different issues, but when organisers sought to bring it all together, there was not as much sharing of ideas or perspectives as had been hoped for.

The key issues for industry stakeholders were identified as follows:

- The need for information/public participation to manage fear about nanotechnology.
- The challenge of separating between the benefits and risks of nanotechnologies—with a feeling that fear was driving the community's response to nanotechnology.
- Education played a role in showing a balance between risks and benefits of nanotechnologies.

Government stakeholders identified

- the difficulty of developing regulation that made sense to a wide range of applications of nanotechnologies in various industries and
- the need to build community trust and the role of information and greater involvement as critical to the acceptance of nanotechnologies in Australia.

Researchers identified

- the challenges of securing funding for specific product developments, while also needing to understand the concerns of the community and identifying the impacts that nanotechnologies might have on communities through the products being developed.

Community group stakeholders identified

- the necessity of regulation,
- the need for more information and public participation in the nanotechnology debate and
- the inherent rights of the public to make informed choices about the products that they use—which may include labelling in many cases.

Change agents identified

- the importance of clear and deliberative processes within technological industries and the value of anticipating the societal needs and concerns within scientific developments [3].

There were also significant differences in what each stakeholder group felt were the engagement processes that would work best, and a general admission that the community were not well linked into engagement activities by any of the stakeholder groups. For instance, industry stakeholders stated that networking and conferences were very helpful for engaging with researchers and government but were uncertain as to how the general public should be best be engaged with, a position similar to that of the research stakeholders. There was a strong sense from both the community sector and change agents that there needed to be an obvious commitment from government to actively involve and engage with them and the wider public in the development of regulation around nanotechnology.

The workshop also identified that the main engagement gaps needing to be addressed were

- limited community input to government policy and regulation,
- limited community input into the social impacts and directions of research and
- poor industry and community dialogue.

And when it came to defining the outcomes of community engagement, there were differences of opinion again, based around stakeholder perspectives. For industry, it was about communicating that nanotechnology was not all about risks, and that the checks and balances of self-regulation were effective. For government stakeholders, it was about regulation keeping up with technological developments and the complexity of nanotechnologies, as well keeping the public debate constructive and free of misinformation.

The research sector identified funding and a need to compete globally, as well as educating an unscientific community, and the community group stakeholders identified the publics' right to know to make an informed decision as well as the need for regulation for worker/family safety. And finally the change agents identified the need for broader public engagement on what type of society we want to live in and what technologies we want to pursue.

In seeking consensus beyond broad principles, it was easier to find agreement on gaps that needed to be addressed, which included

- uncertainties about nanotechnologies and their uses,
- regulatory inquiry into nanotechnology imports and
- need for more dialogue between industry and the community.

However, a subsequent study undertaken through interviews of 14 individuals from industry, researchers and NGOs into what constituted good engagement, or communications, on nanotechnology, found the following key issues that were broadly agreed upon:

- Must discuss what constitutes "nanotechnologies".
- "Accurate" and/or "balanced" information is crucial.
- Address both potential benefits and risks.

- Early forms of engagement is needed.
- Move away from a homogenous approach to focus on specific areas, such as energy, or medical applications.
- More than seeking to "educate" or "inform" publics, towards a "two-way" form of communication or dialogue, in which the wider community is able to provide input into aspects of technological development [25].

The Social Inclusion and Community Engagement workshop ultimately adopted four key principles for social inclusion and community engagement, based on points put forward by the OECD:

(1) Deliberative—emphasises mutual learning and dialogue

(2) Inclusive—involves a wide range of citizens and groups whose views would not otherwise have a direct bearing on policy deliberation

(3) Substantive—topics selected that are appropriate to exchange

(4) Consequential—makes a material difference to the governance of new technologies [24]

Using the learnings from this workshop has led to a larger multi-stakeholder engagement process, which was conducted over a year, where each key stakeholder group spent a day together clarifying their issues and positions—with the general public spending two days—and then representatives of each group came together to develop principles for a framework on public engagement, as outlined in Section 12.6.

Developing agreed-upon principles is an important stage, but how exactly you apply these is an ever harder question to obtain agreement on. A useful model that can help understand the different approaches to engagement, aligned with different purposes and examples of each, has been developed by the International Association for Public Participation.

(1) **Inform**. The goal of the first level of engagement is to provide the public with balanced and factual information, to assist them in understanding the problems, alternatives, opportunities and or solutions. The promise that is made to the public is, we will keep you informed, and examples of this include fact sheets and websites. There is, however, no real scope for two-way engagement to occur.

(2) **Consult**. The purpose of consultation is to obtain public feedback on analysis, alternatives and/or decisions. The promise that is made to the public is that they will be kept informed, they will be listened to and their concerns will be acknowledged. Feedback will also be provided on how their input has influenced decision making. Examples of Consultations include public meetings, focus groups and surveys.

(3) **Involve**. The purpose of involvement is to work directly with the public through a process, to ensure that their concerns and aspirations are consistently understood and considered. The promise that is made to the public under this style of engagement is that they will be worked closely with, and that their concerns and aspirations will be directly reflected in the alternatives developed. Examples of this include workshops and deliberative polling.

(4) **Collaborate**. The purpose of collaboration is to partner with the public in aspects of the decision making, including the development of alternatives and the identification of preferred solutions. The promise made to the public is that they will be looked to for direct advice in developing solutions to problems and their advice and recommendation will be incorporated into the decisions made to the maximum extent possible. Examples of this include citizens advisory committees and participatory decision-making activities.

(5) **Empower**. The purpose of empowering is to place the final decision making into the hands of the public. The types of promises that are made to the public are that whatever they decide will be implemented, and examples of this include citizen juries and delegated decision making [15].

An alternative model is Arnstein's Ladder of Participation, which is based on an eight rung approach that progresses from non-participation up to citizens being empowered (Fig. 12.4) [2].

There is no one "one-size-fits-all" model of engagement techniques as there is no "one-size-fits-all" set of principles that could be used, as each suits different needs and different audiences. However, the fact that there are alternative yet similar models available can assist in engaging with stakeholders to come to some agreement on which frameworks and engagement techniques to use—as even these can be a source of division if not adopted through some form of consultation or engagement.

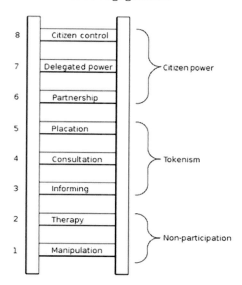

Figure 12.4   Arnstein's ladder of participation.

In relation to nanotechnology public engagement, Kyle and Dodds have recommended that Consult and Involve are the most appropriate forms of engagement in Australia. They argue that these forms of engagement encourage public attitudes and values to play a significant role in the nanotechnologies research and development process, which is an important part of promoting public trust [18].

But social scientists criticise many government or regulatory models of engagement as destroying trust, due to them being overly expert driven, with closed decision making and being divorced from public inputs at their early stages [28,34].

Another key issue highlighted by Kyle and Dodds is when is the appropriate time to engage with the public. Most advocates of public engagement would argue that "upstream" or early engagement is ideal, but with some nanotechnology applications this may need to be reconsidered. Certainly, involving the public too late in the development of any technology is unlikely to result in trust or mutual learnings, as was seen with the release of GM foods, but there might be an argument for more "midstream engagement" on nanotechnology, as at the early development stages there may not be enough information or clarity of applications to draw upon to expect sensible decisions.

12.10   Public Perception Barriers to Good Engagement

In addition to the barriers to good engagement thrown up by the key stakeholders, there are many intrinsic barriers that are inherent in members of the general public, when attempting to communicate complex issues to them. These include

(1) When information is complex, people make decisions based on their values and beliefs.

(2) People seek affirmation of their attitudes (or beliefs)—no matter how fringe—and will reject any information or facts that are counter to their attitudes (or beliefs).

(3) Attitudes that were not formed by logic are not influenced by logical arguments.

(4) Public concerns about contentious science or technologies are almost never about the science—and scientific information therefore does little to influence those concerns.

(5) People most trust those whose values mirror their own.

There are also some competing communication paradigms that impact upon successful engagement that include

(1) If the public are to be more involved in decision making, they need some level of knowledge.

(2) Lay knowledge should be valued, but is only one type of knowledge.

(3) Useful discussions cannot really be held without some scientific knowledge.

(4) Scientific information is often too complicated for the general public to understand, and misinformation and emotional information leads to a distorted understanding [17].

12.11   Examples of Engagement

Rowe and Frewer have listed more than 100 examples of engagement in practice, ranging from Action Planning to Citizens'

Juries, to Community dinners to Computer-Based techniques, to Hotlines and Open Houses and Study Circles [29]. It can, of course, make a significant difference to the outcomes as to which example is chosen, and history shows they tend to be chosen to best suit the organisers' preferred outcomes rather than for the participants' outcomes.

Another analysis of models of engagement, by Abels and Bora, defined seven different types of models and looked at how representative they were and who they favoured (Table 12.1) [1].

According to Dr Gabriele Abels, under most models, one group or another holds a key position. In the consensus conference, it is the lay persons. In the public hearing it is the administrator. In participatory technology assessment models it is often the scientific experts. However, in two models all participating groups enjoy equal rights. He cites these as the voting conference and scenario workshop, which he deems "balanced" [1].

It can be useful to examine some different case studies of nanotechnology engagement and assess how well they met their aims.

Case 1: NanoJury UK: A Citizens' Jury on Nanotechnologies (2005)

The pros and cons of a blockbuster event

NanoJury UK was a citizens' jury on nanotechnologies, organised by the Cambridge University Nanoscience Centre, Greenpeace UK, The Guardian newspaper, and the Politics, Ethics and Life Science Research Centre (PEALS) at Newcastle University. NanoJury ran for five weeks in June and July of 2005 and sought to present a non-specialist perspective on nanotechnologies, and to provide an opportunity for citizens to have a voice on the issues [20].

Twenty randomly chosen British citizens were recruited who heard evidence about a wide range of possible futures and the role that nanotechnologies might play in them. They were informed about nanotechnologies by a group of experts from different fields and heard several witnesses selected by an oversight panel, and a science advisory panel. In the final session they wrote recommendations for the future development of nanotechnologies in the United Kingdom.

An analysis of the NanoJury process made by Jasber Singh and by Tom Wakeford *et al.* at PEALS stated [31]:

Table 12.1 Typology of participatory technology assessment procedures

Procedure types	Criteria for selecting participants	Participants				Key feature
		Policy makers	Interest groups	Scientist/ experts	Lay people	
1 Dialogue Procedure	Representative; partly affected groups		X			Interest group procedure
2 Participative technological assessment (narrow sense)	Representative		X	X		Expert stakeholder procedure
3 Legal hearing	Everybody; those who feel affected			X	X	Decision oriented procedure involving those concerned
4 Consensus conference	Citizens (representative & 'lottery'); experts–deliberate selection by lay people			X	X	Lay people–expert procedure
5 Extended consensus conference	Citizens (representative & 'lottery'); experts–deliberate selection by lay people; interest groups: co-operation.		X	X	X	Lay people, interest groups and experts
6 Voting conference	Citizens (representative & 'lottery'); experts & policy makers; representatives	X	(X)	X	X	Voting-oriented procedure
7 Scenario workshop	Representativess	X		X	X	Procedure involving those affected, experts and policy makers

Procedure types	Form	
	Procedural rules	Social roles
1 Dialogue Procedure	Dialogue/discourse between interest groups; transparency regarding interests involved; understanding for different perspectives	Participants enjoy equal procedural rights
2 Participative technological assessment (narrow sense)	Discourse between scientific experts and interest groups	Experts = key position
3 Legal hearing	Legal decision; affected person have an advisory role	Decision-maker (administrator) = key position. Citizens give arguments, experts deliberate
4 Consensus conference	Questioning of experts by lay people	Lay people = key position, experts deliver knowledge.
5 Extended consensus conference	Participating groups often deliberate separately; interest groups deliver opinion, which is evaluated by lay people	Lay people = key position, experts deliver knowledge (esp. for dialogue with interest groups)
6 Voting conference	Evaluation of different scenarios handed in by stakeholders; voting on scenarios	Participating groups enjoy equal procedural rules
7 Scenario workshop	Evaluation of different scenarios; participating groups deliberate separately as well as in joint sessions	Participating groups enjoy equal procedural rules

(Continued)

Table 12.1 *(Continued)*

Procedure types	Expected achievement	Primary tasks and objectives	Target group	Major issues
		Function of participation		
1 Dialogue Procedure	Disclose divergent perspectives of concerned interest groups; overcome inertia; feed-back into interested associations; settling alternative options for policy-makers	Initiate dialogue among opposing groups; interactive exploration or goals; identification of areas of consensus and lack of agreement	Policy-makers; interest groups; general public	Technology assessment and planning
2 Participative technological assessment (narrow sense)	Risk assessment based on technological state-of-the-art; identify uncontested knowledge as a basis for decision	Resolving status of scientific knowledge by experts and counter-experts; in so doing, clarify political options; legitimise political decisions	Policy-makers; general public	Technology in general
3 Legal hearing	Inform citizens and administration; representation and legal protection of citizen's interests; foster acceptance and legitimacy of administrative decision	Deliberation in a strict sense, i.e. influence decisions by good arguments	Public administration, decision-maker	Specific cases
4 Consensus conference	Typical opinion of informed lay person; also agenda setting	Communication between lay people and experts; fostering and enlightening of public debate	Policy-makers; interest groups; general public	Technology in general
5 Extended consensus conference	Exploration of objectives; typical opinion of informed lay person	Fostering and enlightening of public debate	Policy-makers; interest groups; general public	Technology in general
6 Voting conference	Filter for competing policy options	Fostering and enlightening of public debate; reveal perspectives of different groups	Policy-makers; interest groups; general public	Technology in general
7 Scenario workshop	Disclose divergent perspectives of participating groups; agenda setting; political legitimacy; overcome deadlock	Planning process; dialogue between all involved groups of actors; foster understanding for divergent perspectives	Policy-makers; interest groups; general public	Technology in general

Procedure types	Empirical examples	Underlying model of democracy	
		Typical procedure	Model
1 Dialogue Procedure	Dutch Gideon project, German discourse on Ag-biotechnology	Mediation-oriented stakeholder discourse	Pluralist, but with deliberative elements
2 Participative technological assessment (narrow sense)	PTA on herbicide resistant plants—Berlin	Discursive PTA in a more narrow sense	Not specified, rather deliberative
3 Legal hearing	German law on atomic energy	Public hearing as part of administrative decision making	Formally participatory, actually deliberative
4 Consensus conference	UK Consensus Conference on Plant Biotechnology, Consensus Conference on GM food in Australia.	Consensus conference, citizen's jury	Deliberative
5 Extended consensus conference	UK Citizen Foresight Project on GM food. Citizens Jury on GM crops in India	Modified consensus conference	Deliberative-pluralist
6 Voting conference	Danish Voting Conference on Drinking Water	Voting conference	Deliberative with some pluralist elements
7 Scenario workshop	Danish Scenario Workshop on Urban Ecology; futures Search Conference Traffic Copenhagen.	Scenario workshop (Danish style)	Participatory-deliberative with pluralist elements

As a two-way street process, it has highlighted key questions about the science-society divide and how it leaves people's problems unheard and thus further alienates people from the developments of science. With the growing momentum of upstream engagement, and its likely outcome of debating science independently of people's daily life experience, it seems that the Nanojury should serve as a gentle reminder to challenge the corporate-government tendency to undermine attempts at technology democracy through upstream engagement. For us, the Nanojury analysis is a call to move out of the polluted stream towards science that is incorporated into community development.

However, the Nanojury was also criticised by commentators as having a high profile that did not actually make any significant impact upon policy development [31].

Case 2: Nanotechnology Engagement Group (NEG): Understanding public engagement with nanotechnologies (2005)

Big outcomes—Small audiences

The Nanotechnology Engagement Group (NEG) was convened by the UK not-for-profit group Involve, which specialises in undertaking and understanding public engagements. The collaborating partners were the UK Office of Science and Innovation (OSI) and the Universities of Cambridge and Sheffield. The purpose of the engagement was to document the learning from six UK public engagement projects on nanotechnologies that included: NanoJury UK, Small Talk, Nanodialogues, Nanotechnology, Risk and Sustainability, Citizen Science @ Bristol, and Democs. The study sought to examine stakeholders' expectations of public engagement and then to identify lessons learned and how these might relate to the range of new engagement activities being undertaken.

The NEG undertook a two-year programme of desk research, interviews, meetings with group members, and a workshop for scientists, project organisers, public participants, NGOs, and policy-makers in 2006. The final report, Democratic Technologies?, summarised the experiences of public engagement on nano-technologies that had taken place in the United Kingdom and in other countries, and concluded that while the outcomes were of note, similar processes needed to find ways to involve larger numbers of people in public deliberations about science and technology [12].

Case 3: DEEPEN: Deepening Ethical Engagement and Participation in Emerging Nanotechnologies (2006–09)

A more multi-disciplinary approach

The DEEPEN (Deepening Ethical Engagement and Participation in Emerging Nanotechnologies) project was a three-year project for integrating an understanding of "the ethical challenges posed by emerging nanotechnologies in real world circumstances, and their implications for civil society, for governance, and for scientific practice". Led by the Institute for Hazard and Risk Research (IHRR) at Durham University in the United Kingdom, the project also included researchers based at Darmstadt University of Technology (Germany), the Centre for Social Studies at the University of Coimbra (Portugal), and the University of Twente (the Netherlands) [20].

The project's aim was to deepen ethical understanding of issues on emerging nanotechnologies through an interdisciplinary approach using insights from philosophy, ethics, and social science. The project undertook a series of deliberative forums in which citizens, stakeholders, experts, and decision-makers could develop both convergent and divergent understandings of the social and ethical ramifications of nanotechnology. Focusing on nanosensors and nanomedicines, the project was delivered through nine "integrated work packages" over four phases: surveying of ethical and societal issues of concern; integration; experiments in new deliberative processes; and dissemination [20].

The multi-disciplinary approach of DEEPEN allowed for the outcomes of different approaches to be brought together into a consolidated learning, and one of the key findings was: "the time has come to move away from open-ended conversation on what nanotechnology may provide for our society, and to promote concrete deliberation on possible developments of nanotechnology. Instead of identifying concerns regarding speculative futures, public engagement exercises should focus on current or emerging research directions and technological developments in order to critically assess their possible impacts and their normative implications" [7].

Case 4: The Special Commission for Public Debate (2009–10)

Overwhelmed by Protests

Undertaken in France in 2009 and 2010, the Special Commission for Public Debate sought to hold public debates on nanotechnology issues with an audience of the general public and panellists from scientific, health, and environmental organisations. The series of debates started well, but from late 2009 started attracting more organised protestors in the towns of Grenoble, Rennes and Lyon. This culminated in the public debate in Marseilles being shut down by organisers after protestors clapped, whistled, shouted, threw paper and raised banners with slogans such as "Nano—it's not green, it's totalitarian".

The protestors accused the debates of being government-controlled, rather than independent and were not providing a balanced view of nanotechnologies. Friends of the Earth, who pulled out of their involvement, claimed that the important questions on nanotechnology were not being posed, which included nano-particle toxicity, identification of nanotechnologies that encroach on private life, and military uses of nanotechnologies [11].

Case 5: NICNAS Community Engagement Forums (2004–)

Community representation or interest group representation?

Australia's National Industrial Chemicals Notification and Assessment Scheme (NICNAS) established a Community Engagement Forum, consisting of a group of individuals who represented the interests of workers, public health and the environment, and provides advice to NICNAS and oversight of its community engagement activities.

Membership is for three years and in late 2012 consisted of representatives of the National Toxics Network, the consumers' organisation Choice and trade unions. The Community Engagement Forum meets three times a year to discuss industrial chemical issues and the types of community engagement activities being undertaken by NICNAS, and it also meets with the industry-government consultative committee once a year. The forum also runs some engagement processes.

Albeit being a group that represents the community's interests more than it actually represents the community, the Forum has proved a useful model for engagement of interest groups by the regulator and industry. Its strengths are:

- A shared commitment to the process;
- Good relationships that have been established over time; and
- Wider links with other community groups and interest groups.

Case 6: The University of Wisconsin—Madison's Nano Cafés (2005–)

Informal conversations with experts

Sponsored by members of the Citizens' Coalition on Nanotechnology, in cooperation with faculty at the University of Wisconsin's Madison Nanoscale Science and Engineering Center and the Nelson Institute for Environmental Studies. The Nano Cafés sought to give citizens access to information on nanotechnology research.

The initiative sprung out of the 2005 Madison Areas Citizens' Consensus Conference on Nanotechnology, when several members of the citizen panel wished to continue engaging with scientists and educate the public about nanotechnology. This led to the formation of the Citizens' Coalition on Nanotechnology, which adopted the European model of a science cafe, which provides a casual atmosphere in which people can listen to, or discuss issues of interest, with experts. The Nano Cafes are held in different parts of the community such as coffee shops, libraries or community centres, and University of Wisconsin-Madison experts explained their work, answered questions and addressed concerns from members of the public [20].

In order for the Nano Cafes to be as democratic and participative as possible, more citizens became involved in organising them. However, as participants self-selected to attend, there is a risk that the nano cafes, like most science cafes, only engage with the already engaged, and need to be supplemented by other activities to best reach wider members of the public.

Analysis of the impacts of the Cafes has also led to questions such as, can academics and others who work within institutions really initiate meaningful engagement with members of the public in a predominantly top-down approach? [27].

Clearly every engagement activity is going to have strengths and weaknesses, and the search for a dream model to base engagement activities on is likely to prove elusive. But that also provides a strategic direction for engagement approaches to gain the best outcomes.

12.12   Public Engagement Models for the Future

Historically public engagement has moved from awareness raising, to education, to participative engagement—with some agencies working in all three spaces, and there is now movement towards new and more effective multiple models. This mirrors the evolution of the closely related field of technology assessment. The first generation of technology assessment was typified by the US Office of Technology Assessment in the 1970s. It was characterised by being expert-based, led by government agencies, and sought to provide strategic analysis of developing technologies.

The second generation of technology assessment was typified by the Danish Board of Technology in the 1990s, which was established by Government, but not operating within Government. The second generation models involved selected citizens and key stakeholders making deliberative assessments on the impacts of new technologies, such as occurs in citizens juries.

The third generation is still evolving, but is based around using multiple models and methods, by involving a diversity of interest groups. In practice it involves a lot of trials (and errors) that might even combine different methodologies. It is also typified by distributed governance of management, knowledge and participation. It also has a tendency to blur the boundaries between participating interest groups and individuals.

Professor Arie Rip, one of the key proponents of the third generation of technology assessment, defines it as having multiple technology assessment models that exist at different places or on different paths [28].

He also says that understanding appropriate levels of public engagement can be tied to understanding technology adoption.

Most new technologies go through an up and down curve that starts with a slow rise, triggered by the development of the technology. It rises quickly to a "peak of inflated expectation", after which there is a sharp drop to the "trough of disillusionment". This is followed by slow climb to the "slope of enlightenment" and finally public acceptance reaches whatever plateau is appropriate for that technology [28]. It can be argued that GM foods are, for the most part, approaching the slope of enlightenment and we will then see where the technology finally plateaus. Nanotechnology, by comparison, has perhaps passed the peak of inflated expectation overall, if we look at public attitude research, and is now moving towards the trough of disillusionment, but this might not be such a similar curve to that of GM foods, depending on advances in nanotechnologies in different fields.

One example of how models of engagement can look is that used by the National Enabling Technologies Strategy in Australia, which uses multiple models across different target audiences, seeking to gain maximum engagement via a wide variety of activities (Fig. 12.5).

Nanotechnology Engagement Activities

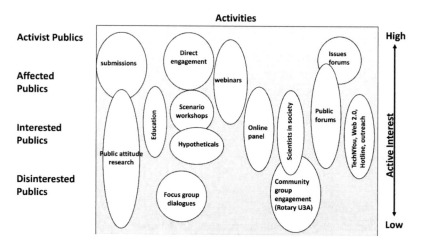

Figure 12.5 The multi-strategy approach to public engagement of the Australian Government's National Enabling Technology Strategy.

12.13 Online Communities and Online Community Engagement

In an era of web 2.0 that is rapidly moving towards web 3.0, online engagement deserves an individual mention. The rapid growth of the Internet and the ability to engage people through popular social network sites may drastically change the way it is possible to engage with members of the community—but many of the fundamental problems and barriers to good engagement are likely to remain.

If 100 people sign up to an online discussion board relating to an aspect of nanotechnology, it is important to know if they represent only an "engaged" public, or other segments of the public too. Likewise, while there is much monitoring of mainstream media for trends in how nanotechnology is being discussed to feed into policy or product development, there has as yet not been as much monitoring of social media.

The development of e-communities, however, will provide a new way to easily reach a target audience, especially with the ability to recruit and develop e-community profiles to match either particular stakeholder or audience segments. It also provides for forums that can cross over information provision, education and engagement, such as using a spokes-and-hub model where a central website can provide information or educative materials that then links to existing social media sites via spokes, such as Facebook, blogs and wikis, providing scope for engagement.

It is not a given though that the e-environment will provide easy ways to reach new publics as there is enormous amount of "competing noise" that has to be overcome to actively engage an audience. Added to this is the concern that Google's search analytics favour similar sites to those you have already searched, in a similar manner to how Amazon.com recommends similar books to those you have purchased, so rather than allow for increased access to varied points of view it tends to support confirmation bias of existing opinions.

One benefit of Internet-based methods of engagement that have been argued is that they allow for a breaking down of the boundaries between experts and non-experts, best typified by web 3.0 practices of citizen-generated content, where the traditional boundaries of who is qualified to be an editor, journalist or commentator are dissolving and content is judged primarily on its appeal.

This may also have a downside, though as content is more likely to be appealing when it is sensational or aligns with existing biases, and while online forums can empower citizens, they can disempower the experts, diluting their voices from public debates. Wikipedia has managed to handle this moderately well in allowing users to edit and re-edit contributor's copy so that in most cases it reaches a plateau of consensus—but with emerging technologies such as nanotechnology, while there might be easy consensus in the scientific spheres, it might be much harder to achieve across wider society in relation to issues where there are polarised views.

There is another engagement conflict here in the online trend towards all opinions having equal weighting in cyberspace, yet a desire to obtain trusted information.

There are also signs of an emergence of a new technological elite, who have the skills and resources to re-engineer web 2.0 tools to suit their own needs or outcomes, manipulating the information flows in new ways. When different stakeholders all start playing in this space, the potential for online engagement to contribute to better engagement risks sliding backwards to a point it is overwhelmed with biased engagement.

For the moment though, the potential is great and the uptake is proving rapid.

12.14 Areas for More Work

While an enormous amount of very good public engagement exercises are being conducted around the globe, there are a few areas that still need to be addressed to obtain better outcomes. Engaging more with the unengaged members of society has already been mentioned, but we also need to look at the nature of the engagements being planned. Many are one off activities in artificially created environments, such as focus groups, or café dialogues with scientists and so on. These can be considered analogous to laboratory experiments for GM foods—they provide useful data, but it is then necessary to transfer the experiment to a glass house trial and then to a field trial to study real-world experience.

Likewise with nanotechnology public engagements we need to find ways to move the experiments we conduct in focus groups and cafes and so on into more real life situations. This is based upon a need to both measure the impact of dialogues in real life situations and to develop engagement models that are more easily transferable to the real world by participants than might be possible from science cafes, or focus groups.

Having done that, there is still a large gap in the research as to what happens to people who take part in engagement activities after the events. Ideally they take their learnings and experiences and share them with their friends and family and colleagues. But more research needs to be done to determine levels of actual topic fade or transference of information.

12.15 So What Does It All Mean?

Public engagement is a bit like nanotechnology in that the more one learns, the more one realises one still has to learn—but this cannot be a reason for indecision. Any amount of "almost good" public engagement is better than no engagement at all. Yet we must acknowledge that the barriers to good engagement are significant and need to be continually worked on. By they are no less surmountable than the barriers to good research outcomes or successful commercialisation. Yes, there are conflicting stake-holders and conflicting principles to get beyond, but if it was easy we would not be committing so many resources to trying to get it right.

Clearly there is no one model that will work best for all occasions, and the contemporary way of thinking is that multiple models need to be employed and the findings of each of them aggregated, as different models appeal to different stakeholders and reach different audiences in different ways.

To put it another way, it is like combining all the different perspectives from all the different cells in a reverse panopticon prison and collectively seeing something new for the first time. But a crucial question then is: who can be trusted to undertake this? For if we are all trapped in the cells of our own perspectives and ideologies, and evidence shows we will simply seek to best position our own interests or perspectives, who is free to provide an independent point of view? Perhaps the answer is in more genuine collaboration between interest groups.

Nanotechnologies are going to have complex impacts upon our societies, not all of them foreseeable, but in order to do justice to the complexity of ways in which the public relate to new technologies we must embrace more complex ways of viewing the public, and more complex ways of viewing new technologies, and more complex ways of viewing the relationships between them. All of which must underpin more complex and diverse engagement practices.

To quote Jose Manuel de Cozar-Escalante [8]:

> In short, we should seek a broader conception of representation for the politics of science and technology, a representation that is better suited to the intricacies of our increasingly technological and globalised world.

References

1. Abels, G. (2005). Forms and functions of participatory technology assessment—or: why should we be more sceptical about public participation? Participatory Approaches in Science and Technology Conference, Edinburgh, June 2006.

2. Arnstein, S. (1969). Ladder of citizen participation, *J. Am. Inst. Planners*, **35**(4), pp. 216–224.

3. Australian Office of Nanotechnology (2008). *Social Inclusion and Community Engagement Report* (Australian Office of Nanotechnology).

4. Binder, A. (2010). Interpersonal amplification of risk? Citizen discussions and their impact on perceptions of risk and benefits of a biological research facility, *Risk Anal.*, **31**, pp. 324–334.

5. Computer Aided Research and Media Analysis (CARMA) (2010). *Nanotechnology, National Enabling Technologies Strategy* (Department of Innovation, Industry, Science and Research).

6. Victorian Department of Innovation, Industry, and Regional Development (DIIRD) (2007). *Community Interest and Engagement with Science and Technology in Victoria—Research Report.*

7. Davies, S., Macnaghten, P., and Kearnes, M. (2009). *Reconfiguring Responsibility: Lessons for Public Policy (Part 1 of the report on Deepening Debate on Nanotechnology)* (Durham: Durham University).

8. de Cozar-Escalante, J. M. (2006). Representation as a matter of agency: a reflection on nanotechnological innovations, Participatory Approaches in Science and Technology Conference, Edinburgh, June 2006.

9. Druckman, J., and Bolsen, T. (2010). *Framing Motivated Reasoning, and Opinions about Emergent Technologies Institute for Policy Research* (Northwestern University, Working Paper Series).

10. Foucault, M. (1979). *Discipline and Punish: the Birth of the Prison* (Penguin, UK).

11. Framing Nano (2010). *Chaos at French Nano Debates*, Framing Nano, March.

12. Gavelin, K., Wilson, R., and Doubleday, R. (2007). *Democratic Technologies? The Final Report of the Nanotechnology Engagement Group* (Involve, UK).

13. Hartz-Karp, J. (2009). International Association for Public Participation 2 (IAP2) International Conference, 22–23 October, Fremantle.

14. Hendriks, C. (2006). When the forums meets interest politics: strategic uses of public deliberation, *Polit. Soc.*, **34**(4), pp. 571–602.

15. International Association for Public Participation 2 (2004) *IAP2 Public Participation Spectrum*. http://www.iap2.org.au/resources/.

16. Ipsos Mori Social Research Institute (2006) *Ingredients for Community Engagement: The Civic Pioneer Experience* (Ipsos Mori Social Research Institute), pp. 12–13.

17. Kuroda, R. (2010). Science in Society: Responsibility of Scientists and Public for 21st Century, Society for the Social Studies of Science Conference, Tokyo, August 25–29.

18. Kyle, R., and Dodds, S. (2009). Avoiding empty rhetoric: engaging publics in debates about nanotechnologies, *Sci. Eng. Ethics*, **15**, pp. 81–96.

19. Lakoff, G. (2004). *Don't Think of an Elephant?* (White River Junction, Vermont: Chelsea Green Publishing).

20. Lafitte, N. B., and Joly, P. B. (2008). Nanotechnology and society: where do we stand in the ladder of citizen participation? *Citizen Participation in Science and Technology Newsletter*, March 2008.

21. Market Attitude Research Services (2011). *Australian Community Attitudes Held about Nanotechnology—Trends 2005–2011* (National Enabling Technologies Strategy, Department of Innovation, Industry, Science and Research).

22. National Enabling Technologies Strategies' Public Awareness and Community Engagement Section, Draft Framework Principles for Multistakeholder Engagement, Department of Innovation, Industry, Science and Research, 2011.

23. National Toxics Network (2008). *Plan of Action,* Available at: http://ntn.org.au/2008/03/25/plan-of-action/.

24. Organization for Economic Cooperation and Development (2008). Conference on Outreach and Public Engagement in Nanotechnology, Delft, Netherlands, 30 October.

25. Petersen, A., Seear, K., and Bowman, D. (2010). *Communicating with Citizens about Nanotechnologies: Views of Key Stakeholders in Australia* (School of Political and Social Inquiry, Faculty of Arts, Monash University).

26. Powell, M., and Collin, M. (2008). Meaningful citizen engagement in science and technology. What would it really take? *Sci. Commun.*, **30**(1), pp. 126–136.

27. Powell, M. (2009). Participatory paradoxes facilitating citizen engagement in science and technology from the top-down?, *Bull. Sci. Technol. Soc.*, **29**(4), pp. 325–342.

28. Rip, A. (2010) Technology assessment of emerging technologies—The next steps, Society for the Social Studies of Science Conference, Tokyo, August 25–29.

29. Rowe, G., and Frewer, L. (2000). Public participation methods: A framework for evaluation, *Sci. Technol. Human Values*, **25**(1), pp. 3–29.

30. Rinie van Est, R., Walhout, B., and Hanssen, L. (2008). *Ten Lessons for a Nanodialogue: How to be Deadly Serious and Still Have Serious Fun.* (The Rathenau Instituut).

31. Singh, J. (2006). Polluted waters: The UK Nanojury as upstream public engagement, http: //www.nanojury.org.uk/pdfs/polluted_waters.pdf/.

32. Wickson, F., Delgado, A., and Kjølberg, L. (2010). Who or what is "the public"?, *Nat. Nanotechnol.*, **5**, pp. 757–758.

33. Workshop conducted at the Australian Academy of Sciences, 1 October 2010.

34. Wynne, B. (2010) Expertise, publics and politics: waving at meanings? Society for the Social Studies of Science Conference, Tokyo, August 25–29, 2010.

Index